Natural Products Desk Reference

Natural Products Desk Reference

John Buckingham
Caroline M. Cooper
Rupert Purchase

CRC Press
Taylor & Francis Group
Boca Raton London New York

CRC Press is an imprint of the
Taylor & Francis Group, an **informa** business

On the front cover: Structures of Morphine (left) and Adriatoxin (right).

On the back cover: Structure of Amphidinolide N.

CRC Press
Taylor & Francis Group
6000 Broken Sound Parkway NW, Suite 300
Boca Raton, FL 33487-2742

© 2016 by Taylor & Francis Group, LLC
CRC Press is an imprint of Taylor & Francis Group, an Informa business

No claim to original U.S. Government works

Printed on acid-free paper
Version Date: 20151019

International Standard Book Number-13: 978-1-4398-7361-8 (Paperback)

Visit the Taylor & Francis Web site at
http://www.taylorandfrancis.com

and the CRC Press Web site at
http://www.crcpress.com

In Memoriam

It is with enormous sadness and regret that I have to report the death of John Buckingham in a car accident while this book was in the last stages of production.

Natural Products Desk Reference *was very much John's project combining two of his favourite subjects of natural product chemistry and taxonomy. His knowledge of natural products, their structures and stereochemistry was encyclopaedic. He wrote authoritatively without apparent effort.*

John corrected the proofs not long before he died, so this book is, I hope, just as he wanted it to be and a fitting memorial to a most valued colleague and friend.

Caroline Cooper, 10th September 2015

Contents

Foreword

The science of natural products is a very broad one, ranging from taxonomy to the isolation and structure elucidation of complex chemical structures to their biology and biosynthesis. It has applications in human medicine and in fields as diverse as agriculture, nutrition and plant pathology. On the medical side, natural products have been the source of or inspiration for 40% of all prescription drugs sold in the United States, including such well-known drugs as Taxol® and penicillin. The importance of natural products to human health was emphasised by the award of the 2015 Nobel Prize in Physiology or Medicine to Professor Satoshi Omura and Dr. William Campbell for the discovery of avermectin from a *Streptomyces* sp. and its development as a treatment for river blindness and other parasitic diseases, and to Professor Tu Youyou of Beijing for her discoveries leading to the development of the antimalarial drug artemisinin from *Artemisia annua*. The structures of natural products are also very diverse, ranging from simple aliphatic carbon chains to high molecular weight proteins.

Because the science of natural products is so broad, it is difficult for any one individual to be familiar with all of its aspects. This is especially true for scientists and others who may encounter natural products as just one aspect of their work, and who may thus need a quick way to look up information on natural product structures and nomenclature, or on taxonomy. This book aims to provide this information in a convenient format. Online resources such as *SciFinder* and the *Dictionary of Natural Products* provide direct access to information on specific natural products, but do not provide the overviews of nomenclature, taxonomy and common natural product skeletons that are found in this book. This handy desk reference was compiled by the same team of experts that produces the *Dictionary of Natural Products*, and it is extensively illustrated with representative chemical structures, making it both authoritative and accessible. It will be a valuable resource to all scientists whose work requires familiarity with natural products in all their diversity.

David G.I. Kingston
Department of Chemistry
Virginia Tech
Blacksburg, Virginia

Preface

Natural products research plays a central role in chemistry and related disciplines. The science of organic chemistry grew out of the study of natural products, and the identification and synthesis of nature's molecules continue two centuries later to set major challenges to organic chemists. The discipline of natural products chemistry is intimately involved in biochemistry and drug research; a large proportion of newly introduced drugs continue to be natural products or molecules based on the structural variation of a natural product model.

The team responsible for the CRC (formerly Chapman & Hall) chemical database has, as one of its main responsibilities, the accurate documentation of all natural products. The results of this endeavour are published as continual updates of the segment of the database forming the *Dictionary of Natural Products* (*DNP*), which, as a separate publication, dates back to 1992.

In 1995, the database team published the *Organic Chemist's Desk Reference* (*OCDR*), a publication containing a mass of information of practical usefulness to organic chemists. A widely welcomed new edition of OCDR appeared in 2011, and a third edition is in preparation. The *Natural Products Desk Reference* (*NPDR*) is similarly compiled as a companion reference source for those with an interest in natural products, not only specialist natural products chemists, but also medicinal chemists, synthetic organic chemists and a wide range of life scientists.

NPDR is more closely based on its relationship with the CRC database than *OCDR*. For general organics, the CRC database is one of many sources of information on which the worker can draw, whilst for natural products, DNP has established itself as the leading edited source of information. We hope that *NPDR* will enable scientists to use DNP effectively but will also provide them with a mass of other useful information which can sometimes be hard to track down. In compiling it, we have drawn on over 20 years of day-to-day experience in the description and classification of all types of natural products.

Acknowledgements

We thank the following people for their contributions to this book:

Professor Bob Hill (University of Glasgow), for advice and also for very valuable work adding to and checking the natural product skeletons in Chapter 6.

Ranjit Munasinghe (consultant, CRC Press), who wrote the section on enzymes.

Steve Walford (CRC Press) and Alison Hodgson (consultant, CRC Press), who checked all the diagrams in Chapter 6.

Janice Shackleton in the London office of CRC Press, who organised the diagrams and typescript.

Trupti Desai, who drew the diagrams, and Chi Huynh, who typed the glossaries.

Finally, Fiona Macdonald at CRC Press in Boca Raton, Florida, who commissioned and supervised the project.

Authors

John Buckingham was a lecturer in organic chemistry at the University of London, United Kingdom. He was involved with the Chapman & Hall/CRC chemical database since its inception in 1980, initially as a Chapman & Hall employee and finally as an editorial consultant. From the database, various editions of the *Dictionary of Organic Compounds* and the *Dictionary of Natural Products* (both of which have been for some years solely electronic) have been produced. In addition, he compiled (with W. Klyne and later with R. A. Hill) two editions and supplements of the *Atlas of Stereochemistry* and co-authored several other specialist dictionaries in the Chapman & Hall/CRC series.

He was also the author of the popular science books *Chasing the Molecule* and *Bitter Nemesis: The Intimate History of Strychnine.*

Caroline M. Cooper completed her BSc in chemistry at King's College London in 1968 and then worked at Glaxo Research in Greenford. She contributes to the *Dictionary of Organic Compounds*, and in 2011, she edited the second edition of *Organic Chemist's Desk Reference*, both published by CRC Press.

Rupert Purchase studied chemistry at the South-East Essex Technical College [Grad. RIC Part II (External), 1967] and the University of York (DPhil, 1972). He is a Fellow of the Royal Society of Chemistry and a Visiting Fellow at the University of Sussex (2014–2017). Dr. Purchase contributes to *The Combined Chemical Dictionary on DVD* published by CRC Press and is a freelance editor for *Science of Synthesis: Houben-Weyl Methods of Molecular Transformations* published by Georg Thieme Verlag, Stuttgart. He edited the Royal Society of Chemistry's (RSC) Environmental Chemistry Group *Bulletin* from 1995 to 2013 and was awarded the RSC's Long Service Award in 2011.

1 The Natural Products Literature
Useful Review Series, Reference Works and Databases

This chapter contains some of the more important reference sources for printed and online information on natural products.

1.1 REVIEW SERIES, REFERENCE BOOKS AND MONOGRAPHS

The following details are provided, where relevant:

- Full review series, reference book, monograph or database title
- CASSI (*Chemical Abstracts Service Source Index*) abbreviated title [in square brackets], obtained via http://cassi.cas.org/search.jsp
- Year(s) of publication
- Change(s) of titles
- Name of the publisher of the current title (*see* Section 2.2 for contact details of the current publishers cited here)

Accounts of Chemical Research [*Acc. Chem. Res.*] (1968–). Publisher: ACS. Wide-ranging review journal with a bias towards interdisciplinary methods and techniques.

ACS Symposium Series [*ACS Symp. Ser.*] (1974–). Publisher: ACS. Ongoing series of books developed from the ACS Divisions Symposia.

Advances in Carbohydrate Chemistry and Biochemistry [*Adv. Carbohydr. Chem. Biochem.*] (Vol. 24–, 1969–). Formerly *Advances in Carbohydrate Chemistry* [*Adv. Carbohydr. Chem.*] (Vols. 1–23, 1945–1968). Publisher: Elsevier.

Advances in Experimental Medicine and Biology [*Adv. Exp. Med. Biol.*] (1967–). Publisher: Springer.

Advances in Marine Biology [*Adv. Marine Biol.*] (Vol. 1–, 1963–). Publisher: Elsevier.

Advances in Phytomedicine [*Adv. Phytomed.*] (Vol. 1–, 2002–). Publisher: Elsevier.

Alkaloids: Chemical and Biological Perspectives [*Alkaloids: Chem. Biol. Perspect.*] (Vols. 1–15, 1983–2001). Publisher: Elsevier. Vols. 1–8.

Bioactive Compounds from Natural Sources, 2nd edn., ed. C. Tringali, 2011. Publisher: CRC Press.

Carotenoids Handbook, eds. G. Britton, S. Liaaen-Jensen and H. Pfander, 2004. Publisher: Birkhäuser-Verlag/Springer. Physico-chemical and spectroscopic properties, isolation procedures and other data for >700 entries.

Chemical Dictionary of Economic Plants, eds. J. B. Harborne and H. Baxter, 2001. Publisher: Wiley.

Chemical Reviews [*Chem. Rev.*] (1924–). Publisher: ACS. Twice-monthly authoritative reviews across the whole of chemistry.

Chemical Society Reviews [*Chem. Soc. Rev.*] (1972–). Successor to *Quarterly Reviews of the Chemical Society* [*Q. Rev. Chem. Soc*] (1947–1971) and *RIC Reviews* [*RIC Rev.*] (1968–1971). Publisher: RSC. Twice-monthly reviews on the chemical sciences.

Chemotaxonomie der Pflanzen: Eine Übersicht über die Verbreitung und die systematische Bedeutung der Pflanzenstoffe (*Plant Chemotaxonomy: A Survey of the Distribution of Plant Constituents and Their Significance for Botanical Classification*), R. Hegnauer, Vols. 1–10, 11a, 11b, 1962–2001. Publisher: Birkhäuser Verlag, Basel.

Chinese and Related North American Herbs: Phytopharmacology and Therapeutic Values, T. S. C. Li, 2002. Publisher: CRC Press.

Chinese Materia Medica: Chemistry, Pharmacology and Applications, Y.-P. Zhu, 1998. Publisher: Harwood Academic Publishers, Amsterdam, the Netherlands.

Comprehensive Natural Products Chemistry [*Compr. Nat. Prod. Chem.*], eds. D. Barton, K. Nakanishi and O. Meth-Cohn, 9 Vols., 1999. Publisher: Elsevier.

Comprehensive Natural Products II: Chemistry and Biology, eds. L. Mander and H.-W. Liu, 9 Vols., 2010. Publisher: Elsevier. The content of the second edition is drawn from the literature covering the decade following the previous edition.

CRC Handbook of Medicinal Herbs, 2nd edn., J. A. Duke, 2002. Publisher: CRC Press. Data on approx. 800 medicinal plant species.

Dictionary of Alkaloids, 2nd edn. with CD-ROM, eds. J. Buckingham, K. H. Baggaley, A. D. Roberts and L. F. Szabo, 2010. Documents chemical information for more than 20,000 alkaloids. Publisher: CRC Press.

Dictionary of Antibiotics and Related Substances, 2nd edn. with CD-ROM, eds. B. W. Bycroft and D. J. Payne, 2013. A compilation reflecting the 10,000 known antibiotics. Publisher: CRC Press.

Dictionary of Carbohydrates, 2nd edn. with CD-ROM, ed. P. M. Collins, 2006. A resource for 26,000 carbohydrate compounds. Publisher: CRC Press.

Dictionary of Flavonoids with CD-ROM, eds. J. Buckingham and V. R. N. Munasinghe, 2015. Lists all known flavonoids (approximately 13,000 compounds). Publisher: CRC Press.

Dictionary of Marine Natural Products with CD-ROM, eds. J. W. Blunt and M. H. G. Munro, 2007. A comprehensive coverage of all known marine natural products with data on 25,000 compounds. Publisher: CRC Press.

Dictionary of Natural Products. See *The Combined Chemical Dictionary on DVD* and *CHEMnetBASE* (Table 1.1).

Dictionary of Plant Toxins, eds. J. B. Harborne, H. Baxter and G. P. Moss, 1996. Publisher: Wiley.

Drugs of Natural Origin: A Treatise of Pharmacognosy, 6th edn., G. Samuelsson and L. Bohlin, 2010. Publisher: Apotekarsocieteten: Swedish Academy of Pharmaceutical Sciences, Stockholm/CRC Press.

Encyclopedia of Biological Chemistry, 2nd edn., eds. W. J. Lennarz and M. D. Lane, 2013. Publisher: Academic Press/Elsevier.

Encyclopedia of Marine Natural Products, 2nd edn., J. M. Kornprobst, 2014. Publisher: Wiley-Blackwell.

Encyclopedia of Traditional Chinese Medicines: Molecular Structures, Pharmacological Activities, Natural Sources and Applications, J. Zhou, G. Xie and X. Yan, 6 Vols., 2011. Publisher: Springer. Describes >23,000 chemical components isolated from approx. 7,000 medicinal plants.

Handbook of Biochemistry and Molecular Biology, 4th edn., eds. R. L. Lundblad and F. M. Macdonald, 2010. Publisher: CRC Press.

Handbook of Chinese Medicinal Plants, W. Tang and G. Eisenbrand, 2011. Publisher: Wiley-VCH.

Handbook of Compounds with Anti-inflammatory and Anti-platelet Aggregation Activities Isolated from Plants, R. M. Pérez Gutiérrez, 2009. Data on 848 compounds classified by their natural product class. Publisher: Nova Science Publishers, Hauppauge, NY.

TABLE 1.1
Some Commercial and Public Online Natural Products Databases[a,b]

Database Name	Number of Entries	Company or Organisation	Website Address
AntiBase 2014: The Natural Compound Identifier	>42,000	Wiley	http://www.wiley-vch.de/publish/en/
Chemical Entities of Biological Interest (ChEBI)[c]	>42,000	European Bioinformatics Institute	http://www.ebi.ac.uk/chebi/
CHEMnetBASE (Combined Chemical Dictionary)	>630,000	CRC Press	http://www.chemnetbase.com/
CHEMnetBASE (Dictionary of Marine Natural Products)	>50,000	CRC Press	http://www.chemnetbase.com/
CHEMnetBASE (Dictionary of Natural Products)	>272,000	CRC Press	http://www.chemnetbase.com/
IBS Database	>50,000	InterBioScreen Ltd	http://www.ibscreen.com/natural.shtml
KEGG ENVIRON Database		Kanehisa Laboratories	http://www.kegg.jp/kegg/drug/environ.html
LipidBank	>7,000	Japanese Conference on the Biochemistry of Lipids	http://lipidbank.jp/
MarinLit		Royal Society of Chemistry	http://pubs.rsc.org/marinlit/
NuBBE Database (Núcleo de Bioensaios Biosíntese e Ecofisiologia de Produtos Naturais). A database of natural products associated with Brazil.			http://nubbe.iq.unesp.br/portal/nubbedb.html
RIKEN Natural Products Encyclopedia (NPEdia)		RIKEN, Japan	http://npd.riken.jp/npedia/
Super Natural Database II	>325,000	Charitè, Medical Faculty of the Humboldt-University	http://bioinf-applied.charite.de/supernatural_new/index.php
The Pherobase		Dr. Ashraf El-Sayed, HortResearch, Lincoln, New Zealand	http://www.pherobase.com/
Traditional Chinese Medicine Database@Taiwan	>61,000	China Medical University, Taiwan	http://tcm.cmu.edu.tw/
ZINC: A free database of commercially-available compounds for virtual screening		Bioinformatics and Chemical Informatics Research Center (BCIRC)	http://zinc.docking.org/vendor0/npd/index.html

[a] Based, in part, on D.-L. Ma, D. S.-H. Chan and C.-H. Leung, Molecular docking for virtual screening of natural product databases, *Chem. Sci.*, 2011, **2**, 1656–1665, and reproduced with permission from The Royal Society of Chemistry.

[b] Natural products databases are reviewed by M. Füllbeck, E. Michalsky, M. Dunkel and R. Preissner, Natural products: Sources and databases, *Nat. Prod. Rep.*, 2006, **23**, 347–356.

[c] J. Hastings et al., The ChEBI reference database and ontology for biologically relevant chemistry: Enhancements for 2013, *Nucleic Acids Research,* 2013, **41**, Database issue D456–D463. doi:10.1093/nar/gks1146.

Handbook of Compounds with Antiprotozoal Activity Isolated from Plants, R. M. Pérez Gutiérrez, 2007. Data on 616 compounds classified by their natural product class. Publisher: Nova Science Publishers, Hauppauge, NY.

Handbook of Compounds with Cytotoxic Activity Isolated from Plants, R. M. Pérez Gutiérrez, 2007. Data on 1752 compounds classified by their natural product class. Publisher: Nova Science Publishers, Hauppauge, NY.

Handbook of Marine Natural Products, eds. E. Fattorusso, W. H. Gerwick and O. Taglialatela-Scafati, 2012. Publisher: Springer.

Handbook of Natural Toxins [*Handb. Nat. Toxins*] (Vols. 1–8, 1983–1995). Publisher: Marcel Dekker, New York.

Handbook of Naturally Occurring Compounds with Antioxidant Activity in Plants, R. M. Pérez Gutiérrez, 2006. Data on 847 compounds classified by their natural product class. Publisher: Nova Science Publishers, Hauppauge, NY.

Handbook of Pharmaceutical Natural Products, G. Brahmachari, 2010. Publisher: Wiley.

Handbook of Porphyrin Science, eds. K. M. Kadish, K. M. Smith and R. Guilard, 35 Vols., 2010–2014. Publisher: World Scientific, Singapore.

Handbook of Secondary Fungal Metabolites, eds. R. J. Cole, M. A. Schweikert and B. B. Jarvis, 3 Vols., 2003. Publisher: Academic Press/Elsevier. Provides a compilation of chemical, physical, spectral and biological data on secondary fungal metabolites.

Handbook of Vitamins, 5th edn., eds. J. Zempleni, J. W. Suttie, J. F. Gregory III and P. J. Stover, 2014. Publisher: CRC Press.

Handbook on Syntheses of Amino Acids, M. A. Blaskovich, 2010. Publisher: ACS/Oxford University Press.

Herbal Drugs: Ethnomedicine to Modern Medicine, ed. K. G. Ramawat, 2009. Publisher: Springer.

Index of Antibiotics from Actinomycetes, ed. H. Umezawa, 2 Vols., 1967–1978. Publisher: University of Tokyo Press.

Kirk-Othmer Encyclopedia of Chemical Technology [*Kirk-Othmer Encycl. Chem. Technol. (5th Ed.)*], 5th ed., 27 vols., 2004–2007. Publisher: Wiley. ***Ullmann's Encyclopedia of Industrial Chemistry,*** 7th ed., 40 Vols., 2011. Publisher: Wiley. 'Ullmann's' was originally published in German (but is now in English) and has a European/Japanese focus; 'Kirk-Othmer' is in English with a North American bias, and both titles contain much pure chemistry, toxicology and pharmacy.

Konstitution und Vorkommen der organischen Pflanzenstoffe (exklusive Alkaloide), 2nd edn., 1976–1985. Publisher: Birkhäuser Verlag, Basel and Stuttgart. Main volume by W. Karrer (1976); 1st supplement by W. Karrer, E. Cherbuliez and C. H. Eugster (1977); 2nd supplement by W. Karrer, E. Cherbuliez, C. H. Eugster and H. Hürlimann (2 Vols., 1981–1985). Comprehensive record of natural products, excluding alkaloids, to the end of 1966.

Leung's Encyclopedia of Common Natural Ingredients, 3rd edn., I. A. Khan and E. A. Abourashed, 2010. Publisher: Wiley.

Mabberley's Plant-book: A Portable Dictionary of Plants, their Classifications, and Uses, 3rd edn., D. J. Mabberley, 2008. Publisher: Cambridge University Press, Cambridge, U.K. provides information on every family and genus of seed-bearing plant (including gymnosperms) plus ferns and club mosses, combining taxonomic details and uses with English and other vernacular names.

Martindale: The Complete Drug Reference, 38th edn., ed. A. Brayfield, 2 Vols., 2014. Publisher: Pharmaceutical Press, London.

Medical Toxicology of Natural Substances: Foods, Fungi, Medicinal Herbs, Plants and Venomous Animals, D. G. Barceloux, 2008. Publisher: Wiley.

Methods in Enzymology [*Methods Enzymol.*] (Vol. 1–, 1955–). Publisher: Elsevier. An ongoing series with over 500 volumes, each devoted to a specific topic in biochemistry. Earlier volumes contain useful properties/procedures for small molecules of biochemical interest.

Natural Product Reports [*Nat. Prod. Rep.*] (1984–). Publisher: RSC. Review series with timely updates on different classes of natural products, though the coverage depends on the availability of a suitable specialist reviewer at any one time. Many issues start with a very useful current awareness section, 'Hot off the Press'.

Natural Product Updates. Online abstracts of developments in natural product chemistry from selected primary journals. Data provided include: structure diagrams, trivial and taxonomic names, molecular formulae, physical and biological properties. Publisher: RSC.

Natural Products from Plants, 2nd edn., eds. L. J. Cseke, A. Kirakosyan, P. B. Kaufman, S. Warber, J. A. Duke and H. L. Brielmann, 2006. Publisher: CRC Press.

Natural Products: Phytochemistry, Botany and Metabolism of Alkaloids, Phenolics and Terpenes, eds. K. G. Ramawat and J.-M. Mérillon, 2013. Publisher: Springer.

Naturally Occurring Quinones, 2nd edn., R. H. Thomson, 1971. Publisher: Academic Press. The second edition was an extensive reworking of the author's 1957 monograph (Butterworth) and was followed by two updates, ***Naturally Occurring Quinones III: Recent Advances***, 1987 and ***Naturally Occurring Quinones IV: Recent Advances***, 1997, both published by Chapman & Hall, London, U.K. (now available from Springer).

Nucleic Acids and Molecular Biology [*Nucleic Acids Mol. Biol.*] (1987–). Review series. Publisher: Springer.

Peptides from A to Z: A Concise Encyclopedia, H.-D. Jakubke and N. Sewald, 2008. Publisher: Wiley.

Phytochemical Dictionary of the Leguminosae, F. Bisby, 2 vols., 1994. Publisher: International Legume Database & Information Service/Chapman & Hall Chemical Database.

Phytochemical Dictionary: A Handbook of Bioactive Compounds from Plants, 2nd edn., eds. J. B. Harborne, H. Baxter and G. P. Moss, 1999. Publisher: Taylor & Francis. Data on approx. 3000 substances.

Phytochemistry Reviews [*Phytochem. Rev.*] (2002–). Publisher: Springer.

Progress in the Chemistry of Organic Natural Products [*Prog. Chem. Org. Nat. Prod.*] (Vol. 37–, 1979–). Formerly *Fortschritte der Chemie Organischer Naturstoffe* [*Fortschr. Chem. Org. Naturst.*] (Vols. 1–36, 1938–1979; publication suspended 1940–1944). Publisher: Springer. This classic series is often referred to as "Zechmeister" after its founder, Laszlo Zechmeister. Review series on various classes of natural products, with one or more topics covered in each volume.

Römpp Encyclopedia Natural Products, eds. B. Fugmann, S. Lang-Fugmann and W. Steglich, 2000. Publisher: Thieme. An update and revision of the 1997 German edition providing details on the chemical structure, stereochemistry, biological properties, physical properties, biological source and medical applications of approx. 6000 secondary metabolites.

Roth Collection of Natural Products Data: Concise Descriptions and Spectra, eds. L. Roth and G. Rupp, 1995. Publisher: Wiley-VCH, Weinheim, Germany.

Specialist Periodical Reports, Publisher: RSC. A series of one-volume updates on developments in particular areas of research. Approx. 10 titles remain current. The more popular appear annually, other titles are sporadic or discontinued. Those of most interest to natural products specialists are *Amino Acids, Peptides and Proteins* [*Spec. Period. Rep.: Amino Acids Pept. Proteins (1994–)*] (Vol. 25–, 1994–) and *Carbohydrate Chemistry* [*Spec. Period. Rep.: Carbohydr. Chem.*] (Vol. 1–, 1968).

Studies in Natural Products Chemistry [*Stud. Nat. Prod. Chem.*] (Vol. 1–, 1988–). Publisher: Elsevier. A series of books edited by Atta-Ur-Rahman, which from Vol. 21 onwards is devoted to bioactive natural products. Vols. 1–14 not online.

Studies in Plant Science [*Stud. Plant Sci.*] (Vols. 1–9, 1991–2002). Publisher: Elsevier.

Substances naturelles d'origine marine, ed. J.-M. Kornprobst, 2 Vols., 2005. Publisher: Lavosier.

The Alkaloids: Chemistry and Biology [*Alkaloids (San Diego, CA, U.S.)*] (Vol. 50–, 1998–). Formerly *The Alkaloids: Chemistry and Physiology* (Vols. 1–20, 1950–1981) and *The Alkaloids: Chemistry and Pharmacology* (Vols. 21–49, 1983–1997). Publisher: Elsevier. The leading review series devoted to alkaloids.

The Combined Chemical Dictionary on DVD, Publisher: CRC Press. Chemical, physical and structural data on more than 630,000 compounds including 272,000 entries in the *Dictionary of Natural Products* and its subsets. Updated twice-yearly. Also available as a web version *via CHEMnetBASE* (Table 1.1).

The Handbook of Natural Flavonoids, 2 Vols., eds. J. B. Harborne and H. Baxter, 1999. Publisher: Wiley.

The Lipid Handbook with CD-ROM, 3rd edn., eds. F. D. Gunstone, J. L. Harwood and A. J. Dijkstra, 2007. Publisher: CRC Press. A large one-volume reference work in two parts; a 780 pages monograph on lipid chemistry, followed by a 617 pages dictionary, which is a reprint of all lipid entries from the Chapman & Hall Database, also searchable on a CD-ROM version in a format uniform with the main database.

The Merck Index, 15th edn., ed. M. J. O'Neil, 2013. Publisher: RSC. A useful one-volume work containing 10,000 monographs on drugs, laboratory chemicals and common natural products.

The Pesticide Manual: A World Compendium, 16th edn., ed. C. MacBean, 2012. Publisher: British Crop Protection Council. One-volume publication containing monographs on pesticides and agrochemicals, current and obsolete.

The Porphyrin Handbook, 20 Vols., eds. K. M. Kadish, K. M. Smith and R. Guilard, 2000-2003. Publisher: Academic Press/Elsevier.

Traditional Chinese Medicines: Molecular Structures, Natural Sources and Applications, 2nd edn., J. Zhou, G. Xie and X. Yan; ed. G. W. A. Mine, 2004. Publisher: Wiley. Data on approx. 9000 chemicals isolated from approx. 4000 natural sources used in Chinese medicine.

Ullmann's Encyclopedia of Industrial Chemistry see under *Kirk-Othmer Encyclopedia of Chemical Technology.*

Wiley Encyclopedia of Chemical Biology, ed. T. P. Begley, 2009. Publisher: Wiley-Blackwell.

1.2   DATABASES

This list gives details of a selection of natural products databases. Website addresses for these and some other public and commercial databases of natural products are provided in Table 1.1.

AntiBase: The Natural Compound Identifier. Maintained by H. Laatsch. Publisher: Wiley. A database, updated annually, of more than 42,000 natural compounds isolated from microorganisms and higher fungi. Includes descriptive information; spectroscopic data (C-13 NMR (CMR), IR, UV and mass spectra); biological data (pharmacological activity, toxicity); information on origin and isolation and a summary of literature sources. Predicted CMR spectra are supplied for those compounds where no measured spectra are available.

Chemical Entities of Biological Interest (ChEBI). Maintained by the European Bioinformatics Institute (part of the European Molecular Biology Laboratory).

A database with more than 42,000 entries, focusing on small molecules, including natural products, and their chemical and pharmacological properties with links to equivalent databases.

CHEMnetBASE. Publisher: CRC Press (formerly Chapman & Hall). This database has sub-structure search capability and includes the *Dictionary of Natural Products* and two subsets: the *Dictionary of Carbohydrates* and the *Dictionary of Marine Natural Products*. Provides descriptive details (chemical structure, molecular formula, mass, elemental composition and CAS Registry Number); physico-chemical data; and information on source, isolation and biological activity; plus relevant literature references.

KEGG ENVIRON Database. Maintained by the Kanehisa Laboratory, Institute for Chemical Research, Kyoto University Part of the KEGG (Kyoto Encyclopedia of Genes and Genomes) collection of online databases, *KEGG ENVIRON* contains structural and efficacy data on: crude drugs, essential oils and medicinal herbs. There are two related resources for natural products: the ***KEGG COMPOUND Database***. It contains details on natural toxins, and the ***KEGG PATHWAY Database***, which includes maps of metabolic pathways and the biosynthesis of secondary metabolites and other natural products.

LipidBank. Maintained by the Japanese Conference on the Biochemistry of Lipids. A free database of natural lipids including fatty acids, carotenoids, glycerolipids, sphingolipids, steroids and fat-soluble vitamins. Contains more than 7000 molecular structures, lipid names (common name and IUPAC), physico-chemical and spectral information. Literature references include those on chemical syntheses.

MarinLit. Established in the 1970s by J. W. Blunt and M. H. G. Munro, University of Canterbury, New Zealand, and from 2013 maintained by the Royal Society of Chemistry. A database of marine natural products and literature, including structure prediction software based on spectroscopic data.

The Pherobase: Database of Pheromones and Semiochemicals. Developed by Dr Ashraf El-Sayed, New Zealand. The browsable search terms of this free database include animal and plant taxa, semiochemicals and functional groups.

1.3 ONLINE TAXONOMY RESOURCES

Details of two large database resources on taxonomy are cited here and some specific taxonomic databases for plants are listed in Table 1.2. Also see Sections 5.1.1 and 5.1.2.

The NCBI Taxonomy Database (http://www.ncbi.nlm.nih.gov/Taxonomy/). Maintained by the National Center for Biotechnology Information (NCBI), National Library of Medicine, Bethesda, MD, USA. The NCBI Taxonomy Database is a curated classification and nomenclature resource for all of the organisms in the public sequence databases. The NCBI taxonomy database is *not a primary source* for taxonomic or phylogenetic information. Furthermore, the database does not follow a single taxonomic treatise but rather attempts to incorporate phylogenetic and taxonomic knowledge from a variety of sources, including the published literature, web-based databases and the advice of sequence submitters and outside taxonomy experts. Consequently, the NCBI taxonomy database is not a phylogenetic or taxonomic authority and should not be cited as such.

Catalogue of Life (http://www.catalogueoflife.org/). Published by Naturalis Biodiversity Center in collaboration with ITIS (Integrated Taxonomic Information System; http://www.itis.gov/). The *Catalogue of Life is a* quality-assured checklist of more than 1.6 million species of plants, animals, fungi and microorganisms, about 70% of all those known to science.

TABLE 1.2

Some Public Online Taxonomic Databases for Plants[a]

Database Name	Organisation	Website Address	Comments
Index Herbariorum	The New York Botanical Garden	http://sciweb.nybg.org/ science2/ IndexHerbariorum.asp	Lists all herbaria of the world.
Index Nominum Genericorum	International Association for Plant Taxonomy and the Smithsonian Institution	http://botany.si.edu/ing/	A compilation of generic names published for all organisms covered by the International Code of Botanical Nomenclature.
International Plant Names Index	The Royal Botanic Gardens, Kew, The Harvard University Herbaria and the Australian National Herbarium	http://www.ipni.org/ index.html	A database of the names and associated basic bibliographical details of seed plants, ferns and fern allies.
Tropicos	Missouri Botanical Garden	http://www.tropicos. org/	A database with >1.2 million scientific names and >4.0 million specimen records.
World Checklist of Selected Plant Families	The Royal Botanic Gardens, Kew	http://apps.kew.org/ wcsp/home.do	A database with information on the accepted scientific names and synonyms of >170 selected plant families.

[a] Based on R. H. J. Erkens, What every chemist should know about plant names, *Nat. Prod. Rep.*, 2011, 28, 11–14, and reproduced with permission from the Royal Society of Chemistry.

2 The Natural Products Literature
Primary Journals

This chapter gives details of specialist natural products journals (not very numerous), plus the more important journals in biochemistry, biology, medicinal chemistry, organic chemistry, pharmacy, pharmacology and toxicology that may contain important information on natural products. Some general chemistry journals have been omitted from the following list although they may sometimes contain information relevant to natural products. A fuller list of chemistry journals is given in the companion *Organic Chemist's Desk Reference*, second edition.

The following items of information are given:

- Full journal title.
- CASSI (the *Chemical Abstracts Service Source Index*) abbreviated title [in square brackets]. CASSI includes details on all literature sources cited in *Chemical Abstracts* since 1907 plus some sources cited in *Beilstein* and *Chemisches Zentralblatt* back to 1830. CASSI gives an abbreviated title for journals, review series and many reference books and monographs, which may be located using the CASSI search tool http://cassi.cas.org/search.jsp. CASSI abbreviations sometimes include the geographical location of the publisher of a journal and the subtitles of those journals which are divided into named parts or sections. These details are omitted here, apart from a few exceptions.
- Years of publication.
- A statement, if applicable, that a journal does not have volume numbers, together with details of when volume numbers were introduced or discontinued. Volume numbers are given for journals which have seen several changes of title.
- Some indication of subject matter where it is not obvious from the title or where a journal is published in two or more parts.
- Changes of journal and superseded titles.
- Translation journals.
- Name of the publisher of the current title (2015) and online (web) archive or of the publisher of the online (web) archive for a former title (Note 1).
- Information on free online access to the full text of chemistry journals on the web (as of 2015) (Note 2).

ACS Medicinal Chemistry Letters [*ACS Med. Chem. Lett.*] (2010–). Publisher: ACS.
Acta Botanica Sinica and *Acta Botanica Sinica* (*English Translation*). See *Journal of Integrative Plant Biology*.
Acta Chemica Scandinavica [*Acta Chem. Scand.*] (1947–1973, 1989–1999). From 1974–1988 (Vols. 29–42) divided into: Series A [*Acta Chem. Scand. Ser. A*] (physical and inorganic chemistry); and Series B [*Acta Chem. Scand. Ser. B*] (organic chemistry and biochemistry). In 1999, absorbed in part by *Journal of the Chemical Society, Dalton Transactions*, *Journal of the Chemical Society, Perkin Transactions 1* and *Journal of the Chemical Society, Perkin Transactions 2*. See *Journal of the Chemical Society*. Free online full-text archive at http://actachemscand.dk/.
Acta Chimica Sinica. See *Chinese Journal of Chemistry* and *Huaxue Xuebao*.

Acta Crystallographica [*Acta Crystallogr.*] (1948–1967). In 1968, divided into: Section A [*Acta Crystallogr. Sect. A*] (1968–) (current subtitle: foundations of crystallography); and Section B [*Acta Crystallogr. Sect. B*] (1968–) (current subtitle: structural science). Later sections added are: Section C [*Acta Crystallogr. Sect. C*] (1983–) (crystal structure communications), formerly *Crystal Structure Communications* [*Cryst. Struct. Commun.*] (1972–1982), Section D [*Acta Crystallogr. Sect. D*] (1993–) (biological crystallography), Section E [*Acta Crystallogr. Sect. E*] (2001–) (structure reports online); and Section F [*Acta Crystallogr. Sect. F*] (2005–) (structural biology and crystallization communications). (Additional CASSI abbreviated subtitles omitted). Some online free access to recent archives for sections A–F *via* PubMed Central (http://www.ncbi.nlm.nih.gov/pmc/). Publisher: International Union of Crystallography http://journals.iucr.org/.

Acta Pharmaceutica [*Acta Pharm.* (*Zagreb, Croatia*)] (1992–). Formerly *Acta Pharmaceutica Jugoslavia* [*Acta Pharm. Jugoslav.*] (1951–1991). Publisher: Croatian Pharmaceutical Society.

Acta Pharmaceutica Fennica. See *European Journal of Pharmaceutical Sciences*.

Acta Pharmaceutica Nordica. See *European Journal of Pharmaceutical Sciences*.

Acta Pharmaceutica Sinica. See *Yaoxue Xuebao*.

Acta Pharmaceutica Sinica B [*Acta Pharm. Sin. B*] (2011–). Open access. Free online full-text archive. Publisher: Elsevier.

Acta Pharmaceutica Suecica. See *European Journal of Pharmaceutical Sciences*.

Acta Pharmacologica Sinica [*Acta Pharmacol. Sin.*] (Vol. 21–, 2000–). In English. Formerly *Zhongguo Yaoli Xuebao* (1980–1999), in English and Chinese. Free online full-text archive 1980–2004. Publisher: Nature Publishing Group.

Acta Phytotherapeutica [*Acta Phytother.*] (Vols. 1–19, 1954–1972). Absorbed by *Quarterly Journal of Crude Drug Research*. See *Pharmaceutical Biology*.

Acta Poloniae Pharmaceutica [*Acta Pol. Pharm.*] (Vols. 1–3, 1937–1939; Vol. 4, 1947–). Publisher: Polish Pharmaceutical Society, Warsaw, Poland.

Acta Poloniae Pharmaceutica (**English Translation**) [*Acta Pol. Pharm.* (*Engl. Transl.*)] (Vols. 20–29, 1963–1972). No longer published.

Agricultural and Biological Chemistry. See *Bioscience, Biotechnology, and Biochemistry*.

American Journal of Botany (*Am. J. Bot.*) (1914–). Selected online free text archive 1997–2009. Publisher: Botanical Society of America.

Amino Acids [*Amino Acids*] (1991–). Publisher: Springer.

Anales de Quimica [*An. Quim.*] (1968–1979, 1990–1995). From 1980–1989, divided into parts including Series C [*An. Quim. Ser. C*] (organic and biochemical). Became *Anales de Quimica International Edition* [*An. Quim. Int. Ed.*] (1996–1998). No longer published.

Analytical Biochemistry [*Anal. Biochem.*] (1960–). Publisher: Elsevier.

Angewandte Chemie [*Angew. Chem.*] (1988–). From 1888–1941, the title was *Zeitschrift fur Angewandte Chemie* [*Z. Angew. Chem.*]. In German, but in 1962 an International Edition in English [*Angew. Chem. Int. Ed. Engl.*] (1962–) was launched, which in 1998 became *Angewandte Chemie, International Edition* [*Angew. Chem. Int. Ed.*] (1998–). The German and English editions have different volume and page numbers. Vol. 1 of the International edition corresponds to Vol. 74 of the German edition. In 1982 and 1983, miniprint supplements were issued. In 1991, *Angewandte Chemie* absorbed *Zeitschrift für Chemie* [*Z. Chem.*] (1961–1990). Publisher: Wiley.

Annals of Applied Biology [*Ann. Appl. Biol.*] (1914–). Publisher: Wiley-Blackwell/Association of Applied Biologists.

Annalen. See *Liebigs Annalen*.

Annalen der Chemie und Pharmazie. See *Liebigs Annalen*.

Annales Pharmaceutiques Françaises [*Ann. Pharm. Fr.*] (1943–). Formed by a merger of *Journal de Pharmacie et de Chemie* [*J. Pharm. Chim.*] (1842–1942) and *Bulletin des*

Sciences Pharmacologiques [*Bull. Sci. Pharmacol*] (1899–1942). Free online full-text archive 1842–1894 from Gallica (Bibliothèque nationale de France) http://gallica..fr/. Publisher: Elsevier.

Annals of the New York Academy of Science [*Ann. N. Y. Acad. Sci.*] (1877–). Irregular. No issue numbers. Publishers: The New York Academy of Sciences and Wiley.

Antibiotics and Chemotherapy (Basel) [*Antibiot. Chemother. (Basel)*] (Vol. 17, 1971–). Formerly *Antibiotica et Chemotherapia* (Vols. 1–16, 1954–1970). Publisher: S. Karger AG, Basel, Switzerland.

Antibiotiki i Khimioterapiya [*Antibiot. Khimioter.*] (1988–). Formerly *Antibiotiki* [*Antibiotiki (Moscow)*] (1956–1984) and *Antibiotiki i Meditsinskaya Bioteknologiya* [*Antibiot. Med. Biotekhnol.*] (1985–1987). Publisher: Media Sphera, Moscow, Russia.

Antimicrobial Agents and Chemotherapy [*Antimicrob. Agents Chemother.*] (Vol. 1, 1972–). Formerly published annually as *Antimicrobial Agents and Chemotherapy* (1961–1970) and *Antimicrobial Agents Annual* (1960). Free online full-text archive 1972–. Publisher: American Society for Microbiology.

Antonie van Leeuwenhoek [*Antonie van Leeuwenhoek*] (1934–). Publisher: Springer.

Applied and Environmental Microbiology [*Appl. Environ. Microbiol.*] (Vol. 31, 1976–). Formerly *Applied Microbiology* [*Appl. Microbiol.*] (Vols. 1–30, 1953–1975). Free online full-text archive 1953–. Publisher: American Society for Microbiology.

Applied Microbiology and Biotechnology [*Appl. Microbiol. Biotechnol.*] (Vol. 19, 1984–). Formerly *European Journal of Applied Microbiology* [*Eur. J. Appl. Microbiol.*] (Vols. 1–4, 1975–1977) and *European Journal of Applied Microbiology and Biotechnology* [*Eur. J. Appl. Microbiol. Biotechnol.*] (Vols. 5–18, 1978–1983). Publisher: Springer.

Archiv der Pharmazie [*Arch. Pharm. (Weinheim, Ger.)*] (1835–). From 1924 to 1971 known as *Archiv der Pharmazie und Berichte der Deutschen Pharmazeutischen Gesellschaft* [*Arch. Pharm. Ber. Dtsch. Pharm. Ges.*]. Publisher: Wiley.

Archives of Biochemistry and Biophysics [*Arch. Biochem. Biophys.*] (1951–). Formerly *Archives of Biochemistry* [*Arch. Biochem.*] (1942–1951). Publisher: Elsevier.

Archives of Insect Biochemistry and Physiology [*Arch. Insect Biochem. Physiol.*] (1983–). Publisher: Wiley.

Archives of Microbiology [*Arch. Microbiol.*] (Vol. 95, 1974–). Formerly *Archiv für Mikrobiologie* [*Arch. Mikrobiol.*] (Vols. 1–13, 1930–1943; 14–94, 1948–1973). Publisher: Springer.

Archives of Pharmacal Research [*Arch. Pharmacal Res.*] (1978–). Publisher: Pharmaceutical Society of Korea/Springer.

ARKIVOC [*ARKIVOC*] (2000–). Electronic journal. Open access. Free online full-text archive from 2000. Publisher: ARKAT USA, Inc.

Arzneimittel-Forschung [*Arzneim.–Forsch.*] (1951–). Drug research. Publisher: Editio Cantor Verlag, Aulendorf, Germany.

Australian Journal of Chemistry [*Aust. J. Chem.*] (1953–). Superseded *Australian Journal of Scientific Research, Series A* [*Aust. J. Sci. Res. Ser. A*] (1948–1952). Publisher: CSIRO Publishing, Melbourne, Victoria, Australia.

Biocatalysis and Biotransformation [*Biocatal. Biotransform.*] (Vol. 12–, 1995–). Formerly *Biocatalysis* [*Biocatalysis*] (Vols. 1–11, 1987–1994). Publisher: Informa Healthcare, London, U.K.

Biochemical and Biophysical Research Communications [*Biochem. Biophys. Res. Commun.*] (1959–). Publisher: Elsevier.

Biochemical Journal [*Biochem. J.*] (1906–). From 1973 to 1983, alternate issues subtitled *Molecular Aspects* and *Cellular Aspects*. Free online full-text archive. Publisher: Portland Press Ltd., Essex, U.K.

Biochemie und Physiologie der Pflanzen [*Biochem. Physiol. Pflanz.*] (Vols. 161–188, 1970–1993). Formerly *Flora (Jena, 1818–1965)* [*Flora (Jena, 1818–1965)*] (Vols. 1–155, 1818–1965), *Flora (Jena), Abteilung A* [*Flora (Jena), Abt. A*] (Vols. 156–160, 1965–1969). Absorbed by *Journal of Plant Physiology*.

Biochemische Zeitschrift [*Biochem. Z.*] (Vols. 1–317, 1906–1944; Vols. 318–346, 1947–1967). Superseded by *European Journal of Biochemistry*.

Biochemical Society Transactions [*Biochem. Soc. Trans.*] (1973–). Replaced a proceedings section formerly included in *Biochemical Journal*. Publisher: Portland Press Ltd., Essex, U.K.

Biochemical Systematics and Ecology [*Biochem. Syst. Ecol.*] (1974–). Formerly *Biochemical Systematics* [*Biochem. Syst.*] (1973). Publisher: Elsevier.

Biochemistry [*Biochemistry*] (1962–). Publisher: ACS.

Biochimica et Biophysica Acta [*Biochim. Biophys. Acta*] (1947–). Issued in different sections. Publisher: Elsevier.

Biochimie [*Biochimie*] (1971–). Formerly *Bulletin de la Société de Chimie Biologique* [*Bull. Soc. Chim. Biol.*] (1914–1970). Publisher: Elsevier.

Biological & Pharmaceutical Bulletin [*Biol. Pharm. Bull.*] (Vol. 16–, 1993–). Formerly *Journal of Pharmacobio–Dynamics* [*J. Pharmacobio–Dyn.*] (Vols. 1–15, 1978–1992). See also *Chemical & Pharmaceutical Bulletin*. Free online full-text archive. Publisher: The Pharmaceutical Society of Japan, Tokyo, Japan.

Biological Chemistry [*Biol. Chem.*] (Vol. 377, 1996–). Superseded *Biological Chemistry Hoppe-Seyler* [*Biol. Chem. Hoppe-Seyler*] (Vols. 366–377, 1985–1996). Formerly *Zeitschrift für Physiologische Chemie* [*Z. Physiol. Chem.*] (1877–1895) and *Hoppe–Seyler's Zeitschrift für Physiologische Chemie* [*Hoppe-Seyler's Z. Physiol. Chem.*] (1895–1984). Publisher: Walter de Gruyter, New York.

Biological Mass Spectrometry. See *Journal of Mass Spectrometry*.

Biomedical and Environmental Mass Spectrometry. See *Journal of Mass Spectrometry*.

Biomedical and Environmental Sciences [*Biomed. Environ. Sci.*] (1988–). Some free online full-text issues. Founded by the Chinese Center for Disease Control and Prevention/Coulston International Corporation, USA. Publisher: Elsevier.

Biomedical Mass Spectrometry. See *Journal of Mass Spectrometry*.

Bioorganic and Medicinal Chemistry [*Bioorg. Med. Chem.*] (1993–). Publisher: Elsevier.

Bioorganic and Medicinal Chemistry Letters [*Bioorg. Med. Chem. Lett.*] (1991–). Publisher: Elsevier.

Bioorganic Chemistry [*Bioorg. Chem.*] (1971–). Publisher: Elsevier.

Bioorganicheskaya Khimia [*Bioorg. Khim.*] (1975–). In Russian. *Bioorganicheskaia Khimia* is an alternative spelling. There is an English-language translation called *Russian Journal of Bioorganic Chemistry* [*Russ. J. Bioorg. Chem.*] (1993–). Formerly *Soviet Journal of Bioorganic Chemistry* [*Sov. J. Bioorg. Chem. (Engl. Transl.)*] (1975–1992). Publisher: Springer/MAIK Nauka/Interperiodica.

Biopolymers [*Biopolymers*] (1963–). Some volumes have the secondary title *Biopolymers: Peptide Science* (no entry in CASSI). Publisher: Wiley.

Bioscience, Biotechnology, and Biochemistry [*Biosci. Biotechnol. Biochem.*] (Vol. 56–, 1992–). Formerly *Bulletin of the Agricultural Chemical Society of Japan* [*Bull. Agric. Chem. Soc. Jpn.*] (1924–1960); and *Agricultural and Biological Chemistry* [*Agric. Biol. Chem.*] (1961–1991). Free online full-text archive. Publisher: Japan Society for Bioscience, Biotechnology and Agrochemistry.

Biotechnology and Applied Biochemistry [*Biotechnol. Appl. Biochem.*] (Vol. 8, 1986–). Formerly *Journal of Applied Biochemistry* [*J. Appl. Biochem.*] (Vols. 1–7, 1979–1985). Publisher (from 2011): Wiley-Blackwell.

Biotechnology and Bioengineering [*Biotechnol. Bioeng.*] (Vol. 4–, 1962–). Formerly *Journal of Biochemical and Microbiological Technology and Engineering* [*J. Biochem. Microbiol. Technol. Eng.*] (Vols. 1–3, 1959–1961). Publisher: Wiley.

BMB Reports [*BMB Rep.*] (Vol. 41–, 2008–). Formerly *Han'guk Saenghwa Hakhoechi* [*Han'guk Saenghwa Hakhoechi*] (Journal of the Biochemical Society of Korea) (Vols. 1–26, 1968–1993), *Korean Biochemical Journal* [*Korean Biochem. J.*] (Vols. 26–27, 1993– 1994), *Journal of Biochemistry and Molecular Biology* [*J. Biochem. Mol. Biol.*] (Vols. 28–40, 1995–2007). Electronic journal. Open access. Free online full-text archive from 2000. Publisher: Korean Society for Biochemistry and Molecular Biology.

Botanica Marina [*Bot. Mar.*] (1959–). Publisher: Walter de Gruyter, New York.

Bryologist [*Bryologist*] (1898–). Publisher: American Bryological and Lichenological Society.

Bulletin de Liaison–Groupe Polyphenols [*Bull. Liaison–Groupe Polyphenols*] (1971–1992). No longer published.

Bulletin de la Société de Chimie Biologique. See *Biochimie*.

Bulletin des Sciences Pharmacologiques. See *Annales Pharmaceutiques Françaises*.

Canadian Journal of Chemistry [*Can. J. Chem.*] (Vol. 29–, 1951–). Continuation of *Canadian Journal of Research* [*Can. J. Res.*] (1929–1935) and its subsequent Section B [*Can. J. Res. Sect. B*] (1935–1950) (chemical sciences). Free online full-text archive 1951–1997. Publisher: NRC Research Press.

Canadian Journal of Microbiology [*Can. J. Microbiol.*] (1954–). Publisher: NRC Research Press.

Canadian Journal of Plant Science [*Can. J. Plant Sci.*] (Vol. 37, 1957–). Formerly *Canadian Journal of Agricultural Science* [*Can. J. Agric. Sci.*] (Vols. 33–36, 1953–1956). Free online full-text archive 1957–2003. Publisher: NRC Research Press.

Carbohydrate Letters [*Carbohydr. Lett.*] (1994–2001). No longer published.

Carbohydrate Polymers [*Carbohydr. Polym.*] (1981–). Publisher: Elsevier.

Carbohydrate Research [*Carbohydr. Res.*] (1965–). Publisher: Elsevier.

Cell Biology and Toxicology [*Cell Biol. Toxicol.*] (1984–). Publisher: Springer.

Cellular and Molecular Life Sciences [*Cell. Mol. Life Sci.*] (1997–). Formerly *Experientia* [*Experientia*] (1945–1996). Publisher: Springer.

Cereal Chemistry [*Cereal Chem.*] (Vol. 1, 1924–). Formerly *Journal of the American Association of Cereal Chemists* [*J. Am. Assoc. Cereal Chem.*] (Vols. 1–8, 1915–1923). Free online full-text archive 1960–1996. Publisher: AACC International.

ChemBioChem [*ChemBioChem*] (2000–). Publisher: Wiley.

Chemical Biology & Drug Design [*Chem. Biol. Drug Des.*] (2006–). Formerly *Journal of Peptide Research*. Publisher: Wiley.

Chemical Communications (Cambridge) [*Chem. Commun. (Cambridge)*] (1996–). Volume numbering commenced with Vol. 46 in 2010. Formerly *Chemical Communications* [*Chem. Commun.*] (1965–1968); *Journal of the Chemical Society* [Part] *D* [*J. Chem. Soc. D*] (1969–1971); and *Journal of the Chemical Society, Chemical Communications* [*J. Chem. Soc. Chem. Commun.*] (1972–1995). See also *Journal of the Chemical Society* and *Proceedings of the Chemical Society, London*. Publisher: RSC.

Chemical & Pharmaceutical Bulletin [*Chem. Pharm. Bull.*] (Vol. 6–, 1958–). Formerly *Pharmaceutical Bulletin* [*Pharm. Bull.*] (Vols. 1–5, 1953–1957). In 1993, biologically oriented papers were transferred to *Biological & Pharmaceutical Bulletin*. Free online full-text archive. Publisher: The Pharmaceutical Society of Japan, Tokyo, Japan.

Chemico–Biological Interactions [*Chem.–Biol. Interact.*] (1969–). Publisher: Elsevier.

Chemistry & Biodiversity [*Chem. Biodiversity*] (2004–). Publisher: Verlag Helvetica Chimica Acta AG, Zürich, Switzerland/Wiley.

Chemistry Central Journal [*Chem. Cent. J.*] (2007–). Free online full-text. Publisher: Springer.

Chemistry and Physics of Lipids [*Chem. Phys. Lipids*] (1966–). Publisher: Elsevier.

Chemistry Letters [*Chem. Lett.*] (1972–). Free online full-text archive 1972–1988 from J-Stage at http://www.jstage.jst.go.jp/browse/_journallist. Publisher: The Chemical Society of Japan.

Chemistry of Natural Compounds. See *Khimiya Prirodnykh Soedinenii.*

ChemMedChem [*ChemMedChem*] (2006–). See also *Farmaco.* Publisher: Wiley.

Chemosphere [*Chemosphere*] (1972–). Publisher: Elsevier.

ChemPlusChem [*ChemPlusChem*] (Vol. 77, 2012–). Publisher: Wiley.

Chinese Chemical Letters [*Chin. Chem. Lett.*] (1991–). Free online full-text archive 1999–2006. Publisher: Elsevier.

Chinese Journal of Chemistry [*Chin. J. Chem.*] (1990–). Formerly *Acta Chimica Sinica (English Edition)* [*Acta Chim. Sin. (Engl. Ed.)*] (1983–1989). *Chin. J. Chem.* and *Huaxue Xuebao* are two separate journals and *Chin. J. Chem.* does not contain translations from *Huaxue Xuebao.* Publisher: Shanghai Institute of Organic Chemistry (Chinese Academy of Sciences) and Wiley-VCH on behalf of the Chinese Chemical Society.

Chinese Traditional and Herbal Drugs. See *Zhongcaoyao.*

Collection of Czechoslovak Chemical Communications [*Collect. Czech. Chem. Commun.*] (Vols. 1–76, 1929–2011). Succeeded by *ChemPlusChem.*

Comparative Biochemistry and Physiology [*Comp. Biochem. Physiol.*] (Vols. 1–37, 1960–1970). In 1971, divided into *Comparative Biochemistry and Physiology, Part A: Molecular & Integrative Physiology* and *Comparative Biochemistry and Physiology, Part B: Biochemistry & Molecular Biology.*

Comparative Biochemistry and Physiology, Part A: Molecular & Integrative Physiology [*Comp. Biochem. Physiol. Part A: Mol. Integr. Physiol.*] (Vol. 38–, 1971–). Formerly *Comparative Biochemistry and Physiology.* Publisher: Elsevier.

Comparative Biochemistry and Physiology, Part B: Biochemistry & Molecular Biology [*Comp. Biochem. Physiol. Part B: Biochem. Mol. Biol.*] (Vol. 38–, 1971–). Formerly *Comparative Biochemistry and Physiology.* Publisher: Elsevier.

Comparative Biochemistry and Physiology, Part C: Toxicology & Pharmacology [*Comp. Biochem. Physiol. Part C: Toxicol. Pharmacol.*] (Vol. 125C–, 2000–). Formerly *Comparative Biochemistry and Physiology, C: Comparative Pharmacology* [*Comp. Biochem. Physiol. C: Comp. Pharmacol.*] (Vols. 50–73C, 1975–1982) and *Comparative Biochemistry and Physiology, Part C: Pharmacology, Toxicology & Endocrinology* [*Comp. Biochem. Physiol. Part C: Pharmacol., Toxicol. Endocrinol.*] (Vols. 74C–124C, 1983–1999). Publisher: Elsevier.

Comparative Biochemistry and Physiology, Part D: Genomics & Proteomics [*Comp. Biochem. Physiol. Part D: Genomics Proteomics*] (Vol. 1D–, 2006–). Publisher: Elsevier.

Crystal Structure Communications. See *Acta Crystallographica.*

Current Bioactive Compounds [*Curr. Bioact. Compd.*] (2005–). Publisher: Bentham Science Publishers Ltd.

Current Protein & Peptide Science [*Curr. Protein Pept. Sci.*] (2000–). Publisher: Bentham Science Publishers Ltd.

Deutsche Apotheker Zeitung [*Dtsch. Apoth. Ztg.*] [DAZ] (Vol. 49, 1934; Vol. 150, 2010). Formerly *Apotheker Zeitung* [*Apoth. Ztg.*] (Vols. 1–48, 1886–1933) and *Standeszeitung Deutscher Apotheker* [*Standesztg. Dtsch. Apoth.*] (Vols. 48–49, 1933–1934). Incorporated and adopted the volume numbering of *Sueddeutsche Apotheker Zeitung* [*Sueddtsch Apoth. Ztg.*] (Vols. 1–90, 1861–1950). Publisher: Deutscher Apotheker Verlag, Stuttgart, Germany.

Environmental Chemistry [*Environ. Chem.*] (2004–). Publisher: CSIRO Publishing, Victoria, Melbourne, Australia.

Environmental Microbiology [*Environ. Microbiol.*] (1999–). Publisher: Society for Applied Microbiology/Wiley-Blackwell.

Environmental Science and Technology [*Environ. Sci. Technol.*] (1967–). Publisher: ACS.

European Journal of Biochemistry. See *FEBS Journal*.

European Journal of Entomology [*Eur. J. Entomol.*] (Vol. 90–, 1993–). Formerly *Casopis Ceské spolecnosti entomologické* (*Acta Societatis Entomologicae Bohemiae*) (1904–1919); *Casopis Ceskoslovenské spolecnosti entomologické* (*Acta Societatis Entomologicae Cechosloveniae*) (1920–1938); *Casopis Ceské spolecnosti entomologické* (*Acta Societatis Entomologicae Bohemiae*) (1939–1944); *Casopis Ceskoslovenské spolecnosti entomologické* (*Acta Societatis Entomologicae Cechosloveniae*) (1945–1952); *Rocenka Ceskoslovenské spolecnosti entomologické* (*Acta Societatis Entomologicae Cechosloveniae*) (1953–1956); *Casopis Ceskoslovenské spolecnosti entomologické* (*Acta Societatis Entomologicae Cechosloveniae*) (1957–1964); and *Acta Entomologica Bohemoslovaca* [*Acta Entomol. Bohemoslov.*] (Vols. 62–89, 1965–1992). Free online full-text for most issues from 1993. Publishers: Institute of Entomology of the Czech Academy of Sciences and the Czech Entomological Society.

European Journal of Lipid Science and Technology [*Eur. J. Lipid Sci. Technol.*] (Vol. 102–, 2000–). Formerly *Fettchemische Umschau* [*Fettchem. Umsch.*] (Vols. 40–43, 1933–1936), *Fette und Seifen* [*Fette Seifen*] (Vols. 43–51, 1936–1944; Vols. 52–54, 1950–1952), *Fette, Seifen, Anstrichmittel* [*Fette Seifen Anstrichm.*] (Vols. 55–88, 1953–1986), *Fett, Wissenschaft Technologie* [*Fett Wiss. Technol.*] (Vols. 89–97, 1987–1995), *Fett/Lipid* [*Fett/Lipid*] (Vols. 98–101, 1996–1999). Publisher: Wiley.

European Journal of Medicinal Chemistry [*Eur. J. Med. Chem.*] (1974–). Formerly *Chimica Therapeutica* [*Chim. Ther*] (1965–73). Publisher: Elsevier.

European Journal of Medicinal Plants [*Eur. J. Med. Plants*] (2011–). Publisher: SCIENCEDOMAIN International.

European Journal of Organic Chemistry [*Eur. J. Org. Chem.*] (1998–) Formed by the merger of *Bulletin de la Société Chimique de France*; *Bulletin des Sociétés Chimique Belges*; *Liebigs Annalen/Recueil*; and *Gazzetta Chimica Italiana*. No volume numbers. Publisher: Wiley.

European Journal of Pharmaceutical Sciences [*Eur. J. Pharm. Sci.*] (1993–). Formed by a merger of *Acta Pharmaceutica Fennica* [*Acta Pharm. Fennica*] (1977–1992) with *Acta Pharmaceutica Nordica* [*Acta Pharm. Nord.*] (1989–1992). In 2000, absorbed *Pharmaceutica Acta Helvetiae*. *Acta Pharmaceutica Nordica* was formed by a merger of: *Acta Pharmaceutica Suecica* [*Acta Pharm. Suec*] (1964–1988) and *Norvegica Pharmaceutica Acta* [*Norv. Pharm. Acta*] (1983–1986). Publisher: Elsevier.

European Journal of Pharmacology [*Eur. J. Pharmacol.*] (1967–). Publisher: Elsevier.

Experientia. See *Cellular and Molecular Life Sciences*.

Experimental Mycology. See *Fungal Genetics and Biology*.

Experimental and Toxicologic Pathology [*Exp. Toxicol. Pathol.*] (Vol. 44–, 1992–). Formerly *Experimentelle Pathologie* [*Exp. Pathol.*] (Vols. 1–18, 1967–1980) and *Experimental Pathology* [*Exp. Pathol.*] (Vols. 19–43, 1981–1991). Publisher: Elsevier.

Extremophiles [Extremophiles] (1997–). Publisher: Springer.

Farmaco [*Farmaco*] (1989–2005) (Drugs). Incorporates *Farmaco, Edizione Scientifica* [*Farmaco Ed. Sci.*] (1953–1988) and *Farmaco, Edizione Pratica* [*Farmaco Ed. Prat.*] (1953–1988). No longer published. Replaced by *ChemMedChem*.

FEBS Journal [*FEBS J.*] (Vol. 272–, 2005–). Formerly *European Journal of Biochemistry* [*Eur. J. Biochem.*] (Vols. 1–271, 1967–2004). [FEBS = Federation of European Biochemical Societies]. Publisher: FEBS/Wiley-Blackwell.

FEMS Microbiology Letters [*FEMS Microbiol. Lett.*] (1977–). [FEMS = Federation of European Microbiological Societies]. Publisher: Wiley-Blackwell (from 2006). Published by Elsevier 1997–2005.

Fett/Lipid. See *European Journal of Lipid Science and Technology*.

Fisheries Science [*Fish. Sci.*] (Vol. 60–, 1994–). Papers in English. Continues, in part, *Nippon Suisan Gakkaishi*. Publisher: Springer.

Fitoterapia [*Fitoterapia*] (1947–). The Journal for the Study of Medicinal Plants. Formerly *Estratti Fluidi Titolati*. Publisher: Elsevier.

Flavour and Fragrance Journal [*Flavour Fragrance J.*] (1985–). Publisher: Wiley.

Food and Chemical Toxicology [*Food Chem. Toxicol.*] (Vol. 20–, 1982–). Formerly *Food and Cosmetics Toxicology* [*Food Cosmet. Toxicol.*] (Vols. 1–19, 1963–1981). Publisher: Elsevier.

Food Chemistry [*Food Chem.*] (1976–). Publisher: Elsevier.

Fundamental and Applied Toxicology. See *Toxicological Sciences*.

Fungal Biology [*Fungal Biol.*] (Vol. 114–, 2010–). Formerly *Transactions of the British Mycological Society* [*Trans. Br. Mycol. Soc.*] (Vols. 1–91, 1896/1897–1988), *Mycological Research* [*Mycol. Res.*] (Vols. 92–113, 1989–2009). Publisher: Elsevier.

Fungal Genetics and Biology [*Fungal Genet. Biol.*] (Vol. 20–, 1996). Formerly *Experimental Mycology* [*Exp. Mycol.*] (Vols. 1–19, 1977–1995). Publisher: Elsevier.

Glycobiology [*Glycobiology*] (1990–). Publisher: Oxford University Press.

Glycoconjugate Journal [*Glycoconjugate J.*] (1984–). In 1994, absorbed *Glycosylation & Disease* [*Glycosylation Dis.*] (Vol. 1, issues 1–4, 1994). Publisher: Springer.

Harmful Algae [*Harmful Algae*] (2002–). Publisher: Elsevier.

Helvetica Chimica Acta [*Helv. Chim. Acta*] (1918–). Publisher: Verlag Helvetica Chimica Acta AG, Zürich, Switzerland/Wiley.

Herba Hungarica [*Herba Hung.*] (1962–1991). No longer published.

Herba Polonica [*Herba Pol.*] (Vol. 11, 1965–). Formerly *Biuletyn Instytutu Roslin Leczniczych*] [*Biul. Inst. Rosl. Lecz.*] (Bulletin of the Institute for Medicinal Plants) (Vols. 3–10, 1957–1964). Publisher: Branch of Medicinal Plants of the Institute of Natural Fibres & Medicinal Plants, Poznań, Poland.

Heterocycles [*Heterocycles*] (1973–). Publisher: The Japan Institute of Heterocyclic Chemistry/Elsevier.

Holzforschung [*Holzforschung*] (1947–). Publisher: Walter de Gruyter, New York.

Hoppe-Seylers Zeitschrift für Physiologische Chemie. See *Biological Chemistry*.

Huaxue Xuebao [*Huaxue Xuebao*] (Journal of Chemistry) (1953–, suspended 1966–1975). In Chinese. Continues, in part, *Journal of the Chinese Chemical Society* (Peking) [*J. Chin. Chem. Soc.* (*Peking*)] (1933–1952; suspended 1937–1939). Has been called *Acta Chimica Sinica* (*Chinese Edition*) [*Acta Chim. Sin.* (*Chin. Ed.*)]. Until 1981, *Chemical Abstracts* named *Huaxue Xuebao* as *Hua Hsueh Hsueh Pao*. Publisher: China International Book Trading Corp., Beijing, China.

Hydrobiologia [*Hydrobiologia*] (1948–). Publisher: Springer.

Hygeia Journal for Drugs & Medicines [*Hygeia: J. Drugs Med.*] (2009–). Publisher: Dr. Madhu C. Divakar, Coimbatore, Tamil Nadu, India. An unrelated publication *Hygeia* appeared from 1923 to 1950.

Indian Drugs [*Indian Drugs*] (1963–). Publisher: Indian Drug Manufacturers' Association, Mumbai, India.

Indian Journal of Natural Products [*Indian J. Nat. Prod.*] (1985–). Publisher: Society of Pharmacognosy (formerly Indian Society of Pharmacognosy).

Indian Journal of Natural Products and Resources [*Indian J. Nat. Prod. Resour.*] (Vol. 1–, 2010–). Formerly *Natural Product Radiance* (Vols. 1–8, 2002–2009). Publisher: National Institute of Science Communication and Information Resources, New Delhi, India.

Indian Journal of Pharmaceutical Sciences [*Indian J. Pharm. Sci.*] (Vol. 40–, 1978–). Formerly *Indian Journal of Pharmacy* [*Indian J. Pharm.*] (Vols. 1–40, 1939–1978). Free online full-text archive 2006–. Publisher: Medknow Publications, Mumbai, India.

Insect Biochemistry and Molecular Biology [*Insect Biochem. Mol. Biol.*] (Vol. 22–, 1992–). Formerly *Insect Biochemistry* [*Insect Biochem.*] (Vols. 1–21, 1971–1991). Publisher: Elsevier.

International Journal of Biochemistry & Cell Biology [*Int. J. Biochem. Cell Biol.*] (Vols. 27–, 1995–). Formerly *International Journal of Biochemistry* [*Int. J. Biochem.*] (Vols. 1–26, 1970–1994). Publisher: Elsevier.

International Journal of Biological Sciences [*Int. J. Biol. Sci.*] (2005–). Electronic journal. Open access. Free online full-text archive from 2005. Publisher: Ivyspring International Publisher, Toronto, Ontario, Canada.

International Journal of Biology and Biotechnology [*Int. J. Biol. Biotechnol.*] (2004–). Publisher: Z–A Scientific Publisher, Karachi, Pakistan.

International Journal of Crude Drug Research. See *Pharmaceutical Biology*.

International Journal of Medicinal Mushrooms [*Int. J. Med. Mushrooms*] (1999–). Publisher: Begell House, Inc., Redding, CT.

International Journal of Peptide and Protein Research [*Int. J. Pept. Protein Res.*] (1972–1996). Formerly *International Journal of Protein Research* [*Int. J. Protein Res.*] (1969–1971). Merged with *Peptide Research* [*Pept. Res.*] (1988–1996) to form *Journal of Peptide Research*. Online archive publisher: Wiley.

International Journal of Peptide Research and Therapeutics [*Int. J. Pept. Res. Ther.*] (2005–). Formerly *Letters in Peptide Science* [*Lett. Pept. Sci.*] (1994–2003). (Not published in 2004). Publisher: Springer.

International Journal of Pharmacognosy. See *Pharmaceutical Biology*.

IUBMB Life [*IUBMB Life*] (1999–). Formerly *Biochemistry International* [*Biochem. Int.*] (1980–) and *Biochemistry and Molecular Biology International* [*Biochem. Mol. Biol. Int.*] (1993–1999). Publisher: Wiley.

Japanese Journal of Antibiotics. See *Journal of Antibiotics*.

Journal de Pharmacie et de Chimie. See *Annales Pharmaceutiques Français*.

Journal of Agricultural and Food Chemistry [*J. Agric. Food Chem.*] (1953–). Publisher: ACS.

Journal of Antibiotics [*J. Antibiot.*] (1948–). English-language translation of the Japanese language journal *Japanese Journal of Antibiotics* [*Jpn. J. Antibiot.*]. From 1953–1967 published as Series A [*J. Antibiot. Ser. A*]. (English language) and Series B [*J. Antibiot. Ser. B*] (Japanese language). Free online full-text archive 2005–2010. Publisher: Japan Antibiotics Research Association, Tokyo, Japan/Nature Publishing Group/Macmillan Publishers Ltd.

Journal of Applied Bacteriology. See *Journal of Applied Microbiology*.

Journal of Applied Chemistry. See *Journal of Chemical Technology and Biotechnology*.

Journal of Applied Entomology [*J. Appl. Entomol.*] (Vol. 101–, 1986–). Formerly *Zeitschrift für Angewandte Entomologie* [*Z. Angew. Entomol.*] (Vols. 1–30, 1914–1944; Vols. 31–100, 1949–1985). Publisher: Wiley-Blackwell.

Journal of Applied Microbiology [*J. Appl. Microbiol.*] (Vol. 82, 1997–). Formerly *Proceedings of the Society of Agricultural Bacteriologists* [*Proc. Soc. Agric. Bacteriol.*] (Vols. 1–7, 1938–1944), *Proceedings of the Society for Applied Bacteriology* [*Proc. Soc. Appl. Bacteriol.*] (1945–Vol. 16, 1953; volume numbering began with volume 13, 1950), *Journal of Applied Bacteriology* [*J. Appl. Bacteriol.*] (Vols. 17–81, 1954–1996). Publisher: Wiley.

Journal of Applied Toxicology [*J. Appl. Toxicol.*] (Vol. 1–, 1981–). Publisher: Wiley.

Journal of Asian Natural Products Research [*J. Asian Nat. Prod. Res.*] (1998–). Publisher: Taylor & Francis.

Journal of Bacteriology [*J. Bacteriol.*] (Vol. 1–, 1916–). Free online full-text archive 1916– (with exclusion of some recently-published papers). Publisher: American Society for Microbiology, Washington, DC.

Journal of Biochemical and Molecular Toxicology [*J. Biochem. Mol. Toxicol.*] (Vol. 12–, 1998–). Formerly *Journal of Biochemical Toxicology* [*J. Biochem. Toxicol.*] (Vols. 1–11, 1986–1996; not published 1997). Publisher: Wiley.

Journal of Biochemistry [*J. Biochem.* (*Tokyo*)] (1922–). Publisher: The Japanese Biochemical Society/Oxford University Press.

Journal of Biochemistry and Molecular Biology. See *BMB Reports.*

Journal of Biological Chemistry [*J. Biol. Chem.*] (1905–). Free online full-text archive. Publisher: The American Society for Biochemistry and Molecular Biology.

Journal of Bioluminescence and Chemiluminescence. See *Luminescence.*

Journal of Bioscience and Bioengineering [*J. Biosci. Bioeng.*] (Vol. 87–, 1999–). Formerly *Journal of Fermentation Technology* [*J. Ferment. Technol.*] (Vols. 55–66, 1977–1988) and *Journal of Fermentation and Bioengineering* [*J. Ferment. Bioeng.*] (Vols. 67–86, 1989–1998) [derived, in part, from *Hakko Kogaku Zasshi* [*Hakko Kogaku Zasshi*] (Vols. 22–54, 1944–1976), previously *Jozogaku Zasshi* [*Jozogaku Zasshi*] (Vols. 1–22, 1923–1944)]. Publisher: Elsevier.

Journal of Biotechnology [*J. Biotechnol.*] (1984–). Publisher: Elsevier.

Journal of Carbohydrate Chemistry [*J. Carbohydr. Chem.*] (1982–). Successor to *Journal of Carbohydrates, Nucleosides, Nucleotides* [*J. Carbohydr. Nucleosides Nucleotides*] (1974–1981), which was divided into *Journal of Carbohydrate Chemistry* and *Nucleosides & Nucleotides* [*Nucleosides Nucleotides*] (1982–1999) later *Nucleosides, Nucleotides & Nucleic Acids* [*Nucleosides Nucleotides Nucleic Acids*]. Publisher: Taylor & Francis.

Journal of Chemical Ecology [*J. Chem. Ecol.*] (1975–). Publisher: Springer.

Journal of Chemical Research [*J. Chem. Res.*] (2004–). Formerly *Journal of Chemical Research, Miniprint* [*J. Chem. Res. Miniprint*] (1977–2003) (a miniprint/microfiche, full-text version) and *Journal of Chemical Research, Synopsis* [*J. Chem. Res. Synop.*] (1977–2003) (a synopsis version). No volume numbers. Publisher: Science Reviews 2000 Ltd., U.K.

Journal of Chemical Technology and Biotechnology [*J. Chem. Technol. Biotechnol.*] (Vol. 36–, 1986–). Formed by the merger of *Journal of Chemical Technology and Biotechnology, Biotechnology* and *Journal of Chemical Technology and Biotechnology, Chemical Technology.* Publisher: Wiley.

Journal of Chemical Technology and Biotechnology, Biotechnology [*J. Chem. Technol. Biotechnol., Biotechnol.*] (Vols. 33B–35B, 1983–1985) and *Journal of Chemical Technology and Biotechnology, Chemical Technology* [*J. Chem. Technol. Biotechnol., Chem. Technol.*] (Vols. 33A–35A, 1983–1985). Formerly *Journal of Applied Chemistry* [*J. Appl. Chem.*] (Vols. 1–20, 1951–1970), *Journal of Applied Chemistry & Biotechnology* [*J. Appl. Chem. Biotechnol.*] (Vols. 21–28, 1971–1978), *Journal of Chemical Technology and Biotechnology (1979–1982)* [*J. Chem. Technol. Biotechnol. (1979–1982)*] (Vols. 29–32, 1979–1982). Merged to form *Journal of Chemical Technology and Biotechnology.* Online archive publisher: Wiley.

Journal of Economic Entomology [*J. Econ. Entomol.*] (1908–). Publisher: The Entomological Society of America.

Journal of Essential Oil–Bearing Plants [*J. Essent. Oil–Bear. Plants*] (1998–). Publisher: Har Krishan Bhalla & Sons, Dehradun, India.

Journal of Essential Oil Research [*J. Essent. Oil Res.*] (1989–). Publisher: Taylor & Francis.

Journal of Ethnopharmacology [*J. Ethnopharmacol.*] (1979–). Publisher: Elsevier.

Journal of Experimental Botany [*J. Exp. Bot.*] (1950–). Publisher: Oxford University Press.

Journal of Experimental Marine Biology and Ecology [*J. Exp. Mar. Biol. Ecol.*] (1967–). Publisher: Elsevier.

Journal of Experimental Zoology [*J. Exp. Zool.*] (Vols. 1–294, 1904–2002). Online archive publisher: Wiley.

Journal of Experimental Zoology Part A: Ecological Genetics and Physiology [*J. Exp. Zool. Part A*] (Vol. 307A–, 2007–). Formerly *Journal of Experimental Zoology, Part A: Comparative Experimental Biology* [*J. Exp. Zool., Part A*] (Vols. 295A–305A, 2003–2006). Continues in part *Journal of Experimental Zoology*. Publisher: Wiley.

Journal of Experimental Zoology, Part B: Molecular and Developmental Evolution [*J. Exp. Zool. Part B*] (Vol. 295B–, 2003–). Continues in part *Journal of Experimental Zoology*. Publisher: Wiley.

Journal of Fermentation and Bioengineering. See *Journal of Bioscience and Bioengineering*.

Journal of Food Science [*J. Food Sci.*] (Vol. 26–, 1961–). Formerly *Food Research* [*Food Res.*] (Vols. 1–25, 1936–1960). Publisher: Wiley.

Journal of Herbs, Spices & Medicinal Plants [*J. Herbs, Spices Med. Plants*] (1992–). Publisher: Taylor & Francis.

Journal of Insect Physiology [*J. Insect Physiol.*] (1957–). Publisher: Elsevier.

Journal of Integrative Plant Biology [*J. Integr. Plant Biol.*] (Vol. 47–, 2005–). Formerly *Zhiwu Xuebao* [*Zhiwu Xuebao*] (Journal of Botany) (Vols. 1–43, 1952–2001; publication suspended 1967–1972), *Acta Botanica Sinica* [*Acta Bot. Sin.*] (Vols. 44–46, 2002–2004). The first issue of volume 15 was also published as an English-language translation *Acta Botanica Sinica (English Translation)* [*Acta Bot. Sin. (Engl. Transl.)*]. Publisher: Institute of Botany, Chinese Academy of Sciences/Wiley.

Journal of Lipid Mediators and Cell Signalling [*J. Lipid Mediators Cell Signalling*] (1994–1997). Formerly *Journal of Lipid Mediators* [*J. Lipid Mediators*] (1989–1993). Merged with *Prostaglandins* [*Prostaglandins*] (1972–1997) to become *Prostaglandins & Other Lipid Mediators*.

Journal of Lipid Research [*J. Lipid Res.*] (1959–). Free online full-text archive. Publisher: The American Society for Biochemistry and Molecular Biology.

Journal of Luminescence [*J. Lumin.*] (1970–). Publisher: Elsevier.

Journal of Magnetic Resonance [*J. Magn. Reson.*] (1997–, 1969–1992). Formerly divided into Series A [*J. Magn. Reson. Ser. A*] (1993–1996); and Series B [*J. Magn. Reson. Ser. B*] (1993–1996). Publisher: Elsevier.

Journal of Marine Biotechnology. See *Marine Biotechnology*.

Journal of Mass Spectrometry [*J. Mass Spectrom.*] (1995–). Formerly *Organic Mass Spectrometry* [*Org. Mass Spectrom.*] (1968–1994). Incorporates *Biological Mass Spectrometry* [*Biol. Mass Spectrom.*] [1991–1994]: formerly *Biomedical Mass Spectrometry* [*Biomed. Mass Spectrom.*] (1974–1985); and *Biomedical and Environmental Mass Spectrometry* [*Biomed. Environ. Mass Spectrom.*] (1986–1990). Publisher: Wiley.

Journal of Medicinal and Aromatic Plant Sciences [*J. Med. Aromat. Plant Sci.*] (Vol. 18–, 1996–). Formerly *Current Research on Medicinal and Aromatic Plants* [*Curr. Res. Med. Aromat. Plants*] (Vols. 1–17, 1979–1995). Publisher: Central Institute of Medicinal and Aromatic Plants, Lucknow, India.

Journal of Medicinal Chemistry [*J. Med. Chem.*] (1963–). Formerly *Journal of Medicinal and Pharmaceutical Chemistry* [*J. Med. Pharm. Chem.*] (1959–1962). Publisher: ACS.

Journal of Medicinal Plant Research. See *Planta Medica*.

Journal of Medicinal Plants Research [*J. Med. Plants Res.*] (2007–). Free online full-text archive. Publisher: Academic Journals, Lagos, Nigeria.

Journal of Medicinal Plants Studies [*J. Med. Plants Stud.*] (2013–). Free online full-text archive. Publisher: Academic Journals, Lagos, Nigeria.

Journal of Microbiology and Biotechnology [*J. Microbiol. Biotechnol.*] (1991–). Free online full-text archive. Publisher: Korean Society for Microbiology and Biotechnology.

Journal of Molecular Biology [*J. Mol. Biol.*] (1959–). Publisher: Elsevier.

Journal of Natural Medicines [*J. Nat. Med.*] (Vol. 60–, 2006–). Formerly *Yakuyo Shokubutsu To Shoyaku* [*Yakuyo Shokubutsu To Shoyaku*] (Journal of Pharmacognostical Society of Japan) (Vols. 3–4/5, 1949–1950/51), *Shoyakugaku Zasshi* [*Shoyakugaku Zasshi*] (Journal of Pharmacognosy) (Vols. 6–47, 1952–1993), *Natural Medicines* [*Nat. Med. (Tokyo, Jpn.)*] (Vols. 48–59, 1994–2005). Publisher: Springer.

Journal of Natural Products [*J. Nat. Prod.*] (1979–). Formerly *Lloydia* [*Lloydia*] (1938–1978). Publisher: ACS.

Journal of Natural Products [*J. Nat. Prod. (Gorakhpur, India)*] (2008–). Electronic journal. Open access. Free online full-text issues at http://www.JournalOfNaturalProducts.com. Not related to the *Journal of Natural Products* published by ACS.

Journal of Natural Remedies [*J. Nat. Rem.*] (2001–). Publisher: Natural Remedies Pty. Ltd., Bangalore, India.

Journal of Natural Toxins [*J. Nat. Toxins*] (1992–2002). No longer published.

Journal of Organic Chemistry [*J. Org. Chem.*] (1936–). Publisher: ACS.

Journal of Peptide Research [*J. Pept. Res.*] (1997–2005). Formed by the merger of *International Journal of Peptide and Protein Research* and *Peptide Research*. Superseded by *Chemical Biology & Drug Design*. Online archive publisher: Wiley.

Journal of Peptide Science [*J. Pept. Sci.*] (1995–). Publisher: European Peptide Society and Wiley.

Journal de Pharmacie et de Chimie. See *Annales Pharmaceutiques Françaises*.

Journal of Pharmaceutical Sciences [*J. Pharm. Sci.*] (Vol. 50–, 1961–). Formerly *Journal of the American Pharmaceutical Association* [*J. Am. Pharm. Assoc.*] (Vols. 1–49, 1912–1960). Publisher: American Pharmacists Association/Wiley.

Journal of Pharmacy and Pharmacology [*J. Pharm. Pharmacol.*] (1929–). From 1929–1948, the title was *Quarterly Journal of Pharmacy and Pharmacology* [*Q. J. Pharm. Pharmacol.*]. Publisher: Pharmaceutical Press.

Journal of Phytopathology [*J. Phytopathol.*] (Vol. 115–, 1986–). Formerly *Phytopathologische Zeitschrift* [*Phytopathol. Z.*] (Vols. 1–114, 1929–1985; not published 1944–1948). Publisher: Wiley.

Journal of Plant Biochemistry and Biotechnology [*J. Plant Biochem. Biotechnol.*] (1992–). Publisher: Society for Plant Biochemistry and Biotechnology, New Delhi, India/IOS Press, Amsterdam, the Netherlands.

Journal of Plant Growth Regulation [*J. Plant Growth Regul.*] (1982–). Publisher: Springer.

Journal of Plant Physiology [*J. Plant Physiol.*] (Vol. 115–, 1984–). Formerly *Zeitschrift für Botanik* [*Z. Bot.*] (Vols. 1–52, 1909–1965; not published 1945–1951), *Zeitschrift für Pflanzenphysiologie* [*Z. Pflanzenphysiol.*] (Vols. 53–114, 1965–1984). Absorbed *Biochemie und Physiologie der Pflanzen*. Publisher: Elsevier.

Journal of Porphyrins and Phthalocyanines [*J. Porphyrins Phthalocyanines*] (1997–). Publisher: World Scientific Publishing Co., Singapore.

Journal of Protein Chemistry. See *Protein Journal*.

Journal of Steroid Biochemistry and Molecular Biology [*J. Steroid Biochem. Mol. Biol.*]. (1990–). Formerly *Journal of Steroid Biochemistry* [*J. Steroid Biochem.*] (1969–1990). Publisher: Elsevier.

Journal of the American Oil Chemists' Society [*J. Am. Oil Chem. Soc.*] (Vol. 24–, 1947–). Publisher: The American Oil Chemists' Society (AOCS), Urbana, IL.

Journal of the American Pharmaceutical Association. See *Journal of Pharmaceutical Sciences*.

Journal of the Chemical Society [*J. Chem. Soc.*] (Vols. 1–32, 1849–1877: 1926–1965). Vol. 1 also assigned to the year 1848. Volume numbers were discontinued in 1925. From 1849–1862, an alternative title was *Quarterly Journal, Chemical Society* [*Q. J. Chem. Soc.*] (1849–1862). From 1878–1925, issued as *Journal of the Chemical Society, Transactions* [*J. Chem. Soc. Trans.*] (Vols. 33–127, 1878–1925) and *Journal of the Chemical Society, Abstracts* [*J. Chem. Soc. Abstr.*] (Vols. 34–128, 1878–1925). (Odd numbered volume numbers only used for the *Transactions*; even numbered volume numbers only used for the *Abstracts*.) In 1966, divided into parts including Part C [*J. Chem. Soc. C*] (1966–1971) (organic). *Chemical Communications* [*Chem. Commun.*] (1965–69) became Part D [*J. Chem. Soc. D*] (1970–1971). In 1972, Parts C and D were superseded by *Journal of the Chemical Society, Perkin Transactions 1* [*J. Chem. Soc. Perkin Trans. 1*] (1972–2002) (organic and bioorganic); and *Journal of the Chemical Society, Chemical Communications* [*J. Chem. Soc. Chem. Commun.*] (1972–1995) (preliminary communications), respectively. In 1996, *Journal of the Chemical Society, Chemical Communications* became *Chemical Communications (Cambridge)*. In 2003, *Journal of the Chemical Society, Perkin Transactions 1* absorbed into *Organic & Biomolecular Chemistry*. Online archive publisher: RSC.

Journal of the Chemical Society of Japan. See *Nippon Kagaku Kaishi*.

Journal of the Chinese Chemical Society (Peking). See *Huaxue Xuebao*.

Journal of the Chinese Chemical Society (Taipei) [*J. Chin. Chem. Soc. (Taipei)*] (1954–). In English. Free online full-text issues from 1988. Publisher: The Chemical Society, Taipei, Taiwan.

Journal of the Pharmaceutical Society of Japan. See *Yakugaku Zasshi*.

Journal of the Royal Netherlands Chemical Society. See *Recueil des Travaux Chimiques des Pays-Bas*.

Journal of the Science of Food and Agriculture [*J. Sci. Food Agric.*] (1950–). Publisher: The Society of Chemical Industry/Wiley.

Journal of Traditional Medicines [*J. Trad. Med.*] (Vol. 21–, 2004–). Formerly *Wakan Iyaku Gakkaishi* [*Wakan Iyaku Gakkaishi*] (Journal of the Society for Chinese and Japanese Medicine and Pharmacy) (Vols. 1–10, 1984–1993), *Wakan Iyakugaku Zasshi* [*Wakan Iyakugaku Zasshi*] (Journal of Chinese and Japanese Medicine and Pharmacy) (Vols. 11–20, 1994–2003). Free online full-text archive from 2004. Publisher: The Office of Medical and Pharmaceutical Society for WAKAN-YAKU, Institute of Natural Medicine, University of Toyama, Toyama, Japan.

Journal of Venomous Animals and Toxins Including Tropical Diseases [*J. Venomous Anim. Toxins Incl. Trop. Dis.*] (Vol. 8–, 2002–). Formerly *Journal of Venomous Animals and Toxins* [*J. Venomous Anim. Toxins*] (Vols. 1–7, 1995–2001). Free online full-text archive. Publisher: Center for the Study of Venoms and Venomous Animals, São Paulo State University.

Journal of Wood Science [*J. Wood Sci.*] (Vol. 44–, 1998–). Continues the English-language content of *Mokuzai Gakkaishi* [*Mokuzai Gakkaishi*] (Journal of the Wood Research Society) (Vol. 1–, 1955–). Publisher: Springer.

Justus Liebigs Annalen der Chemie. See *Liebigs Annalen*.

Khimiko-Farmatsevticheskii Zhurnal (*Khim.-Farm.Zh.*) (1967–). In Russian. There is an English-language translation called *Pharmaceutical Chemistry Journal* [*Pharm. Chem. J.*] (1967–). Translation published by Springer.

Khimiya Prirodnykh Soedinenii [*Khim. Prir. Soedin.*] (1965–). In Russian. There is an English-language translation called *Chemistry of Natural Compounds* [*Chem. Nat. Compd.*] (1965–). Translation published by Springer.

Korean Journal of Pharmacognosy. See *Saengyak Hakhoechi*.

Lichenologist [*Lichenologist*] (1958–). Publisher: Cambridge University Press.

Liebigs Annalen [*Liebigs Ann.*] (1995–1996). Formerly *Annalen der Pharmacie* [*Ann. Pharm. (Lemgo, Ger.)*] (Vols. 1–32, 1832–1839) and *Justus Liebigs Annalen der Chemie*

[*Justus Liebigs Ann. Chem.*] (1840–1978). Sometimes referred to colloquially as *Annalen*. [Other former titles, not adopted by CASSI, are: *Annalen der Chemie und Pharmacie* [*Ann. Chem. Pharm.*] (1840–1873); *Justus Liebigs Annalen der Chemie und Pharmacie* [*Justus Liebigs Ann. Chem. Pharm.*] (1873–1874)]. In 1979, became *Liebigs Annalen der Chemie* [*Liebigs Ann. Chem.*] (1979–1994) which was superseded by *Liebigs Annalen* [*Liebigs Ann.*] (1995–1996). Volume numbers were used until 1972 (Vol. 766). In 1997, merged with *Recueil des Travaux Chimiques des Pays-Bas* to form *Liebigs Annalen/Recueil* [*Liebigs Ann./Recl.*] (1997). No longer published. In 1998, superseded by **European Journal of Organic Chemistry**. Online archive publisher: Wiley.

Letters in Peptide Science. See *International Journal of Peptide Research and Therapeutics*.

Liebigs Annalen/Recueil. See *Liebigs Annalen*.

Life Sciences [*Life Sci.*] (1962–). Publisher: Elsevier.

Limnology and Oceanography [*Limnol. Oceanogr.*] (1956–). Publisher: American Society of Limnology and Oceanography (ASLO). Free on-line full text archive three years after publication.

Lipids [*Lipids*] (1966–). Publisher: American Oil Chemists' Society/Springer.

Lloydia. See *Journal of Natural Products*.

Luminescence (The Journal of Biological and Chemical Luminescence) [*Luminescence*] (Vol. 14–, 1999–). Formerly *Journal of Bioluminescence and Chemiluminescence* [*J. Biolumin. Chemilumin.*] (Vols. 1–13, 1986–1998). Publisher: Wiley.

Magnetic Resonance in Chemistry [*Magn. Reson. Chem.*] (Vol. 23–, 1985–). Formerly *Organic Magnetic Resonance* [*Org. Magn. Reson.*] (1969–1984). Publisher: Wiley.

Marine Biology (Berlin) [*Mar. Biol. (Berlin)*] (1967–). Publisher: Springer.

Marine Biology Letters [*Mar. Biol. Lett.*] (Vols. 1–5, 1979–1984). Absorbed by *Journal of Experimental Marine Biology and Ecology*.

Marine Biotechnology [*Mar. Biotechnol.*] (Vol. 1–, 1999–). Formed by the merger of *Journal of Marine Biotechnology* [*J. Mar. Biotechnol.*] (Vols. 1–6, 1993–1998) and *Molecular Marine Biology and Biotechnology* [*Mol. Mar. Biol. Biotechnol.*] (Vols. 1–7, 1991–1998). Publisher: Springer.

Marine Drugs [*Mar. Drugs*] (2003–). Electronic journal. Open access. Free online full-text archive from 2003. Publisher: Molecular Diversity Preservation International (MDPI), Basel, Switzerland.

MedChemComm [*MedChemComm*] (2010–). Publisher: RSC.

Medicinal Plants – International Journal of Phytomedicines and Related Industries (2009–). No entry in CASSI. Publisher: IndianJournals.com, New Delhi.

Mokuzai Gakkaishi. See *Journal of Wood Science*.

Molecular Marine Biology and Biotechnology. See *Marine Biotechnology*.

Molecular Pharmacology [*Mol. Pharmacol.*] (1965–). Publisher: American Society for Pharmacology and Experimental Therapeutics.

Molecules [*Molecules*] (1996–). Electronic journal. Open access. Free online full-text archive from 1996. Publisher: Molecular Diversity Preservation International (MDPI), Basel, Switzerland.

Mycologia [*Mycologia*] (Vol. 1–, 1909–). Formed by the merger of *Journal of Mycology* (Vols. 1–14, 1895–1980) (no entry in CASSI) and *Mycological Bulletin* (Vols. 2–6, 1904–1908) (Formerly *Ohio Mycological Bulletin* (Vol. 1, 1903) (no entries in CASSI). Publisher: Mycological Society of America.

Mycological Research. See *Fungal Biology*.

Mycopathologia [*Mycopathologia*] (Vols. 1–4, 1938–1949; Vol. 55–, 1975–). Published as *Mycopathologia & Mycologia Applicata* (Vols. 5–54, 1950–1974). Publisher: Springer.

Mycotoxin Research [*Mycotoxin Res.*] (1985–). Publisher: Springer.

Mycotoxins [*Mycotoxins*] (1975–) (Adopted volume numbering in 2000 starting from Vol. 50). Early issues named *Maikotokishin (Tokyo)* [*Maikotokishin (Tokyo)*] or *Proceedings of the Japanese Association of Mycotoxicology* [*Proc. Jpn. Assoc. Mycotoxicol.*]. Free online full-text archive from J-Stage at http://www.jstage.jst.go.jp/browse/myco. Publisher: Japanese Association of Mycotoxicology (Maikotokishin Kenkyukai), Tokyo, Japan.

Natural Medicines. See *Journal of Natural Medicines*.

Natural Product Communications [*Nat. Prod. Commun.*] (2006–). Publisher: Natural Product, Inc., Westerville, OH.

Natural Product Letters. See *Natural Product Research*.

Natural Product Research [*Nat. Prod. Res.*] (2003–2006). From 2006, issued as Part A [*Nat. Prod. Res. Part A*] and Part B [*Nat. Prod. Res. Part B*]. Formerly *Natural Product Letters* [*Nat. Prod. Lett.*] (1992–2002). Publisher: Taylor & Francis.

Natural Product Sciences [*Nat. Prod. Sci.*] (1995–). Publisher: Korean Society of Pharmacognosy, Seoul, South Korea.

Natural Products: An Indian Journal [*Nat. Prod.*] (2005–). Publisher: Trade Science Inc.

Natural Product Research [*Nat. Prod. Res.*] (2003–). Some issues from 2006– named Part A and Part B. Formerly *Natural Product Letters* [*Nat. Prod. Lett.*] (1992–2002). Publisher: Taylor & Francis.

Natural Product Research and Development. See *Tianran Chanwu Yanjiu Yu Kaifa*.

Natural Products and Bioprospecting [*Nat. Prod. Bioprospect.*] (2011–). Publisher: Springer.

Natural Products Chemistry & Research [*Nat. Prod. Chem. Res.*] (2013–). Publisher: OMICS Publishing Group.

Natural Products Journal [*Nat. Prod. J.*] (2011–). Publisher: Bentham Science.

Natural Toxins [*Nat. Toxins*] (1992–1999). No longer published.

Nature [*Nature (London)*] (1869–). Publisher: Nature Publishing Group/Macmillan Publishers Ltd.

Nature Chemical Biology [*Nat. Chem. Biol.*] (2005–). Publisher: Nature Publishing Group/Macmillan Publishers Ltd.

Nature Chemistry [*Nat. Chem.*] (2009–). Publisher: Nature Publishing Group/Macmillan Publishers Ltd.

Nature Medicine [*Nat. Med. (N. Y.)*] (1995–). Publisher: Nature Publishing Group/Macmillan Publishers Ltd.

Naturwissenschaften [*Naturwissenschaften*] (1913–). Publisher: Springer.

Naunyn-Schmiedeberg's Archives of Pharmacology [*Naunyn-Schmiedeberg's Arch. Pharmacol.*] (Vol. 272–, 1972–). Formerly *Archiv für Experimentelle Pathologie und Pharmakologie* [*Arch. Exp. Pathol. Pharmakol.*] (Vols. 1–109, 1873–1925), *Naunyn-Schmiedebergs Archiv für Experimentelle Pathologie und Pharmakologie* [*Naunyn-Schmiedebergs Arch. Exp. Pathol. Pharmakol.*] (Vols. 110–203, 1925–1944; 204–253, 1947–1966), *Naunyn-Schmiedebergs Archiv für Pharmakologie und Experimentelle Pathologie* [*Naunyn-Schmiedebergs Arch. Pharmakol. Exp. Pathol.*] (Vols. 254–263, 1966–1969), *Naunyn-Schmiedebergs Archiv für Pharmakologie* [*Naunyn-Schmiedebergs Arch. Pharmakol.*] (Vols. 264–271, 1969–1971). Publisher: Springer.

Nippon Suisan Gakkaishi [*Nippon Suisan Gakkaishi*] (Bulletin of the Japanese Society of Fisheries). Papers in Japanese and in English. (Vol. 1–, 1932–). Free online full-text archive 1932– (with exclusion of some recently published papers) from J-Stage at http://www.jstage.jst.go.jp/browse/suisan. See also *Fisheries Science*.

Norvegica Pharmaceutica Acta. See *European Journal of Pharmaceutical Sciences*.

Nucleosides, Nucleotides & Nucleic Acids [*Nucleosides Nucleotides Nucleic Acids*] (Vol. 19–, 2000–). Formerly *Journal of Carbohydrates, Nucleosides, Nucleotides* [*J. Carbohydr.*

Nucleosides Nucleotides] (1974–1981) which was divided into *Nucleosides & Nucleotides* [*Nucleosides Nucleotides*] (Vols. 1–18, 1982–1999) and *Journal of Carbohydrate Chemistry* [*J. Carbohydr. Chem.*]. Publisher: Taylor & Francis.

Nucleic Acids Research [*Nucleic Acids Res.*] (1974–). Free online full-text archive via publisher's website or PubMed Central (http://www.ncbi.nlm.nih.gov/pmc/). Publisher: Oxford University Press.

Open Medicinal Chemistry Journal [*Open Med. Chem. J.*] (2007–). Electronic Journal. Open access. Free online full-text archive from 2007. Publisher: Bentham Open.

Open Natural Products Journal [*Open Nat. Prod. J.*] (2008–). Electronic Journal. Open access. Free online full-text archive from 2008. Publisher: Bentham Open.

Organic & Biomolecular Chemistry [*Org. Biomol. Chem.*] (Vol. 1–, 2003–). Formed by a merger of *Journal of the Chemical Society, Perkin Transactions 1* and *Journal of the Chemical Society, Perkin Transactions 2*. See also *Journal of the Chemical Society*. Publisher: RSC.

Organic Letters [*Org. Lett.*] (1999–). Publisher: ACS.

Organic Magnetic Resonance. See *Magnetic Resonance in Chemistry*.

Organic Mass Spectrometry. See *Journal of Mass Spectrometry*.

Organic and Medicinal Chemistry Letters [*Org. Med. Chem. Lett.*] (2011–2015). Merged with *Chemistry Central Journal*.

Peptide Chemistry. See *Peptide Science*.

Peptide Research. See *International Journal of Peptide and Protein Research*.

Peptide Science [*Pept. Sci.*]. Proceedings of the Japanese Peptide Symposium. (35th–, 1998–). Formerly *Peptide Chemistry* [*Pept. Chem.*] (14th–34th, 1976–1996). Publisher: Protein Research Foundation, Osaka, Japan.

Peptide Science. See *Biopolymers*.

Peptides [*Peptides (Amsterdam, Neth.)*] (2007–). Previous CASSI abbreviations include *Peptides (N. Y.)* (1980–2006). Publisher: Elsevier.

Pesticide Biochemistry and Physiology [*Pestic. Biochem. Physiol.*] (1971–). Publisher: Elsevier.

Pest Management Science [*Pest Manage. Sci.*] (Vol. 56–, 2000–). Formerly *Pesticide Science* [*Pestic. Sci.*] (Vols. 1–55, 1970–1999). Publisher: Society of Chemical Industry/Wiley.

Pharmaceutical Biology [*Pharm. Biol. London, U.K.*] (Vol. 36–, 1998–). Formerly *Quarterly Journal of Crude Drug Research* [*Q. J. Crude Drug Res.*] (Vols. 1–19, 1961–1981), *International Journal of Crude Drug Research* [*Int. J. Crude Drug Res.*] (Vols. 20–28, 1982–1990), *International Journal of Pharmacognosy* [*Int. J. Pharmacogn.*] (Vols. 29–35, 1991–1997). Publisher: Taylor & Francis.

Pharmaceutical Bulletin. See *Chemical and Pharmaceutical Bulletin*.

Pharmaceutical Journal [*Pharm. J.*] (Vols. 55–81, 1896–1908; Vol. 131–, 1933–). (Vols. 55–81 also called 4th series, Vols. 1–27). Formerly *Pharmaceutical Journal and Transactions* (Vols. 1–54, 1841–1895) (Vols. 19–29 also called 2nd series, Vols. 1–11; and Vols. 30–54 were called 3rd series, Vols. 1–25). (No entry in CASSI). From 1909–1933, named *Pharmaceutical Journal and Pharmacist* [*Pharm. J. Pharm.*] (Vols. 82–131). Publisher: Royal Pharmaceutical Society of Great Britain, London, U.K.

Pharmacognosy Magazine [*Pharmacogn. Mag.*] (2005–). Publisher: Medknow Publications, Mumbai, India.

Pharmacological Research [*Pharmacol. Res.*] (Vol. 21–, 1989–). Formerly *Pharmacological Research Communications* [*Pharmacol. Res. Commun.*] (Vols. 1–20, 1969–1988). Publisher: Elsevier.

Pharmazeutische Zeitung [*Pharm. Ztg.*] (1855–). See also *PZ Prisma*. Publisher: Govi-Verlag Pharmazeutischer Verlag, Eschborn, Germany.

Pharmaceutica Acta Helvetiae [*Pharm. Acta Helv.*] (1926–2000). Absorbed by *European Journal of Pharmaceutical Sciences.*

Pharmazie [*Pharmazie*] (1946–). Publisher: Govi-Verlag Pharmazeutischer Verlag, Eschborn, Germany.

Phycological Research [*Phycol. Res.*] (Vol. 43–, 1995–). Formerly *Japanese Journal of Phycology* [*Jpn. J. Phycol.*] (Vols. 26–42, 1978–1994). Publisher: Wiley-Blackwell.

Physiologia Plantarum [*Physiol. Plant.*] (1948–). Publisher: Wiley-Blackwell.

Phytochemical Analysis [*Phytochem. Anal.*] (1990–). Publisher: Wiley.

Phytochemistry [*Phytochemistry*] (1961–). Publisher: Elsevier.

Phytochemistry Letters [*Phytochem. Lett.*] (2008–). Publisher: Phytochemical Society of Europe/Elsevier.

Phytochemistry Reviews [*Phytochem. Rev.*] (2002–). Publisher: Springer.

Phytomedicine [*Phytomedicine*] (1994–). Publisher: Elsevier.

Phytopathologische Zeitschrift. See *Journal of Phytopathology.*

Phytopathology [*Phytopathology*] (1911–). Publisher: The American Phytopathological Society.

Phytotherapie [*Phytotherapie*] (Vol. 1–, 2003–). Supersedes *Cahiers de Phytotherapie.* Publisher: Springer.

Phytotherapy Research [*Phytother. Res.*] (1987–). Publisher: Wiley-Blackwell.

Planta [*Planta*] (1925–). Publisher: Springer.

Plant Cell [*Plant Cell*] (1989–). Publisher: American Society of Plant Biologists.

Plant Cell Reports [*Plant Cell Rep.*] (Vol. 1–, 1981–). Publisher: Springer.

Plant Physiology [*Plant Physiol.*] (1926–). Publisher: American Society of Plant Biologists.

Plant Science (***Lucknow, India***) [*Plant Sci. (Lucknow, India)*] (1969–). Publisher: Association for Advancement of Plant Sciences (India).

Plant Science (***Shannon, Ireland***) [*Plant Sci. (Shannon, Irel.)*] (Vol. 38–, 1985–). Formerly *Plant Science Letters* [*Plant Sci. Lett.*] (Vols. 1–37, 1973–1985). Publisher: Elsevier.

Planta Medica [*Planta Med.*] (1953–). Sometimes referred to as *Journal of Medicinal Plant Research: Planta Medica* [*J. Med. Plant Res.: Planta Med.*]. Publisher: Thieme.

Planta Medica Letters [*Planta Med. Lett.*] (2014–). Electronic journal. Open access. Publisher: Thieme.

Proceedings of the Chemical Society, London [*Proc. Chem. Soc, London*] (1885–1914, 1957–1964). Superseded by *Chemical Communications* [*Chem. Commun.*] (1965–1969). From 1915–1956 there was a Proceedings section in *Journal of the Chemical Society.* See also *Chemical Communications (Cambridge).*

Proceedings of the National Academy of Sciences of the United States of America [*Proc. Natl. Acad. Sci. U.S.A.*] (1863–). Free online full-text archive (online issues available six months after the print publication; some current full-text content also free online). Publisher: National Academy of Sciences, USA.

Prostaglandins & Other Lipid Mediators [*Prostaglandins Other Lipid Mediators*] (1998–). Formed by a merger of *Prostaglandins* [*Prostaglandins*] (1972–1997) and *Journal of Lipid Mediators and Cell Signalling* [*J. Lipid Mediators Cell Signalling*] (1994–1997). Publisher: Elsevier.

Protein Journal [*Protein J.*] (Vol. 23–, 2004). Formerly *Journal of Protein Chemistry* [*J. Protein Chem.*] (Vols. 1–22, 1982–2003). Publisher: Springer.

Protein & Peptide Letters [*Protein Pept. Lett.*] (1994–). Publisher: Bentham Science Publishers Ltd.

Pure and Applied Chemistry [*Pure Appl. Chem.*] (I960–). Free online full-text archive. Publisher: IUPAC.

PZ Prisma [*PZ Prisma*] (Vol. 1–, 1994–). Formerly *PZ Wissenschaft* [*PZ Wiss.*] (1988–1993), which replaced a 'Scientific Edition' section of *Pharmazeutische Zeitung* published 1983–1987. Publisher: Govi-Verlag Pharmazeutischer Verlag, Eschborn, Germany.

Quarterly Journal of Crude Drug Research. See *Pharmaceutical Biology*.

Records of Natural Products [*Rec. Nat. Prod.*] (2007–). Electronic journal. Open access. Free online full-text archive from 2007. Publisher: Academy of Chemistry of Globe, Turkey.

Recueil des Travaux Chimiques des Pays-Bas [*Recl. Trav. Chim. Pays-Bas*] (1882–1996). Also known as *Journal of the Royal Netherlands Chemical Society* [*J. R. Neth. Chem. Soc.*]. From 1897–1919, the title was *Recueil des Travaux Chimiques des Pays–Bas et de la Belgique* [*Recl. Trav. Chim. Pays-Bas Belg.*], and from 1980–1984, the title was *Recueil: Journal of the Royal Netherlands Chemical Society* [*Recl.: J. R. Neth. Chem. Soc.*]. No longer published. Merged with *Chemische Berichte* [*Chem. Ber.*] to form *Chemische Berichte/Recueil* and with *Liebigs Annalen* [*Liebigs Ann.*] to form *Liebigs Annalen/Recueil*.

Regulatory Peptides [*Regul. Pept.*] (1980–). Publisher: Elsevier.

Research Journal of Medicinal Plant [*Res. J. Med. Plant*] (2007–). Free online full-text archive. Publisher: Academic Journals Inc., New York.

Research Journal of Phytochemistry [*Res. J. Phytochem.*] (2007–). Free online full-text archive. Publisher: Academic Journals Inc., New York.

Revista Brasileira de Farmacognosia [*Rev. Bras. Farmacogn.*] (1986–). Free online full-text archive. Publisher: Elsevier.

Russian Journal of Bioorganic Chemistry. See *Bioorganicheskaya Khimia*.

Saengyak Hakhoechi [*Saengyak Hakhoechi*] (Journal of the Society of Pharmacognosy) (1970–). Publisher: The Korean Society of Pharmacognosy.

Science [*Science (Washington, DC)*] (1883–). Publisher: American Association for the Advancement of Science http://www.sciencemag.org/.

Scientia Pharmaceutica [*Sci. Pharm.*] (1930–). Free online full-text archive from 2006. Publisher: The Austrian Journal of Pharmaceutical Sciences.

Shoyakugaku Zasshi. See *Journal of Natural Medicines*.

Soviet Journal of Bioorganic Chemistry. See *Bioorganicheskaya Khimia*.

Spectrochimica Acta [*Spectrochim. Acta*] (1939–1966). From Vol. 23, divided into parts including Part A [*Spectrochim. Acta Part A*] (1967–) (molecular spectroscopy; from 1995, subtitle is molecular and biomolecular spectroscopy.

SpringerPlus [*SpringerPlus*] (2012–). Chemistry and Materials Science Section. Publisher: Springer.

Steroids [*Steroids*] (1963–). Publisher: Elsevier.

Tetrahedron [*Tetrahedron*] (1957–). From 1958 to 1962, more than one volume number was issued each year: 1957, Vol. 1, 1958; Vols. 2–4, 1959; Vols. 5–7, 1960; Vols. 8–11, 1961; Vols. 12–16, 1962; Vols. 17–18, 1963; Vol. 19 *et seq.* 2011, Vol. 67. Publisher: Elsevier.

Tetrahedron: Asymmetry [*Tetrahedron: Asymmetry*] (1990–). Publisher: Elsevier.

Tetrahedron Letters [*Tetrahedron Lett.*] (1959–). In the print edition, volume numbers were first used in 1980 (Vol. 21). In the online edition, volume numbers were assigned from 1959/60 (called volume 1). The 48 issues for 1959–1960 were paginated separately, and the issue numbering differs between the print and online editions: print edition, issues 1–21 (1959); issues 1–27 (1960); online edition, issues 1–21 (1959); issues 22–48 (1960). Publisher: Elsevier.

The Bryologist. See *Bryologist*.

The International Journal of Biochemistry & Cell Biology. See *International Journal of Biochemistry & Cell Biology.*

The Protein Journal. See *Protein Journal.*

Tianran Chanwu Yanjiu Yu Kaifa [*Tianran Chanwu Yanjiu Yu Kaifa*] (Natural Products Research and Development) (1989–). Publisher: Tianran Chanwu Yanjiu Yu Kaifa Bianjibu, Chengdu, China.

Toxicological and Environmental Chemistry [*Toxicol. Environ. Chem.*] (Vol. 3, issue 3/4–, 1981–). Formerly *Toxicological and Environmental Chemistry Reviews* [*Toxicol. Environ. Chem. Rev.*] (Vols. 1–3, 1972–1980). Publisher: Taylor & Francis.

Toxicological Sciences [*Toxicol. Sci.*] (Vol. 41–, 1998–). Formerly *Fundamental and Applied Toxicology* [*Fundam. Appl. Toxicol.*] (Vols. 1–40, 1981–1997). Publisher: Oxford University Press.

Toxicology [*Toxicology*] (1973–). Publisher: Elsevier.

Toxicology Letters [*Toxicol. Lett.*] (1977–). Publisher: Elsevier.

Toxicon [*Toxicon*] (1962–). Publisher: Elsevier.

Toxin Reviews [*Toxin Rev.*] (Vol. 24–, 2005–). Formerly *Journal of Toxicology, Toxin Reviews* [*J. Toxicol. Toxin Rev.*] (Vols. 1–23, 1982–2004). Publisher: Taylor & Francis.

Transactions of the British Mycological Society. See *Fungal Biology.*

Wakan Iyaku Gakkaishi. See *Journal of Traditional Medicines.*

Wakan Iyakugaku Zasshi. See *Journal of Traditional Medicines.*

Weed Research [*Weed Res.*] (1961–). Publisher: Wiley-Blackwell.

Weed Science [*Weed Sci.*] (1968–). Publisher: Weed Science Society of America.

World Journal of Microbiology & Biotechnology [*World J. Microbiol. Biotechnol.*] (Vol. 6–, 1990–). Formerly *MIRCEN Journal of Applied Microbiology and Biotechnology* [*MIRCEN J. Appl. Microbiol. Biotechnol.*] (Vols. 1–5, 1985–1989). [MIRCEN = Microbiological Resources Centres]. Publisher: Springer.

Xenobiotica [*Xenobiotica*] (1971–). Publisher: Taylor & Francis Group.

Yakugaku Zasshi [*Yakugaku Zasshi*] (Journal of Pharmacy) (1881–). Also known as *Journal of the Pharmaceutical Society of Japan.* In Japanese. No English-language translation is available. Free online full-text archive from 1881. Publisher: The Pharmaceutical Society of Japan, Tokyo, Japan.

Yaoxue Xuebao [*Yaoxue Xuebao*] (Pharmaceutical Journal) (1953–). Also known as *Acta Pharmaceutica Sinica.* Not to be confused with *Acta Pharmaceutica Sinica B.* Publisher: Yaoxue Xuebao Bianjibu.

Yunnan Zhiwu Yanjiu [*Yunnan Zhiwu Yanjiu*] (Yunnan Botanical Research) (1979–). Also known as *Acta Botanica Yunnanica.* Publisher: Yunnan Plant Research Bianjibu.

Zeitschrift für Angewandte Chemie. See *Angewandte Chemie.*

Zeitschrift für Chemie. See *Angewandte Chemie.*

Zeitschrift für Naturforschung [*Z. Naturforsch.*] (1946). In 1947, divided into Teil A [*Z. Naturforsch. A*] (1947–) (physical sciences); and Teil B [*Z. Naturforsch. B*] (1947–) (chemical sciences); to which was later added Teil C [*Z. Naturforsch. C*] (1973–) (biosciences— previously included in Teil B). (Additional CASSI abbreviated subtitles omitted). Free online full-text archive 1946–2011. Publisher: Verlag der Zeitschrift für Naturforschung, Tübingen, Germany.

Zeitschrift für Pflanzenphysiologie. See *Journal of Plant Physiology.*

Zeitschrift für Physiologische Chemie. See *Biological Chemistry.*

Zhiwu Xuebao. See *Journal of Integrative Plant Biology.*

Zhongcaoyao [*Zhongcaoyao*] (Chinese Herbal Medicine) (Vol. 11–, 1980–). Formerly *Chung Ts'ao T'ung Hsin.* Publisher: Zhongcaoyao Zazhi Bianjibu.

Zhongguo Yaoli Xuebao. See *Acta Pharmacologica Sinica.*

2.1 ELECTRONIC SOURCES FOR CHEMISTRY JOURNALS

With very few exceptions, all the current printed chemistry journals are available online on the web, and some recent additions to the chemistry literature are only available electronically. For the majority of titles, access to the online full text of a journal and its archive is by paid subscription, but for an increasing number of chemistry journals, free online open access to current issues and full-text archives, either partial or complete, is now allowed. For some authors, open access of published papers on the web may be a condition of funding of their research.

Tables of contents for the current and archival issues of chemistry journals are free online, and usually abstracts are also provided by the publisher. Search engines give details of the internet addresses (URLs) for chemistry journals, and there are also websites which provide hyperlinks to most of the chemistry journals currently online, for example Cambridge University's Department of Chemistry website http://www.ch.cam.ac.uk/c2k/. Free full-text chemistry journals on the web are listed on the Belarusian State University website http://www.abc.chemistry.bsu.by/current/fulltext. htm. (The content and permanence of any website cannot be guaranteed and internet addresses are subject to modification.)

2.2 LEADING PUBLISHERS OF CHEMISTRY JOURNALS AND CHEMICAL INFORMATION

Publisher	Abbreviated Name	Internet Address
American Chemical Society, Washington, DC	ACS	http://pubs.acs.org/
Elsevier, Oxford, U.K. (and elsewhere)	Elsevier	http://www.sciencedirect.com/science
John Wiley & Sons Inc., Hoboken, NJ. (Includes VCH; Wiley-Blackwell; Wiley InterScience; Wiley-VCH)	Wiley or Wiley-Blackwell	http://eu.wiley.com/WileyCDA/ Section/index.html
MAIK Nauka/Interperiodica		http://www.maik.rssi.ru/
Royal Society of Chemistry, Cambridge, U.K.	RSC	http://www.rsc.org/
Springer Publishing Company, New York and Berlin	Springer	http://www.springerpub.com/
Taylor & Francis Group, London, U.K. (includes CRC Press)	Taylor & Francis	http://www.taylorandfrancisgroup. com/
Thieme Publishing Group, Stuttgart, Germany	Thieme	http://www.thieme-chemistry.com/

3 Nomenclature

The complexity of natural product nomenclature means that precise rules cannot be given, only guiding principles and examples. Descriptions of the nomenclature of individual types of natural products are given throughout Chapter 6. This chapter deals with general points. See also IUPAC Commission on *Nomenclature of Organic Chemistry*, revised Section F, Natural Products and Related Compounds, available from http://www.cchem.qmul.ac.uk/iupac/section F/.

A fuller treatment of general organic nomenclature is given in the companion *Organic Chemist's Desk Reference*, 2nd edn., Caroline Cooper, ed., Chapters 3–6, and references therein.

3.1 GENERAL

Types of names that can be used for natural products comprise *functional parent names*, *trivial names*, *systematic names*, *semisystematic names* and *semitrivial names*, which are described in the following text. These categories overlap.

IUPAC basic principles should always be followed whenever possible when naming natural products. A few deviations are customarily made, however, to simplify nomenclature and bring related compounds together. For example, compounds containing lactone and carboxylic groups are often named in breach of the IUPAC principle that a compound should have only one principal functional group.

3β,14β-Dihydroxy-5β-card-20-enolid-19-oic acid (Cannogeninic acid)

Some other breaches of good IUPAC nomenclature practice are found in the literature. For example, natural anthraquinones were often numbered by authors so as to make the usual carbon substituent invariably 2-, overriding the correct IUPAC numbering (not recommended). Other examples are highlighted in Chapter 6.

Chrysophanol
4,5-Dihydroxy-2-methylanthraquinone
(older literature)
1,8-Dihydroxy-3-methylanthraquinone
(correct IUPAC name)

3.2 FUNCTIONAL PARENT NAMES

Some well-defined groups such as carbohydrates and peptides are named on the basis of stem names containing functional groups (e.g. glucose, alanine). These are called *functional parents* and are used universally. They are not described in detail here. CAS gives precedence to these names and may often name compounds using functional parent names in preference to other systematic names if the molecule contains the functional parent structure.

Valine   2-Amino-3,3-dimethylbutanoic acid
= 3-Methylvaline (CAS)

3.3 TRIVIAL NAMES

Early natural products could only be named trivially. Trivial names are those which convey no structural information. For complex molecules they may be preferable to either systematic or semi-systematic names.

Allocation of a trivial name to a new natural product is recommended. Failure to allocate one can make the retrieval of subsequent information more difficult as it may be lost sight of amongst the documentation of large numbers of synthetic and semisynthetic analogues. Severe problems can also be caused if the structure is subsequently revised.

Most researchers follow this policy, which deviates from the IUPAC recommendation that a trivial name should be introduced only when the structure is unknown. Many new complex natural products now just receive a trivial name followed in due course by a CAS systematic name.

3.3.1 DERIVATION

Trivial names are mostly derived from the Latin binomial of the producing species, and this is the route recommended by IUPAC.

A small early step towards systematisation was the universal adoption of the suffix *-in(e)* to denote an alkaloid. However, the -in(e) suffix is not confined to alkaloids.

IUPAC has made recommendations to extend the regularisation of suffixes, e.g. -une to denote a natural product of unknown structure and for which the identity of the functional groups has not been established. However, these recommendations are hardly ever followed in practice.

Some natural product names are not derived from binomials:

Morphin(e), from Morpheus, Greek god of dreams (1816)
Baiyunoside, from the Chinese crude drug Bai-Yun-Shen (1985)
Almazole A, from an unidentified seaweed collected at Almadies, Senegal (1994)
Louisfieserone, after organic chemist Louis Fieser (1980)
Musettamycin, Rudolfomycin and *Marcellomycin*, after characters in *La Boheme* (1978)

Trivial names very often contain affixes or suffixes indicating what functional groups they contain, but they should *not* contain fragments, which describe functions that they do *not* contain.
For example, the following are confusing:

Theobromine. From *Theobroma* spp. (1841). Contains no bromine.
Epibatidine. From *Epipedobates anthonyi* (1992). Not an epimer. Name sometimes given as *epi*-Batidine, but there is no such compound as Batidine.
Spirotropone. From *Spirotropis longifolia* (2012). Not a spiro compound, nor a tropone.

Series of compounds from the same source are often differentiated by letter and/or number sequences, for example Exemplarines A, B, C, ... or Exemplarines 1, 2, 3, ... Practice differs as to whether the first of the series is called Exemplarine and the second member Exemplarine A, or whether the first is called Exemplarine A and the second Exemplarine B. The latter is preferable. Numerical suffixes are often used, as in Gibberellins A_1–A_{131} (successive members of this series have been added by various workers over a long period). Suffixes are normally allocated in alphanumeric order of characterisation, but there are exceptions. The suffixes in Usambaridines/Dihydrousambaridines Br, Vi and Vc are their colour reactions.

3.3.2 TERMINAL -E AND OTHER TYPOGRAPHY

In German and some other languages, the terminal -e of, e.g. alkaloid names is not used, for example, the German name for Strychnine is Strychnin. The presence or absence of this -e is trivial and two names found in the literature, one with and one without the -e are considered duplicates. The -e may be substituted by -a in some other languages, e.g. Italian.
Non-alphanumeric typography has occasionally appeared:

Adouétines (various suffix letters), presumably from French *adoucissement*, alleviation of pain (1963)
Palau' amine (Hawaiian; from a sponge collected at Palau (Belau) (1993)

These names can be simplified to Adouetines and Palauamine.

3.3.3 DUPLICATION OF TRIVIAL NAMES

There are a considerable number of duplications and higher multiples of trivial names in the literature.

Densiflorine	Alkaloid from *Fumaria densiflora* (1983)
	Another alkaloid from *Fumaria densiflora* (1984)
Odoratin(e)	Alkaloid from *Viola odorata* (1961)
	Sesquiterpenoid from *Hymenoxys odorata* (1968)
	Nortriterpenoid from *Cedrela odorata* (1972)
	Chalcone from *Eupatorium odoratum* (1973)
	Isoflavonoid from *Dipteryx odorata* (1974)
	Another isoflavonoid from *Dipteryx odorata* (1979)

Exemplarine/Exemplarine A should be considered a duplication if the names refer to different substances.

3.3.4 PREMATURE PUBLICATION

Some published names refer to names assigned by the authors at the isolation stage, before they went on to show that the compounds were, in fact, identical with known (often very well-known) compounds. These names should not be published. This partly accounts for the fact that Oleanolic acid, for example has at least 18 trivial synonyms.

3.3.5 SUBSEQUENT MODIFICATION

Once published, a trivial name should not be subsequently modified. An example is Bryostatin 3. The original work assigned incorrect stereochemistry at C-20. When this error was corrected (1991), the original compound was renamed 20-Epibryostatin 3 and the newly-isolated C-20 epimer was called Bryostatin 3. This is confusing.

3.4 SYSTEMATIC NAMES

All natural products whose structures are known can be named systematically using IUPAC principles. However, it is possible to arrive at more than one valid name for all but the simplest molecules, for example using different available software packages. CAS nomenclature is much more rigorous. Every natural product should have one unique CAS name, but except for relatively simple natural products systematic names, including CAS names, are too cumbersome for convenient use, a fact that is recognized in the IUPAC recommendations.

A much fuller treatment of systematic nomenclature is given in the companion *Organic Chemist's Desk Reference*.

3.5 SEMISYSTEMATIC NAMES

These are names based on semisystematic parents, e.g. Labdane, Ergoline.

Labdane

For naming many natural products, semisystematic names have several advantages over systematic names:

1. The numbering of side chains such as labdane C-11 to C-16 is not covered by systematic numbering.
2. Introduction of different functional groups, e.g. an *OH* group into Labdane, will in general change the numbering in a systematic name.
3. Biogenetically related skeletons can be numbered in a consistent way to bring out underlying biosynthetic relationships.

The boundary between systematic and semisystematic names is one of historical precedence and convenience. Compounds derived from common skeletons such as cholestane could be named systematically (as cyclopentenophenanthrenes) but cholestane is universally treated in CAS and everywhere else as a stereoparent.

Cholestane (5α- shown)

3.5.1 CHOICE OF SEMISYSTEMATIC PARENT

The 1999 IUPAC recommendations give directions for arriving at a semisystematic parent, together with a list of examples of those in current use. This list can be extended by authors in the field. A semisystematic parent should contain as many common features as possible present in the known members of the group, including stereochemistry. For natural products that depart from the main systematic parent, modifications can then be introduced as outlined in Section 3.5.4.

A newly introduced semisystematic parent should not normally contain functionality such as an oxygen atom. However, existing parents containing such functionality (e.g. Spirostan) are allowed (the ending -an indicating a nonhydrocarbon). A stereoparent may be fully saturated or partly unsaturated; the level of unsaturation should be chosen so as to minimize the need for hydro- or dehydroprefixes for the majority of compounds in the group.

Spirostan (5α, 25R- shown)

Dehydro- positions need not be contiguous.

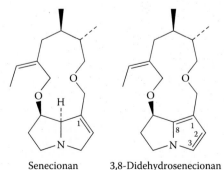

Senecionan   3,8-Didehydrosenecionan

3.5.2 NUMBERING OF SEMISYSTEMATIC PARENTS (IUPAC REFERENCE RF-3.5)

Where a numbering system has already evolved in the literature for a particular group, IUPAC recommends its adoption, provided it covers the numbering of all skeletal atoms. If no scheme pre-exists, IUPAC recommends the adoption of the systematic numbering scheme for the senior fundamental parent of the ring system and that the numbering should start at the same locant as in the systematic skeleton.

However, there are two major complications:

1. Natural product chemists normally number all carbons of a skeleton including atoms at fusion positions. Therefore, although numbering may start at the same locus, the schemes rapidly diverge.
2. In the natural products literature it is customary to follow a biogenetic numbering scheme. Biosynthesis may involve the loss of carbon atoms, fission of rings, rearrangements, etc., so the biogenetic scheme may contain discontinuous numbering. It is possible to deal with these new skeletons using the IUPAC scheme of modifiers described below, but authors often prefer to introduce a completely new semisystematic parent skeleton.

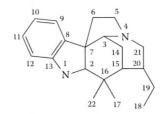

Akuammicine indole alkaloids
Example of discontinuous numbering
by biogenetic analogy

These schemes assume that the biosynthetic origin of the group is known or can be readily inferred. When the biosynthesis of a new group is unknown, several different schemes may exist from different workers.

3.5.3 LIST OF FUNDAMENTAL PARENT STRUCTURES (IUPAC)

IUPAC has listed fundamental parent structures (see IUPAC recommendations for Natural Products and Related Structures 1999). However, many more are found in the literature as shown in Chapter 6.

3.5.4 APPROVED MODIFIERS (AFFIXES)

IUPAC gives detailed guidance for the modification of semisystematic parents. For a full documentation, see the 1999 recommendations, sections RF-2 to RF-10. The following diagrams briefly illustrate the approved possibilities:

Modifications to the skeletal structure

1. Removal of skeletal atoms without affecting the number of rings

Germacrane   13-Norgermacrane

2. Addition of skeletal atoms without affecting the number of rings

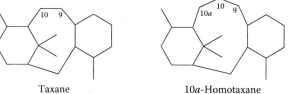

Taxane

10*a*-Homotaxane
(N.B. insertion of new
carbon after highest numbered possible atom)

3. Bond formation

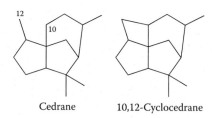

Cedrane   10,12-Cyclocedrane

4. Bond cleavage

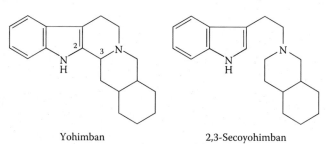

Yohimban   2,3-Secoyohimban

5. Bond migration

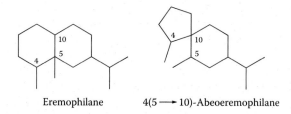

Eremophilane   4(5 ⟶ 10)-Abeoeremophilane

6. Removal of a terminal ring

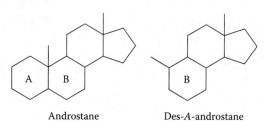

Androstane   Des-*A*-androstane

7. Fusion of additional rings

Naphtho[2',1':2,3]androstane

3.5.5   COMBINATION AND SORTING OF PREFIXES

Prefixes can be combined to describe multiple modifications to a notional parent, e.g. 8(14)-Seco-8(14)homopregnane. See IUPAC recommendations section RF-4.7. In some cases, it is possible to arrive at a structure by more than one route using different combinations of prefixes, and section RF-4.7 includes rules for making a choice, e.g. homo/nor- is preferred over cyclo/seco-. However, the IUPAC rules for making this choice and for placing the prefixes in correct order are complex. It is recommended not to introduce new names based on such combinations unless it is known that the new name is uniform with CAS practice.

There are considerable variations found in the literature concerning the italicisation and alphabetical ordering of these prefixes.

IUPAC (rule RF-4.7.3) recommends that prefixes should be presented in the following order; atomic replacement prefixes (oxa, aza etc.) followed by bond rearrangement prefixes (abeo, cyclo, retro, seco etc.), followed by those that indicate addition or removal of skeletal atoms (apo, de, des, nor). These prefixes modify the parent structure and thus come after ordinary substituents, e.g. nitro-. The ordering of prefixes within each subgroup is alphabetical except for the first group where by long-established practice, oxa- precedes aza.

Thus, 12-Bromo-11-(tribromomethyl)-15-aza-3,5-cyclo-19-norcholestane.

IUPAC recommends that these prefixes are not italicised, however, they are often found italicised in the literature. On the other hand, IUPAC recommends that stereochemical prefixes such as epi- should be italicised.

This policy is complex and prone to errors and inconsistencies. DNP adopts the policy that all such prefixes are nonitalicised and are sorted alphabetically alongside substituents.

Thus, 15-Aza-12-bromo-3,5-cyclo-19-nor-11-(tribromomethyl)cholestane.

3.6   SEMITRIVIAL NAMES

These are defined here as names in which a trivial parent is modified by one or more systematically derived modifiers. IUPAC (gold book) defines semitrivial name as a synonym for semisystematic name (a name in which at least one part is used in a systematic sense). The definition given here distinguishes between the two. In a semisystematic name (as defined above and by IUPAC), the parent structure is semisystematic, e.g. labdane. In a semitrivial name as here defined, the parent structure is trivial, e.g. Zeatin. It should be noted that the IUPAC list of natural product fundamental structures contains not only semisystematic parents (e.g. labdane) but also trivial parents, e.g. Cephalotaxine, Vincaleucoblastidine, modification of which produces semitrivial names according to the above definition.

Examples:

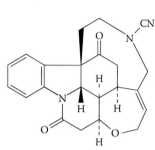

2-(1,1-Dimethylallyl)-5,7-diprenyltryptophan
Could be named in several other ways,
trivial name Echinine is preferable

N-Cyano-*sec*-pseudostrychnine
Could be named in several other ways,
does not have a trivial name

This class of name is moderately widespread in the literature but is problematical for the following reasons:

1. Many of the systematic modifiers have more than one form.
2. Different authors may base the name on different trivial parents.
3. Severe difficulties may be generated if there is a subsequent structure revision.

An extreme example of the proliferation of semitrivial names in the literature for a single compound is the following, where all the semitrivial names listed can be found in the literature:

6-(3-Methylbut-2-enyl)aminopurine
N^6-(δ^2-Isopentenyl)adenine
6-(γ,γ-Dimethylallylamino)purine
N^6-(3-Methylbut-2-enyl)adenine
N^6-Prenyladenine
Deoxyzeatin
I^6-Ade
6-(3-Methyl-2-pentenylamino)purine (erroneous)

3.7   STRUCTURE REVISION

According to most estimates, a large proportion of newly published natural product structures are incorrect. (Maier, M.E., *Nat. Prod. Rep.*, 2009, **26**, 1105–1124 [revision of all classes of natural product by synthesis]; Nicolaou, K.C. et al., *Angew. Chem. Int. Ed.*, 2005, **44**, 1012–1044 [ditto]; Suyama, T.L. et al., *Biorg. Med. Chem.*, 2011, **19**, 6675–6701 [revision of marine natural product structures by all methods].)

When the structure of a natural product is revised, its systematic, semisystematic and/or semi-trivial names change and this can cause problems in the retrieval of subsequent information. Trivial names remain unchanged.

1. *Example of structure revision of a natural product assigned a trivial name*
   Hortensin. Original assignment: 2,4'-Dihydroxy-6,7-dimethoxyflavone. *First revision*: 4',5-Dihydroxy-6,7-dimethoxyflavone (1992). *Second revision*: 5,7-Dihydroxy-4',6-dimethoxyflavone (1994).
2. *Example of structure revision of a natural product assigned only a semitrivial name*
   Terpenoid $C_{37}H_{52}O_6$ from *Maprouana africana*. Assigned an ursane structure and named 7β-Hydroxymaprounic acid 3-*p*-hydroxybenzoate (1983). *First revision*; following structure revision of Maprounic acid to a taraxerane identical with the known Aleuritolic acid, name changed to 7β-Hydroxyaleuritolic acid 3-*p*-hydroxybenzoate (1989). *Second revision*: based on revised placement of acyloxy group, name changed to 1β-Hydroxyaleuritolic acid 3-*p*-hydroxybenzoate (1995).

   Case (1) is clearly simpler and less error-prone.

3.8   CHEMICAL ABSTRACTS NOMENCLATURE POLICY

CAS aims to generate systematic names according to CAS nomenclature rules for all substances of known structure. However, for compounds of new and complex structural type, including many natural products, this may not be possible within the short timeframe in which indexing takes place. Therefore, a trivial or semitrivial name is often published in the interval before a systematic name can come into use depending on CAS resources available.

Bicyclomycin (structure established 1972)

- Trivial name: Bicyclomycin used until 13CI.
- Replaced with systematic name: 2-Oxa-7,9-diazabicyclo[4.2.2]decane-8,10-dione, 6-hydroxy-5-methylene-1-(1,2,3-trihydroxy-2-methylpropyl)- in 14CI.

Parsonsianine (structure established 1990)

- Semitrivial name: 22-Norparsonsine, 16-hydroxy- used until 12CI.
- Not indexed in 13CI or 14CI.
- Systematic name: 7*H*-[1,5,10]Trioxacyclotetradecino[7,8,9-*gh*]pyrrolizine-2,5,9(8*H*)-trione, 8-ethyl-3,4,11,13,15,16,16*a*,16*b*-octahydro-3,4,8-trihydroxy-7-methyl-3-(1-methylethyl) introduced in 15CI.

3.8.1   CAS Registry Numbers and Structure Revision

A CAS number normally applies to the *substance*, not the *structure*. When the structure of a natural product is revised, CAS normally uses the registry number (RN) already assigned to the substance for the revised structure and the CAS name for the revised structure is now assigned to this number in the registry. Examples are Maxonine [125797-63-9], Nazlinine [136945-81-8], Flaciformin [102275-31-0] and Glabrene [60008-03-9].

Shown below are the original and revised structures of Glabrene:

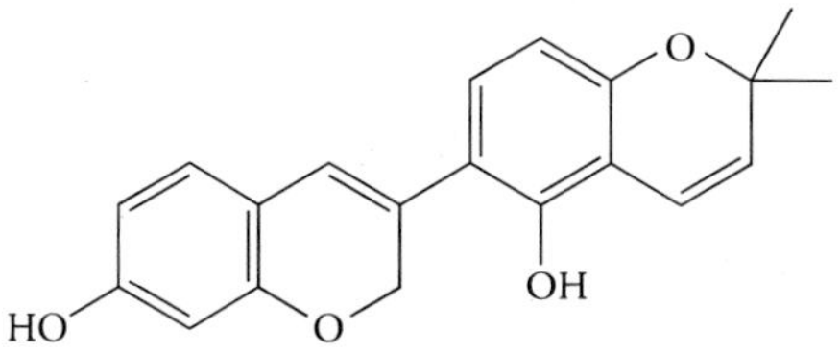

Originally proposed structure
2',2'-Dimethyl[3,6'-bi-2*H*-1-
benzopyran]-5',7-diol

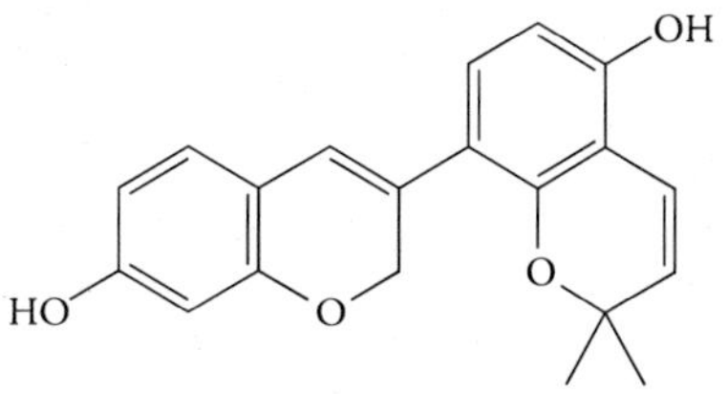

Revised structure (1996)
2',2'-Dimethyl[3,8'-bi-2*H*-1-
benzopyran]-5',7-diol

However, there are always cases current where CAS has not yet resolved the identity and two RNs may therefore be in use.

Examples: Lobophytol [77449-20-8] and [134175-15-8]; Thalflavine [30274-27-4] and [125617-79-0], Ottensin [917078-10-5] and [1040736-77-3].

Complications may arise, for example, because the new structure may already have been known as a synthetic compound, or because of tautomerism, or because different isolations or syntheses report the structure with incomplete levels of stereochemical specification. These factors affect chemistry generally and CAS policy for dealing with them is described in more detail in CAS documentation.

In other cases, new CAS numbers may be introduced and the original one suppressed even though it corresponds to the correct structure for a natural product. For example, Bruceajavanin C was originally allocated the CAS number [1169810-06-3]. Later its stereochemistry was revised (change of *C*21-configuration from 21*S*- to 21*R*-), and the new natural product having the same structure with 21*S*-configuration was isolated and given the trivial name Brujavanone D. The 21*R*-compound now corresponding to Bruceajavanin C was given the new CAS number [1417426-14-2] and the 21*R*-epimer corresponding to Brujavanone D the new number [1417715-15-1]. The number [1169810-06-3] is now reported in the CAS database as a deleted number.

3.8.2   CAS POLICY ON RACEMATES

Until the 14th collective index period, CAS policy was to allocate a separate CAS number to the racemic, i.e. (±)-form, as well as a general number. From the 14th collective index period onwards, the separate (±)-numbers were in principle suppressed and data referring to the racemate, or to isolates having unspecified chirality, were referred back to the general number. This needs to be borne in mind when accessing the pre-2002 literature. For example; Lactic acid; unspecified number [50-21-5] remains but (±)-number [598-82-3] is no longer used. The same policy has been applied to substances with more than one chiral centre, for example the number [36294-30-3] referring to (*R**,*R**)-(±)-2,3-Dihydroxybutanoic acid is replaced by that for (*R**,*R**)-2,3-Dihydroxybutanoic acid [759-06-8].

If the (±)-number predated the general number, it was given precedence.

GLOSSARY OF MISCELLANEOUS TERMS USED
IN NATURAL PRODUCT NOMENCLATURE

Cross references to other terms in the glossary are italicised.

abeo-: IUPAC-approved structure-modifying prefix. Indicates that a bond has migrated. For example, in a 5(4 →3)-abeoterpenoid, the 5–4 bond has been replaced by a 5–3 bond contracting ring A from six to five members. See IUPAC section RF-4.5. IUPAC now recommends that abeo- be not italicised.

Normal triterpene   5(4 ⟶3)-Abeotriterpenoid

Acetals: Diethers of *gem*-diols $R_2C(OR)_2$ (R can be the same or different). Often named as derivatives of aldehydes or ketones. Thus, acetaldehyde dimethyl acetal is $H_3CCH(OMe)_2$. It is now more usual to name them as dialkoxy compounds, e.g. 1,1-dimethoxyethane. The term 'acetal' is sometimes extended to compounds containing heteroatoms other than oxygen, as in *N,O*-acetals $R_2C(OR)(NR_2)$. Derivatives of ketones can be called ketals. This term was abandoned by IUPAC but is now reinstated.

Acetogenins: A class of long-chain polyketide metabolites found in plants (Annonaceae).

Acetonides: Cyclic acetals derived from acetone and diols. Better described as isopropylidene derivatives.

Glycerol acetonide =
1,2-Isopropylideneglycerol

aci-: The acid form of (prefix).

Aglycones (aglycons): Nonsugar compounds remaining after hydrolysis of the glycosyl groups from *glycosides*.

Aldaric acids: Dibasic acids derived from *aldoses* by oxidation at both ends of the chain, e.g. Glucaric acid.

Aldonic acids: Monobasic acids derived from *aldoses* by oxidation at the aldehyde end of the chain, e.g. Gluconic acid.

Aldoses: General name for sugar polyhydroxyaldehydes and their relatives. See Chapter 7.

Alduronic acids: Monobasic acids derived from *aldoses* by oxidation at the terminal (CH_2OH) end of the chain, for example, Glucuronic acid.

allo- (Greek 'other'): (1) A configurational prefix used in carbohydrate nomenclature. (2) A general prefix to denote close relationship, e.g. alloaromadendrene. (3) The more stable of a pair of geometric isomers, e.g. allomaleic acid = fumaric acid (obsolete).

Angucyclines: Class of polyketide-derived antibiotics containing a nonlinear tetracarbocyclic (benz[*a*]anthracene) skeleton, isomeric with tetracyclines.

anhydro: A subtractive prefix denoting the loss of the elements of water within one molecule.

D-Gulonic acid   2,3-Anhydro-D-gulonic acid

Anhydrosulfide: An analogue of an anhydride in which the oxygen atom connecting the two acyl residues has been replaced by a sulfur atom. Anhydroselenides are the Se analogues.

Anteiso acids: Fatty acids with chain branching at the antepenultimate carbon atom, e.g. $H_3CCH_2CH(CH_3)(CH_2)_{10}COOH$ = 12-Methyltetradecanoic acid = Anteisopentadecanoic acid.

Anthocyanins: Flavonoid glycosidic pigments. On hydrolysis they give anthocyanidins, which are oxygenated derivatives of flavylium salts.

apo (Greek 'from'): (1) Preceded by a locant indicates that all of the molecule beyond the carbon atom corresponding to that locant has been replaced by a hydrogen atom. Oxidative degradation products of carotenoids can be named as apocarotenoids; IUPAC-approved modifier for semisystematic skeletal names (carotenoids only); see IUPAC section RF-4.4.2. The prefix diapo indicates removal of fragments from both ends of the molecule. (2) In general means 'derived from', e.g. apomorphine.

Apolignans: Subclass of *lignans* in which a further carbocyclic cyclisation has taken place to produce a naphthalenoid structure. Obsolescent.

ar-: Abbreviation for 'aromatic', used as a locant to indicate an attachment at an unspecified position on an aromatic ring. Thus, in *ar*-methylaniline, the methyl group is attached to the aromatic ring and not to the amine *N* atom.

Ar: Denotes an unspecified aryl group.

Ascorbigens: Condensation products of ascorbic acid found in plants.

Aurones: Group of flavonoid natural products based on the 2-benzylidenefuranone nucleus.

aza-: Denotes replacement of a carbon atom or other heteroatom in a semisystematic skeleton by nitrogen.

Azaphilones: A family of unsaturated fungal natural products based on the 6*H*-2-benzopyran-6-one nucleus or its aza- analogues. (Gao, J.-M. et al., *Chem. Rev.*, 2013, **113**, 4755–4811.)

Bacteriocins: Peptides produced by some bacteria, e.g. Lactobacilli, with antibacterial activity. Of antibiotic and food industry interest.

Betaines: Originally meant zwitterionic compounds related to betaine having structure $R_3N^+CH_2COO^-$ (betaine itself has R = CH_3). Extended into a general class name for compounds containing a cationic centre and an anionic centre. They can also be called 'inner salts' or 'zwitterionic compounds'. Named as 'hydroxide, inner salts' in CAS.

Betalains: Class of zwitterionic alkaloidal plant pigments restricted to the Centrospermae, where they are responsible for flower colours in place of anthocyanins.

Bilanes (bilins, bilichromes): General term for linear tetrapyrrolic pigments such as Biliverdin.

Bisnor: See *nor*.

carba-: Replacement prefix indicating that a carbon atom has replaced a heteroatom, e.g. 5-carbaglucopyranose (a cyclohexane).

Carbapenems: Antibiotics containing the 1-azabicyclo[3.2.0]heptane nucleus. Related to *penicillins*.

-carbolactone: Suffix denoting the presence of a lactone ring fused to a ring system.

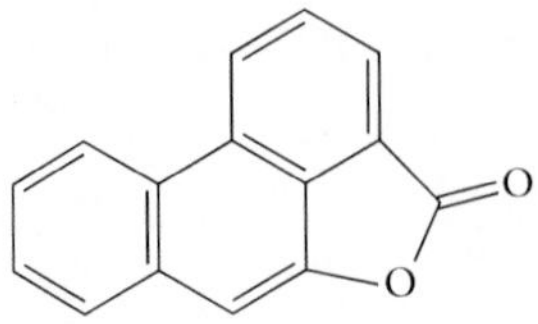

1,10-Phenanthrenecarbolactone

-carbonitrile: Suffix denoting –C≡N.

Carotenoids: Tetraterpenes arising from head to head coupling of two geranylgeranylpyrophosphate units. They are red to orange pigments found in plant and animal tissue.

Cephalosporins: Antibiotics containing the 5-thia-1-azabicyclo[4.2.0]octane skeleton. Related to *penicillins*.

Ceramides: *N*-Acylated *sphingoids*.

Chalcones (chalkones): Substituted derivatives of the parent compound 1,3-diphenyl-2-propen-1-one, PhCH=CHCOPh. The simplest group of flavonoids from which the other classes are derived.

Chlorins: A subclass of cyclic tetrapyrroles.

Clavams: Antibiotics containing the 1-aza-4-oxabicyclo[3.2.0]heptane nucleus. Related to *penicillins*.

Corrinoids: Compounds containing the corrin nucleus. A subclass of cyclic tetrapyrroles.

Coumarins: Compounds based on the 2*H*-1-benzopyran-2-one skeleton, including many natural products.

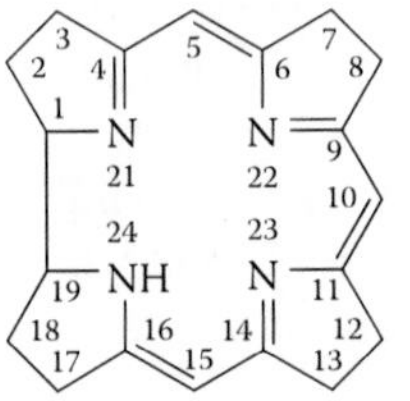

The number 20 is omitted when numbering the corrin nucleus so that the numbering system will correspond to that of the *porphyrin* nucleus. (*Pure Appl. Chem.*, 1976, **48**, 495.)

Coumestans: Flavonoid plant products based on the benzofuro[3,2-*c*]benzopyran nucleus.

Cyclitols: Cycloalkanepolyols in which an OH group is attached to each ring carbon. See Chapter 7.

cyclo-: IUPAC-approved modifier for semisystematic skeletons. (IUPAC section RF-4.3) Indicates that an additional ring has been formed by means of a direct link between atoms of the fundamental skeleton.

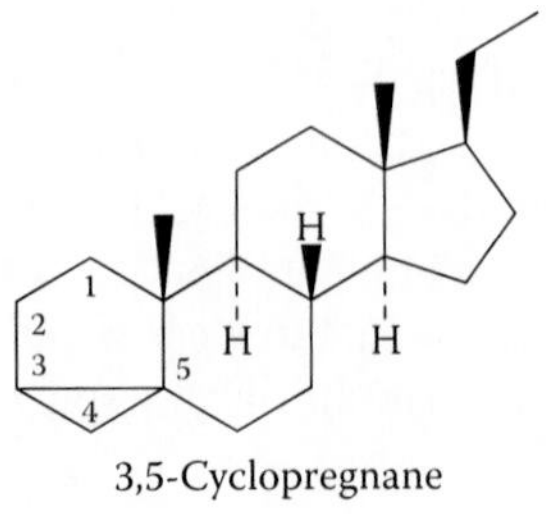

3,5-Cyclopregnane

Cyclodextrins: Cyclic oligosaccharides consisting of α-D-(1 → 4)-linked D-glucose residues.

Cyclotides: A class of plant cyclopeptides. (Shim, Y.Y. et al, *J. Nat. Prod.*, 2015, **78**, 645–652.)

Dalbaheptides: General name introduced for glycoside antibiotics of the Vancomycin–Ristocetin group.

de-: The prefix de- followed by the name of a group or atom denotes replacement of that group or atom by hydrogen. Thus in de-*N*-methylmorphine, the *N*-Me group of morphine has been replaced by *N*-H. Can be used to denote the loss of a complex group, e.g. in steroid nomenclature to denote the loss of an entire ring as in de-*A*-cholestane. *Des-* should be preferred for this usage.

dehydro-: Prefix denoting loss of two hydrogen atoms. Thus, 7,8-didehydrocholesterol is cholesterol with an additional double bond between atoms 7 and 8. In common usage, 'dehydro' is sometimes used instead of 'didehydro' but didehydro is preferred. Has also been used for removal of H_2O not H_2 as in Dehydromorphine. (Obsolete, not recommended).

Depsides: Esters formed from two or more molecules of the same or different phenolic acids. Widespread in lichens.

Depsidones: Lichen products formed by lactonic cyclisation of *depsides*.

Depsipeptides: Compounds containing amino acids and hydroxy acids (not necessarily α-hydroxy acids) and having both ester and peptide bonds.

des-: Prefix denoting loss of a complex substituent, e.g. Des(acetoxy)matricarin. Should be preferred to *de-* in this situation. Occasionally used in steroid/terpenoid nomenclature to describe loss of an entire ring, as in Des-*A*-cholestane. See IUPAC section RF-4.6.

Desmoside: Descriptor of the number of glycoside chains present in a saponin. E.g. Hederagenin bisdesmosides are glycosides of hederagenin glycosylated at C23 and C25.

diapo-: See *apo-*.

Didehydro: See *dehydro*.

Dimeth(yl)allyl: Obsolete descriptor for prenyl $(H_3C)_2C=CHCH_2–$

Dinor: See *nor*.

Diterpenoids: Terpenes having a C_{20} skeleton or modified C_{20} skeleton.

Dithio: –S–S–.

Dithioacetals: Sulfur analogues of acetals $R_2C(SR)_2$ (R is the same or different).

Ecdysteroids: Steroidal insect moulting and sex hormones.

Eicosanoids: See *icosanoids*.

Ellagitannins: *Tannins* in which at least two galloyl units are C–C coupled to each other.

Epi (Greek 'upon'): Denotes the opposite configuration at a chiral centre (e.g. 4-epiabietic acid). In carbohydrate chemistry, it denotes an isomer differing in configuration at the α-carbon.

Etio (aetio) (Greek *aitia*, 'cause'): Denotes a degradation product, e.g. Etiocholanic acid. Obsolescent.

Eupeptide: Peptide containing only normal –CONH- linkages involving the α-NH_2 groups of the amino acids.

Fatty acids: Carboxylic acids derived from animal or vegetable fat or oil. The term is sometimes used to denote all acyclic aliphatic carboxylic acids. See Chapter 6.

Flavaglines: Plant products based on the cyclopenta[*b*]benzofuran skeleton.

Flavanones: Flavonoid plant products based on the dihydro-2-phenylbenzopyran-4-one nucleus.

Flavans: Flavonoid plant products based on the dihydro-2-phenylbenzopyran nucleus.

Flavones: Flavonoid plant products based on the 2-phenylbenzopyran-4-one nucleus.

Flavonoids: A large group of natural products that are widespread in higher plants, derived by cyclisation of a *chalcone* precursor.

Flavonotannins (flavotannins): *Tannins* and tannin-like substances containing flavonoid (catechin-like) residues.

friedo-: In a (usually triterpenoid) name, denotes that a methyl group has migrated from one position to another.

Normal triterpene   *D*-friedo-

D:A-friedo-   *D:B*-friedo-

D:C-friedo-

Furanoses: Cyclic acetal or hemiacetal forms of saccharides in which the ring is five-membered. See Chapter 7.

Gallotannins: *Tannins* in which galloyl (3,4,5-trihydroxybenzoyl) or galloylgalloyl units are bound to a variety of polyols or terpenoids.

***gem(inal)*:** Used to denote that two groups are attached to the same atom as in *gem*-diol and *gem*-dimethyl groups.

Geoporphyrins (petroporphyrins): Degraded porphyrins found in oil shales and other geological deposits. Not true natural products; artefacts of slow decomposition.

Glucosinolates (glycosinolates): Naturally occurring mustard oil glycosides, typically R-C(SR')=N-OSO$_3$H (R' = glucose in all currently characterised natural products, potentially other sugars).

Glycals: Alkenic sugars with a double bond between positions 1 and 2.

Glycan: (1) The carbohydrate portion of a *glycopeptide* or *glycoprotein*. (2) Synonym for *polysaccharides* (obsolete).

Glycaric acids: Another name for *aldaric acids*.

Glycerides: Esters of glycerol with fatty acids. Subdivided into mono-, di-, and triglycerides.

Glycolipids: Fatty acid glycosides important in cellular recognition.

Glycopeptides, glycoproteins: Substances in which one or more carbohydrate components are covalently linked to the amino acid side-chain(s) of a peptide or protein. Important in cell recognition and many have antibiotic properties.

Glycosides: Compounds in which the anomeric oxygen atom of a sugar (or N atom of an amino-sugar, etc.) is alkylated with one of a very wide variety of aglycones.

Hemiacetals: Compounds $R^1CH(OH)OR^2$ or $R^1R^2C(OH)OR^3$ formed by condensation of an alcohol with a carbonyl compound.

Hemiketal: A hemiacetal derived from a ketone.

Hemimercaptals, hemimercaptoles: Compounds $R^1R^2C(SH)(SR^3)$. Sulfur analogues of hemiacetals and hemiketals.

Hemiterpenoids: C_5 terpenoids formed of a single isoprene unit and their relatives.

Hetero (Greek *heteros*, 'other'): Prefix meaning 'different', e.g. heteroxanthine, heterocycle.

Heterodetic: Peptide chain containing bonds other than *eupeptide* bonds, e.g. disulfide, ester.

Hexofuranose: Six-carbon sugar in the five-membered (furanose) ring form.

Hexopyranose: Six-carbon sugar in the six-membered (pyranose) ring form.

Hexose: Sugar with six carbon atoms. See Chapter 7.

homo-: IUPAC-approved skeletal modification to semisystematic names (see IUPAC section RF-4.2). Denotes incorporation of CH_2 as an additional member in a ring or chain. The additional carbon atom is given an *a*- suffix and the highest available number, in this case 4*a*-. Plural=dihomo, etc.

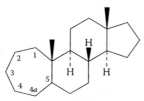

A-Homoandrostane

Homodetic: Peptide chain consisting only of *eupeptide* bonds.

hydro-: Denotes an added hydrogen atom. Thus dihydro denotes saturation of one double bond.

Hydrodisulfides: Compounds R–S–SH.

Hydroperoxides: Compounds R–O–OH.

Hypo (Greek 'under'): Indicates a lower state of oxidation, e.g. hypoxanthine.

Icosanoids (eicosanoids): Natural products produced in animals which are unsaturated C_{20} fatty acids and their oxidation products. A major subclass of *oxylipins*. They perform important biochemical signalling roles. The spelling icosa- is IUPAC preferred.

Inner salts: CAS considers compounds such as *betaines* to be formed by the loss of water from the corresponding hydroxides and names them using the expression 'hydroxide, inner salt', e.g. $Me_3N^+CH_2COO^-$ (Betaine) = Carboxy-*N,N,N*-trimethylammonium hydroxide inner salt (CAS).

Inosamines: Aminodeoxyinositols, i.e. 6-amino-1,2,3,4,5-cyclohexanepentols.

Inositols: 1,2,3,4,5,6-Cyclohexanehexols. See Chapter 7.

Inososes: 2,3,4,5,6-Pentahydroxycyclohexanones.

Iridoids: Type of cyclopentanoid sesquiterpenoids.

iso-: Prefix denoting isomerism, especially carbon chain branching (isohexanoic acid = 4-methylpentanoic acid). In the old literature it can be treated as a separable prefix, e.g. *iso-propyl*; in the modern literature it is usually treated as an inseparable prefix, e.g. isopropyl.

Isoacids: Fatty acids with chain branching at the penultimate position of the chain, e.g. $(H_3C)_2CH(CH_2)_{12}COOH$ = Isopalmitic (14-methylpentadecanoic) acid.

Isocyanates: Compounds RNCO. Thus, methyl isocyanate is MeNCO.

Isocyanides: Compounds RNC. Thus, methyl isocyanide is MeNC.

Isoflavans: Flavonoid plant products based on the 2,3-dihydro-3-phenylbenzopyran-4-one nucleus.

Isoflavones: Flavonoid plant products based on the 3-phenylbenzopyran-4-one nucleus.

Isonitriles: See *isocyanides*.

Isopentenyl: Alternative name for prenyl $(H_3C)_2C=CHCH_2-$, not recommended.

Isopeptide: Peptide containing amide bonds formed from non-α NH_2 groups, normally the *C*-6 NH_2 of lysine.

Isoprenoids: Compounds, especially terpenoids, derived formally from isoprene units. Isoprene is 2-methyl-1,3-butadiene, $H_2C=C(CH_3)CH=CH_2$.

Isoprenyl: Alternative name for prenyl $(H_3C)_2C=CHCH_2-$; confusing, not recommended.

Isothiocyanates: Compounds RNCS. Thus, methyl isothiocyanate is MeNCS.

-ium: Suffix denoting a positively charged species.

Ketals: Acetals derived from ketones.

Keto: Oxo, O= Now used only in a generic sense as in 'ketoesters', etc.

Knottins: Polypeptides with a secondary structure forming a knot topology based on the disulfide linkages of cysteine or some rarer modified amino acids.

Lactams: Compounds containing a group of –CO–NH– as part of a ring. β-Lactams have four-membered rings, γ-lactams have five-membered rings, etc.

γ-Butyrolactam or 4-Butanelactam

Lactims: Tautomers of lactams containing a group –C(OH)=N– as part of a ring.

Lactones: Intramolecular cyclic esters of hydroxyacids. They contain a group –CO–O– as part of a ring. β-Lactones have four-membered rings, γ-lactones have five-membered rings, etc.

D-Glucono-1,4-lactone

Lambda (λ): Italicised prefix indicating abnormal (higher) valency of a ring heteroatom, e.g. S, Se, Te. The 2006 CAS nomenclature changes introduced this system to replace Roman numeral valency suffixes used earlier.

1,2,3-Thia-(*S IV*)diazole, 9CI → 1λ⁴δ²-Thiadiazole

Lantibiotics: Peptide antibiotics (*bacteriocins*) produced by Gram-positive bacteria and containing the characteristic amino acid Lanthionine.

leuco- (Greek 'white'): Prefix denoting usually the reduced colourless form of a coloured compound, e.g. a flavonoid.

Leucoanthocyanidins: Flavan-3,4-diols.

Lignans: Class of plant natural products formed from phenylpropionate (cinnamate) residues by various oxidative linkages and other cyclisations, etc. In traditional lignans, the primary linkage is α,α' (8,8'-in lignan numbering). See also *neolignans*.

Lipopeptides: Peptides containing lipid residues.

Macrolides: Macrocyclic lactones.

Mercaptals: Dithioacetals.

Mercaptans: An old name for thiols. Thus, ethyl mercaptan is ethanethiol, EtSH.

Mercaptoles: Mercaptals derived from ketones.

Meroterpenoids: Terpenoids in which a substantial portion of the main skeleton of the molecule is derived from alternative biochemical pathways, e.g. polyketide.

Mes(yl)ate: A salt or ester of methanesulfonic acid MeSO₃H.

meta-: Denotes 1,3-substitution on a benzene ring.

Methine: =C–.

Methiodide, methobromide, methochloride, methoperchlorate, methopicrate, etc.: Indicates a base quaternised with methyl iodide, etc.

Monoterpenoids: Terpenoids having a C₁₀ skeleton.

Mustard oils: Old term for isothiocyanates.

n-: Abbreviation for normal (unbranched) as in *n*-butane.

Nactins: Polyketide natural products which are oligomers of nonactic acid and closely related units produced by streptomycetes.

Nonactic acid

Naphtho: The ring fusion prefix derived from naphthalene.

Neo (Greek 'new'): (1) A newly characterised stereoisomer (e.g. neomenthol). (2) A quaternary branched hydrocarbon. (3) In terpenes, the prefix neo indicates the bond migration that converts a *gem*-dimethyl grouping directly attached to a ring carbon into an isopropyl group.

Neoflavonoids: *Flavonoid* plant products based on the 4-phenyl-2-benzopyranone nucleus.

Neolignans: *Lignans* in which the linkage between the phenylpropanoid residues is other than α,α'-(8,8'-in lignan numbering) as in 'traditional' lignans. A structurally diverse and rapidly growing type of natural product. The term neolignan now has little usefulness and should probably be discontinued.

Neosteroids: Occasionally used to refer to ring B aromatic steroids.

Nitriles: Compounds RCN. The suffix '-nitrile' denotes a –CN group at the end of an aliphatic chain. Thus, butanenitrile is $H_3CCH_2CH_2CN$. Nitriles can also be named as cyano-substituted compounds.

2-Furancarbonitrile or 2-Cyanofuran

Nonadrides: Polyketide fungal products typically based on a nine-membered carbocyclic ring with attached furandione (maleic anhydride) rings.

nor-: Used mainly in naming steroids and terpenes. Denotes elimination of one CH_2 group from a chain or contraction of a ring by one CH_2 unit. Where there is a choice, the atom is considered as removed that retains as far as possible the original skeletal numbering or has the highest possible numbering, in this case 4.

19-Norpregnane   A-Norandrostane

The form *A*-nor- etc. to denote removal of a –CH_2 from a ring is no longer recommended by IUPAC but is still used by CAS.

The plural form should be *bisnor* when two carbon atoms are lost from the same site and *dinor* where they are lost from different sites, but in practice the terms are used interchangeably. IUPAC prefers dinor-.

In older usage, particularly for monoterpenes, 'nor' denotes loss of all methyl groups attached to a ring system, e.g. norbornane, norpinane. (IUPAC disallows all except norbornane).

Nucleoside: Monomeric fragment of nucleic acids consisting of a nitrogenous base (e.g. adenine) covalently linked to a sugar, e.g. ribose. See Chapter 7.

Nucleotide: Monomeric fragment of nucleic acids consisting of a *nucleoside* linked to a phosphate group.

o-: Abbreviation of *ortho-*.

-olide: Suffix denoting a lactone.

5-Pentanolide   4-Pentanolide   3-Pentanolide

oligo-: Prefix meaning 'a few' as in oligosaccharides, oligopeptides.

Oligosaccharides: Sugars obtained by glycosidic condensation of three or more, but a defined number, of monosaccharides.

-olium: Denotes a positively charged species derived from a base with a name ending in -ole, e.g. pyrrolium.

-onium: Indicates a positively charged species such as ammonium, phosphonium, sulfonium, oxonium, etc.

Orbitides: A class of plant cyclopeptides. (Shim, Y.Y. et al, *J. Nat. Prod.*, 2015, **78**, 645–652.)

ortho-: Denotes 1,2-substitution in a benzene ring (abbreviated to *o*-).

Orthoesters: Compounds $R^1C(OR^2)_3$, esters of the hypothetical orthoacids $R^1C(OH)_3$. Thus, ethyl orthoacetate is $H_3CC(OEt)_3$; orthoacetic acid is $H_3CC(OH)_3$.

oxa-: Denotes replacement of a carbon atom or other heteroatom in a semisystematic skeleton by oxygen.

Oxylipins: Widespread natural products formed from fatty acids by oxidation. Occur in plants, animals and fungi and include the *eicosanoids*. Many are physiologically and medicinally important, such as the *prostaglandins*.

p-: Abbreviation of *para-*.

para- **(Greek 'beside', beyond'):** Denotes 1,4-substitution in a benzene ring.

Peltogynins: Group of flavonoids based on the [2]benzopyrano[4,3-*b*][1]benzopyran skeleton, resulting from formation of an additional pyran ring in a flavone.

Penicillins: Antibiotics containing the 4-thia-1-azabicyclo[3.2.0]heptane nucleus.

Penitrems: Class of tremorigenic microbial toxins incorporating indole alkaloid and terpenoid fragments.

Pentofuranose: A 5-carbon sugar in the 5-membered ring (furanose) form.

Pentopyranose: A 5-carbon sugar in the 6-membered ring (pyranose) form.

Pentose: A 5-carbon sugar.

Peptaibols: Fungally produced oligopeptides and polypeptides containing unusual amino acids, especially 2-aminobutanoic acid, and with a terminal $-CH_2OH$ group.

Peptolides: Synonym for *depsipeptides*.

per: (1) The highest state of oxidation, e.g. perchloric acid. (2) Presence of a peroxide ($-O-O-$) group, e.g. perbenzoic acid. (3) Exhaustive substitution or addition, e.g. perhydronaphthalene.

Peroxyacids: Acids containing the group $-C(O)OOH$. Thus, peroxypropanoic acid is $H_3CCH_2C(O)OOH$ (also named as propaneperoxoic acid).

Petroporphyrins: See *geoporphyrins*.

Phlorotannins: Oligomers of phloroglucinol (1,3,5-benzenetriol) found mostly in the cell walls of brown algae. They can be considered as *tannins* according to some definitions.

Phosphatidic acids: Derivatives of glycerol in which one primary OH group is esterified with phosphoric acid and the other two OH groups are esterified with fatty acids.

Phospholipids: Fatty acid diglycerides linked to a phosphorus-containing aliphatic residue, most frequently choline. Biochemicals important in cell membrane structure.

phyto-: Prefix meaning 'plant', e.g. phytobilins = *bilins* from plants.

Phytoecdysteroids: *Ecdysteroids* produced by plants as defence mechanisms against insect predators.

Picrate: An ester, salt or addition compound of picric acid (2,4,6-trinitrophenol). Formerly used extensively to characterise alkaloids, etc., now infrequently used.

Polyketides: Large category of natural products based on Claisen-type oligomerisation of acetate, propionate and higher acyl residues followed by a wide range of further cyclisations, condensations, etc.

Polysaccharides: Carbohydrate polymers produced by condensation of an indefinite number of monosaccharide residues. May be linear or chain-branched and may consist of one (homopolysaccharide) or more than one (heteropolysaccharide) type of sugar.

Porphyrins: Tetrapyrrolic pigments comprising a numerically limited number of important biochemicals including the chlorophylls and bile pigments.

21H,23H-Porphine
or Porphyrin

Fischer numbering
(obsolete)

Prenyl: 3-Methyl-2-butenyl, $(H_3C)_2C=CHCH_2-$. A very common substituent in natural product structures. Has also been called isoprenyl (misleading; avoid), γ,γ-Dimethylallyl, 3,3-Dimethylpropenyl, Dimethallyl, Isopentenyl and β,β-Dimethylacrylyl. Prenyl is now universally preferred.

Proanthocyanidins: 3-Hydroxyflavans.

Prostaglandins: Metabolically important group of *icosanoids* incorporating a cyclopentane ring.

Proteins: Polypeptides of high molecular weight (above 10,000).

Pseudo (Greek 'false'): Prefix indicating resemblance to, especially isomerism with, e.g. pseudocumene or ψ-cumene.

Pterocarpanoids: Type of modified flavonoid based on formation of additional pyranoid and furanoid rings from an isoflavonoid precursor.

Quercitols: Deoxyinositols, i.e. 1,2,3,4,5-cyclohexanepentols.

retro-: (1) (In carotene names) The prefix *retro-* and a pair of locants denotes a shift, by one position, of all single and double bonds delineated by the pair of locants. The first locant cited is that of the carbon atom that has lost a proton, the second that of the carbon atom that has gained a proton. See IUPAC section RF-4.5. (2) (In peptide names) When used with a trivially named peptide, *retro-* denotes that the amino acid sequence is the reverse of that in the naturally occurring compound. (3) Has been used in steroid nomenclature to denote the unusual (9β,10α)-configuration (not recommended).

Rotenoids: Type of flavonoid characterised by formation of an additional pyran ring of an isoflavonoid.

S-: Denotes sulfur as a locant. Also an important stereochemical descriptor (sinister-).

Saponins: Obsolescent term for hydrolysable *glycosides* from plants.

Schardinger dextrins: Another name for *cyclodextrins*.

sec-: Abbreviation of 'secondary' as in *sec*-butyl.

seco-: IUPAC-approved modifier for semisystematic skeletons, see IUPAC section RF-4.4. Denotes fission of a ring with addition of a hydrogen atom at each terminal group thus created.

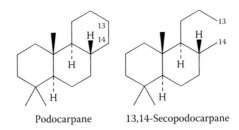

Podocarpane   13,14-Secopodocarpane

Sesquaterterpenoids: C_{35} terpenoids formed of seven isoprenoid units (rare).

sesqui-: Numerical prefix meaning 1.5 as in *sesquiterpenoids*.

Sesquiterpenoids: C_{15} terpenoids formed of three isoprenoid units.

Sester: Numerical prefix meaning 2.5 as in sesterterpenoids.

Sesterterpenoids: Terpenes having a C_{25} skeleton formed of five isoprenoid units and their relatives.

Siderophores: A loosely defined group of iron-complexing metabolites produced mainly by bacteria but also by plants. The majority are nitrogenous and their complexing ability is due to hydroxamate and/or catechol residues.

Sphingoids (sphingolipids): Refers to sphingamine (D-*erythro*-2-amino-1,3-octadecanediol), its homologues, stereoisomers and derivatives. Biochemicals important in cellular recognition and nerve cell function.

Styphnates: Complexes with 2,4,6-trinitro-1,3-benzenediol (styphnic acid), formerly used to characterise alkaloids, etc. (cf. *picrates*).

Tannins: An important class of plant polyphenolic compounds mostly containing galloyl (3,4,5-trihydroxybenzoyl) or modified galloyl residues. They are difficult to define in precise structural terms. Historically they were defined functionally as substances having tanning properties so they are structurally diverse; many of the compounds now considered tannins have no tanning properties. See also *gallotannins*; *ellagitannins*; *flavonotannins*. (Khanbabee, K. and Van Ree, T., *Nat. Prod. Rep.*, 2001, **18**, 641–649.)

Terpenoids: A large class of organic compounds, the common structural feature of which is a carbon skeleton of repeating isoprene units.

tert-: Abbreviation of tertiary as in *tert*-butyl.

Tetracyclines: Class of polyketide antibiotic substances containing a linear 4-ring carbocyclic (naphthacene) skeleton.

Tetramic acids: Natural products based on 2,4-pyrrolidinedione. (Royles, B.J.L., *Chem. Rev.*, 1995, **95**, 1981–2001.)

Tetraterpenoids: Terpenes having a C_{40} skeleton formed of eight isoprenoid units and their relatives. Principally comprising the *carotenoids*.

thia-: Denotes replacement of a carbon atom or other heteroatom in a semisystematic skeleton by sulfur.

-thial: Suffix denoting –CHS at the end of an aliphatic chain. Thus, hexanethial is $H_3C(CH_2)_4CHS$.

Thio: Denotes replacement of oxygen by sulfur as in thiophenol, thiourea. Also, the multiplying radical –S–. Dithio is –S–S–, trithio is –S–S–S–, etc.

Thioacetals: Sulfur analogues of *acetals*.

Thioaldehydes: Sulfur analogues of aldehydes, RCHS.

Thiocarboxylic acids: Compounds RC(S)OH, RC(O)SH and RC(S)SH, sulfur analogues of carboxylic acids.

Thiocyanates: Compounds RSCN. Thus, methyl thiocyanate is MeSCN.

Thioketones: Sulfur analogues of ketones.

-thiol: Suffix denoting –SH. 'Thiols' are compounds RSH.

-thione: Suffix denoting a thioketone. Thus, 2-butanethione is $H_3CC(=S)CH_2CH_3$.

Thromboxanes: Group of *icosanoids* closely related to prostaglandins.

Tosylate (tosate): An ester of *p*-toluenesulfonic acid (4-methylbenzenesulfonic acid).

Triterpenoids: Terpenes having a C_{30} skeleton formed of six isoprenoid units and their relatives.

Tropolones: Compounds containing the 2-hydroxycyclohexa-2,4,6-trienone ring system, showing quasi-aromatic properties.

Tropones: Compounds containing the cyclohexa-2,4,6-trienone ring system.

Ulosaric acids, ulosonic acids, ulosuronic acids: Acids derived from the oxidation of ketoses.

-ulose: Denotes a ketose; '-ulofuranose' and '-ulopyranose' denote a ketose in the cyclic hemiacetal form having five- and six-membered rings, respectively.

Uronic acids: Monocarboxylic acids derived by oxidation of the terminal CH_2OH of aldoses. Names are formed by replacing the '-ose' ending of the aldose name by '-uronic acid'.

Vicinal: Neighboring.

Zwiitterionic compounds: General term for compounds containing both positive and negative charges. See also *betaine*.

4 Stereochemistry

This chapter deals only with topics specific to natural products. A much fuller account of stereochemistry is given in the companion *Organic Chemist's Desk Reference* and in standard works.

4.1 TRIVIALLY NAMED COMPOUNDS

Allocation of a trivial name to a natural product traditionally implies a certain relative stereochemistry in the case of molecules with two or more chiral centres or other stereogenic features, even though at the time of allocation of the name the configurations may not be known. Known diastereomers may be given different trivial names or may be distinguished by prefixes such as epi-, allo- and epiallo-. Traditionally, such prefixes were italicised and separable (*epiallo*-Corynantheine, indexed under C). Modern practice is for them to be inseparable: for example Epiallocorynantheine, indexed under E (see Section 4.6).

There are a few examples in the literature of two enantiomers being isolated separately, e.g. Ipomeamarone (1943), later found to be the enantiomer of Ngaione (isolated 1925). Retention of the separate names for the two enantiomers is unsatisfactory since the racemate can have either of two names.

4.2 SEMISYSTEMATICALLY NAMED COMPOUNDS

A parent semisystematic structure should include as much stereochemistry as possible that is common to the relevant class of natural products, and the name of the fundamental parent implies the normal absolute configuration at all chiral centres as well as double bond configurations, except in some cases where certain stereogenic centres are variable and have to be specified. Thus in the name $(5\alpha,9\alpha)$-pregnan-3α-ol, the three alphas perform different functions. 3α specifies the configuration of a substituent, 9α specifies an unusual configuration (normal pregnane implies 9β-) and 5α- is necessary because the 5-configuration in pregnanes is variable.

Configurations at stereocentres in planar ring systems are denoted by the symbols α- (below the plane), β- (above the plane) and ξ- (unknown). Use of these symbols implies an unambiguous orientation for the ring system, either as specified by IUPAC, or as in common use in the literature. Ambiguity may arise where the stereogenic centre is attached to a part of the molecule that is free to rotate, or has been rotated to a nonstandard position, e.g. Vitamin D_2, or where there is a bridged ring system, e.g. the 19- and 20-positions in the Atidane skeleton. In such cases the $(R,S$-) system should be introduced.

TABLE 4.1

Skeleton	Parent Name	Variation
Sesquiterpenoids	Cadinanes–Muurolanes–Bulgaranes–Amorphanes	Epimerism at stereogenic centres
	Cadinanes–Calamenenes–Cadalenes	Dehydrogenation
	Bicyclogermacranes–Lepidozanes	Epimerism
	Acoranes–Alaskanes	Enantiomerism
Diterpenoids	Clerodanes–Neoclerodanes (kolavanes)	Enantiomerism
	Pimaranes–Isopimaranes	Enantiomerism
	Kauranes–Phyllocladanes	Epimerism
	Bifloranes–Serrulatanes	Dehydrogenation
Triterpenoids	Ursanes–Taraxastanes	Epimerism
Steroids	Ergostanes–Campestanes	Epimerism
	Stigmastanes–Poriferastanes	Epimerism

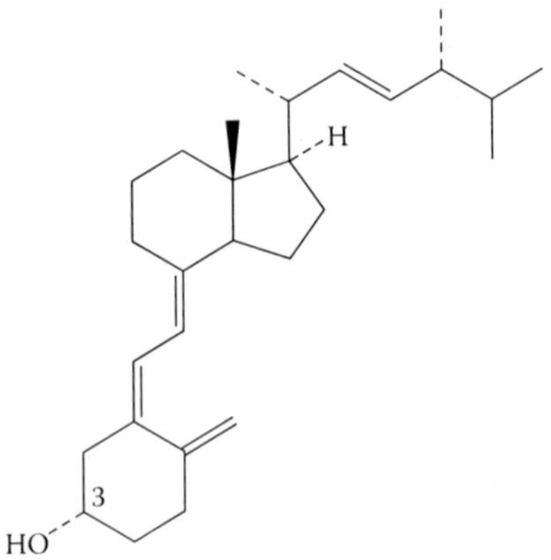

Vitamin D₂ =
(3S,5Z,7E,22E)-
9,10-Secoergosta-5,7,10(19),22-tetraen-3-ol

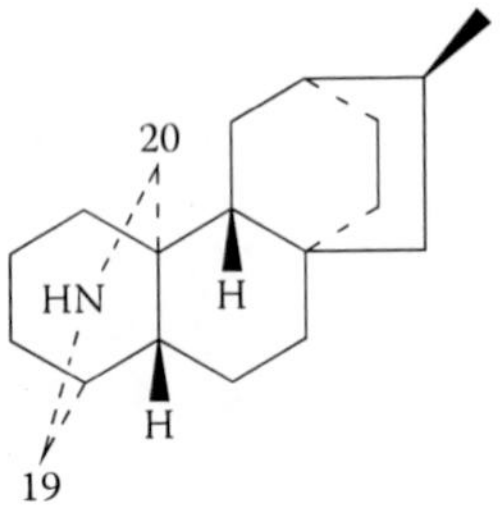

Atidane skeleton

There are a number of cases among terpenoids and steroids in which different parent skeletal names are in use where the only variation between them is stereochemical, either epimerism, enantiomerism or dehydrogenation leading to fewer centres of chirality (Table 4.1).

Most of these nomenclatural variants are obsolescent and it is recommended to use the most prevalent name throughout (e.g. name all muurolanes, bulgaranes, amorphanes, calamenenes and cadalenes as cadinanes, using appropriate stereodescriptors and unsaturation labels).

4.3 THE *ent-* CONVENTION

This is used for natural products, mostly terpenoids but also some others, e.g. catechin flavonoids. Use of the *ent-* prefix implies inversion of configuration at *all* chiral centres in the molecule so that, for example *ent*-6α,13R-Dihydroxy-5β-labdan-15-al is the enantiomer of 6α,13R-Dihydroxy-5β-labdan-15-al. Use of the convention can easily introduce errors, especially when describing transformations. For example, going from *ent*-5β-Labdan-3α-ol to *ent*-5β-Labdane-3α,13R-diol requires addition of a 13S-hydroxy group.

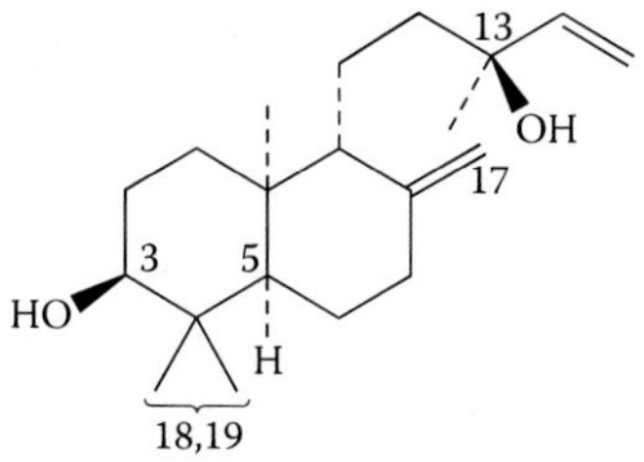

ent-5β-Labdane-3α,13*R*-diol

Particular care is needed with kauranes (and some other smaller terpenoid groups) where CAS (on the historical grounds that nearly all natural kauranes are in fact *ent*-kauranes on the biogenetic scheme) calls *ent*-Kaurane, Kaurane.

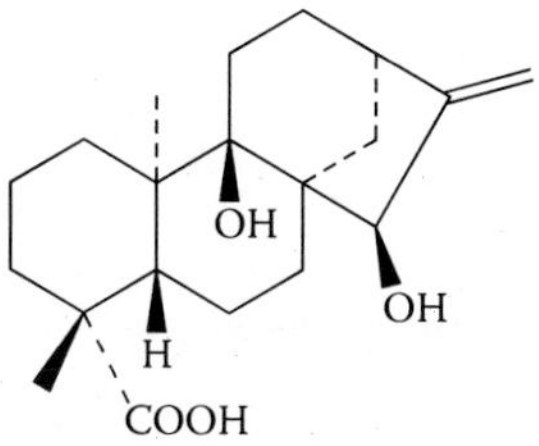

ent-9,15α-Dihydroxykaur-16-en-19-oic acid
= Kaur-16-en-18-oic acid, 9,15-
dihydroxy (4α,15β)(CAS)

4.4   CAS POLICY ON STEREOCHEMICAL LABELS

Until recently, CAS stereochemical labelling functioned on the principle that all stereochemistry in a given molecule had to be expressed by a single descriptor, which could often be complex and non-intuitive. This is no longer required and recent labels are more intuitive.

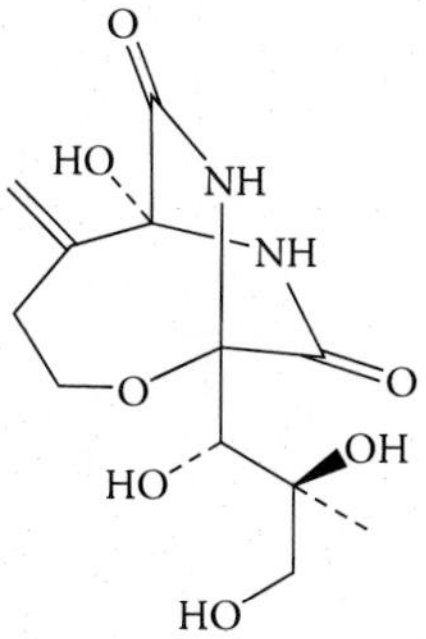

Bicyclomycin

For example, bicyclomycin (above) is named in the 14th and 15th collective indexes as follows:

2-Oxa-7,9-diazabicyclo[4.2.2]decane-8,10-dione, 6-hydroxy-5-methylene-1-(1,2,3-trihydroxy-
2-methylpropyl)-, [1S-[1α,1(1R*,2R*,6β)]]-, 14CI

2-Oxa-7,9-diazabicyclo[4.2.2]decane-8,10-dione, 6-hydroxy-5-methylene-1-[(1S,2S)-1,2,3-
trihydroxy-2-methylpropyl]-, (1S,6R)-, 15CI

CAS differs from most of the literature for compounds substituted at a *gem*-dimethyl site. Usual practice is to differentiate between the methyl carbons in the numbering scheme, with the α-Me invariably being assigned the lower number. CAS practice is to treat the two Me groups when unsubstituted as indistinguishable, assigning the substituent to the lower-numbered Me group and adding a stereodescriptor to indicate the configuration at the newly generated stereogenic feature. Particular care needs to be taken with *ent*- compounds.

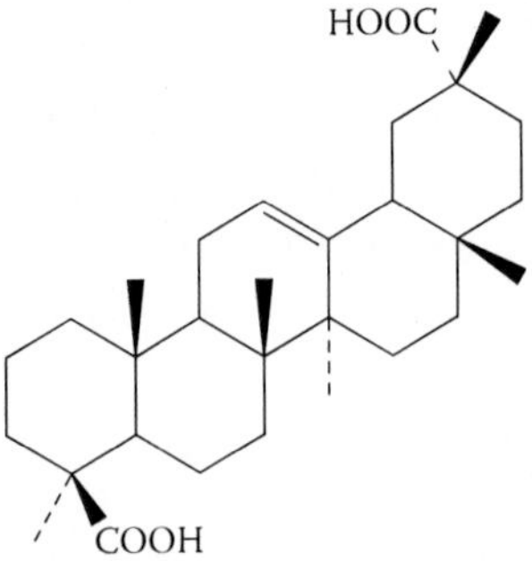

Olean-12-ene-24,29-dioic acid =
Olean-12-ene-23,29-dioic acid
(4β,29α)(CAS)

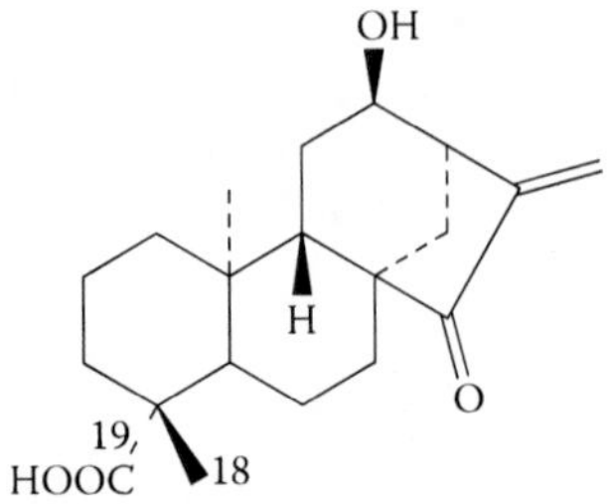

Tremulin
= *ent*-12α-Hydroxy-15-oxo-16-kauren-19-oic acid
=12-Hydroxy-15-oxokaur-16-en-18-oic acid (4α,12β)(CAS)

4.5 DICTIONARY OF NATURAL PRODUCTS (DNP) POLICY

In *Dictionary of Natural Products* (DNP) all stereoisomeric forms, including *ent*-forms, of a given constitutional structure are collected together as variants within the same entry. For example, the entry for 8(17)-Labdadien-13-ol contains information on the four variants (13R)-form (Manool), (13*S*)-form (13-Epimanool), (*ent*-13R)-form (*ent*-Manool) and (*ent*-13S)-form (*ent*-13-Epimanool).

For important natural products with one chiral centre (e.g. amino acids), the properties of the racemate (and frequently the non-natural enantiomer also) are given.

If only one stereoisomer of a natural product with several stereogenic features occurs naturally, the structure diagram shows that stereoisomer and (except in the case of well-established categories such as steroids) the diagram is normally labelled 'Absolute configuration' or 'Relative configuration'. If more than one stereoisomer occurs naturally, the diagram is labelled appropriately. The stereoisomer shown is normally the first one documented in the entry, except in a few cases, e.g. the common amino acids where the *S*-form is the usual and most important stereoisomer.

Data on the racemates of such natural products is only given for a small number of commercially important compounds, or where the (±)-form has been separately characterised as a natural product. For many other natural products the physical properties are not given for the racemate but the CAS registry numbers for the racemate and other non-natural stereoisomers are often given.

5 The Natural World and Sources of Natural Products

This chapter gives a brief overview of the natural world and its classification, focussing on the organisms that are the richest sources of natural products, and with reference to the organisation of data in the *Dictionary of Natural Products* (DNP). The main groups of microorganisms, plants, animals and fungi are briefly reviewed.

5.1 USEFUL LITERATURE SOURCES

Also see Chapter 1, in particular Section 1.2.

5.1.1 CATALOGUE OF LIFE (HTTP://WWW.CATALOGUEOFLIFE.ORG)

This database is an international cooperative effort that is planned to become a complete record of the world's species, and currently contains over 1.5 million, about 70% of the reported total. It is compiled from sectors provided by 134 taxonomic databases from around the world, and its most important technical feature is that software (wrapper programs) has been developed to integrate and reconcile the taxonomy from these different sources. The database gives authorities and the status of each name, e.g. accepted name, provisionally accepted name or synonym. Clicking on the taxon name gives information about taxonomic position, species distribution and other data. Coverage is near-comprehensive for many important groups of organisms, but may be limited for smaller groups where no database currently exists, or where the owners of the database have not yet joined the consortium. DNP species names are hyperlinked to CoL.

5.1.2 OTHER TAXONOMIC DATABASES

Perusal of the web will show that there are a considerable number of other databases which are taxonomically centred, some of them giving more detailed information than *Catalogue of Life* (CoL) about individual species or groups (such as a photographic record). Many are contributing partners to CoL in which case their home database contains this additional information which is not shared in the CoL project. It should be borne in mind that some other databases may adhere to a different taxonomic view compared with CoL, especially where they are geographically based (e.g. describing the fauna of a particular region) (Blunt et al., 2008; Brummitt, 1992; Cimino et al., 2001; Erskens, 2011; Kornprobst, 2010).

5.2 TAXONOMIC HIERARCHIES

The term 'species' can be defined in various ways and is not a precisely defined term. Taxonomy does not regulate whether or not the species itself is a recognised entity, but through the regulatory bodies seeks to make precise rulings about which *name* is valid for any one claimed taxon.

The nineteenth century high-level classification of organisms into plant and animal kingdoms has been largely abandoned since about the 1960s with the development of cladistic analysis. Since then it has become increasingly clear that certain groups of organisms, some of them previously little studied, such as the cyanobacteria or cyanophytes (so-called blue-green algae; more closely related to bacteria), the chromista (including the brown algae) and the archaebacteria or archaea, show greater differences in genetics and fundamental biochemistry from each other and from the so-called higher organisms, than higher plants and animals show from each other. It has been proposed that the most fundamental division should be between the prokaryotes and the eukaryotes, a classification that has now been generally accepted, but subsequently modified to include the discovery of the archaea in the 1970s.

There are some differences between botanical and zoological (and microbiological) taxonomic protocols, but for plants the descending order of taxonomic groups is *phylum, class, order, family, genus* and *species*. There may also be intermediate levels such as subclass and superfamily. All schemes depend on a particular taxonomic view, and this is particularly true of taxa intermediate between phylum and genus, for which numerous schemes may exist in the literature. Modern taxonomy attempts to name and classify only *clades*, that is groups that can be shown (by genetics) to descend from a common ancestor. For most groups, there are too many sub-branching clades to be handled by the traditional levels of hierarchy.

5.3   REGULATION OF TAXONOMIC NOMENCLATURE

5.3.1   REGULATING BODIES

Plants: *International Code of Nomenclature for algae, fungi, and plants* (ICN). Replaced in 2011, the *International Code of Botanical Nomenclature* (ICBN). Its terms of reference now include various nonphotosynthetic groups such as slime moulds (www.iapt-taxon.org).

Cultivated plants: *International Code of Nomenclature for Cultivated Plants* (ICNCP), also known as the *Cultivated Plant Code*, 8th Edition, 2009, published by the International Society for Horticultural Science, available at http://www.ishs.org/sci/icracpco.htm.

Animals: *International Code of Zoological Nomenclature* (ICZN), 4th Edition, 2000, published by the International Commission on Zoological Nomenclature, available at http://www.nhm.ac.uk/hosted-sites/iczn/code.

5.3.2   AUTHORITIES

The complete expression of a binomial name requires as an addendum the authority for that name which may be simple (e.g. *Taraxacum absurdum* Soest) or more complex, e.g. *Taraxacum albiflorum* (Makino) Koest., where the species has been subject to a taxonomic revision. There are some differences in authority presentation between the botanical and zoological literatures. *Catalogue of Life* gives complete authority information. *Dictionary of Natural Products* does not include authorities.

5.3.3   REVISION OF TAXONOMIC NAMES

Revision of taxonomic names is a frequent occurrence. Revisions are increasingly based on genetic research but may still be based on nongenetically based expert opinion by a taxonomist specialising in a particular group. The new name once properly published (publication is now allowed electronically) becomes part of the authority. A name may also be superseded if it is shown to be a synonym for another (by comparison of the type specimens).

The *Catalogue of Life* is recommended as the most complete literature source. E.g. *Stephanotis mucronata* Blanco (Merr.) is reported to be a synonym for *Jasminanthes mucronata* (Blanco) W.D. Stevens and P.T. Li.

5.3.4   DUPLICATION OF TAXONOMIC NAMES

Duplication of binomial names within the separate botanical and zoological domains should not be possible, but, unfortunately, the two codes function entirely separately (also separately from fungi until the recent extension of the botanical code to include fungi) so there may be duplication of genus names between the two codes.

Examples:

Tandonia (Animalia, Gastropoda)
Tandonia (Plantae, Monocotyledons, Euphorbiaceae)
Tandonia (Fungi, Ascomycetes)

5.3.5 HERBAL EXTRACT NAMES

Latin-derived names for herbal extracts and other natural drugs are used in pharmacy and pharmacognosy. The orthography is not well controlled but the names should not be italicised, to avoid confusion with Latin binomials.

Examples:

Fructus Sophorae (Latin *fructus*, fruit); dried fruits of *Sophora japonica*
Radix puerariae or Puerariae Radix (Latin *radix*, root), root of *Pueraria lobata*
Cortex fraxin (Latin *cortex*, bark), bark of *Fraxinus* spp.

5.3.6 TAXONOMIC DATA IN THE *DICTIONARY OF NATURAL PRODUCTS*

The binomial names given in DNP are in general those found in the primary literature. New genus names appearing on the DNP database are validated, corrected where necessary and assigned to a position in the taxonomic hierarchy by inspection on a six-monthly basis.

Genus names are validated using the *Catalogue of Life*. Where the name reported in the primary literature is not yet in CoL, the genus name is cross-checked with other internet resources. If a genus name cannot be validated or appears unreliable, this is noted in the entry.

5.4 THE MAIN TAXONOMIC GROUPS

The Z code numbers given with each group below are the Type of Organism codes in the *Dictionary of Natural Products*, by which searches can be performed (e.g. find all compounds with a furan ring produced by lichens). The Type of Organism codes are broadly based (at the level of family to phylum) and the Z code scheme was evolved to facilitate this sort of search rather than to be taxonomically precise.

5.4.1 ARCHAEA (ARCHAEBACTERIA) (ZA)

The archaea are prokaryotic organisms inhabiting extreme environments, both marine and terrestrial, such as hydrothermal vents, and also highly saline regions. There are three generally recognised groups; thermophiles (heat-tolerant), halophiles (tolerant of highly saline media such as the Dead Sea; some species are also extremely alkali tolerant, growing in media up to pH 12) and methanogens (some species of which are also highly thermotolerant). It is also convenient to recognise a group of 'psychrophiles' tolerant of cold Arctic and Antarctic conditions. It now appears that they are the most numerous bacteria in the marine environment. They show major differences from other prokaryotes in their genome, and these are carried through into fundamental differences in their membrane structure and biochemistry. Their cell walls do not contain the glycopeptides found in the eubacteria. Stabilisation of the membrane structure is effected by esters of glycerol with characteristic branched-chain terpenoid fatty acids, a role which in prokaryotes is performed by carotenoids and/or hopanoids, and in eukaryotes by sterols.

5.4.2 Eubacteria (ZB0001)

The eubacteria are characterised by their cell wall structure, which is based on a glycoprotein formed of *N*-Acetylglucosamine and *N*-Acetylmuramic acid, cross-linked by peptide side chains containing unusual amino acids which render different bacterial strains biochemically and immunologically distinct. In Gram-positive bacteria, the glycoprotein coat forms the outermost layer; in Gram-negative bacteria, there is a further outer membrane coat which prevents this layer being stained by the reagent. Bacteria may be photosynthetic or nonphotosynthetic, and the photosynthetic bacteria may be anaerobic (sulfur bacteria) or aerobic (which includes the cyanobacteria). The former group utilises the bacteriochlorophylls as photosynthetic pigments.

The classification of bacteria is a complex and specialised subject, and there is no official classification scheme. The nearest equivalent is the scheme being evolved for Bergey's Manual (Garrity et al., 2004), which is work in progress and available for inspection online. Further major subdivisions such as alpha-, beta-, gamma- and deltaproteobacteria have been delineated according to various schemes, but the overall picture is complex. Within the context of natural products, the majority of investigations have been into two major bacterial subcategories, the actinomycetes and the cyanobacteria (see below). The code ZB0001 is used in DNP for all eubacteria which do not fall into one of these two major groups (Garrity et al., 2004; König et al., 2006; Lepage et al., 1992; Moore, 2005; Piel, 2004).

5.4.3 Cyanobacteria (Cyanophytes, Blue-Green Algae, Myxophyceae) (ZB1000)

The older term blue-green algae is now considered a misnomer, and the cyanobacteria are considered a subdivision of the photosynthetic eubacteria. They are unicellular organisms, which are both marine and terrestrial; some marine species also inhabit fresh water. Some are truly monocellular, but when found unassociated with other organisms, many species adhere via their mucilaginous coats into filaments or tufts visible to the naked eye, and are sometimes found as large colonies known as stromatolites. These are well documented in the Precambrian fossil record and thus place the cyanobacteria among the earliest known organisms. Schemes for the subclassification of cyanobacteria are based on their mode and degree of such association, or alternatively by the type of spores formed. Attempts have also been made to classify them chemotaxonomically. About 7500 species have currently been described. According to one view, as few as 200 of these may be taxonomically distinct, but conversely according to recent chemical studies, a colony appearing to consist of a single species may comprise many genetically distinct strains. Cyanobacteria are responsible for frequent algal blooms, the toxicity of which is associated with their high level of secondary metabolites.

Their cell wall structures contain some sterols, but more characteristically conjugated hopanoids such as Bacteriohopanetetrol. The fundamental chemotaxonomic distinction between prochlorophytes and cyanobacteria lies in their photosynthetic pigments; in cyanobacteria there is chlorophyll a but no chlorophyll b, which is replaced by the phycobilins, phycoerythrobilin and phycocyanobilin. These pigments are also found in the red algae, which show other chemical similarities to the cyanobacteria, notably in their polysaccharides. Cyanobacteria contain characteristic xanthophylls such as Myxoxanthophyll. Marine cyanobacteria participate in a wide variety of commensal and symbiotic arrays.

The most characteristic secondary metabolites are nitrogenous. The known metabolites are also characterised by a high degree of halogenation.

5.4.4 Actinomycetes (ZB5000)

These are a particular class of Gram-positive eubacteria showing filamentous growth and some similarities to fungi. In the past, they have often been classified as filamentous fungi and are sometimes called 'higher bacteria'. They also merit special treatment biochemically speaking because of their vast production of different types of natural product, many of them with strong antibiotic or other pharmacological activity. The most important genera by far in terms of natural products are *Streptomyces* and *Actinomyces*; according to which definition is used, the actinomycetes can also include the important pathogenic genera *Nocardia* and *Mycobacterium* (Burja et al., 2001; Gerwick et al., 2001; Moore, 2005).

5.4.5 PROTOZOA (ZD0001)

The term protozoa is difficult to define taxonomically and is subject to ongoing modification in the light of biochemical studies. It was formerly used as a blanket term to describe almost any kind of unicellular organism, but it is now known, for example, that the dinoflagellates (see ZH7000) are more closely related to the brown algae than to other unicellular organisms. In DNP, the code ZD0001 is used for all unicellular organisms that cannot be placed elsewhere. The ciliates are non-photosynthetic organisms but can often harbour photosynthetic algae as symbionts.

5.4.6 ALGAE (GENERAL)

The algae can be described as lower, mostly multicellular plants of a simple body plan, lacking well-defined differentiation into roots, stems and leaves. The classification of algae has undergone a number of changes in recent decades and there is no definitive overall plan. The most fundamental division is between the brown algal branch and the green algal branch, two groupings which show large biochemical differences. The green branch comprises not only the green algae proper (Chlorophyta), but also the red algae (Rhodophyta), which are now considered more closely related to the green algae than either of them are to the brown algae and their relatives.

5.4.6.1 Chlorophytes (Green Algae) (ZE)

About 7000 species are recognised, freshwater and marine. In their fundamental biochemistry (photosynthetic pigments, storage polysaccharides, etc.), they resemble the higher plants. Some members are unicellular, sometimes as endophytes in other species of green algae.

Green algae photosynthesise using the common carotenoids α- and β-carotenes, and contain a range of relatively common xanthophylls. The most common storage polysaccharides are amylose and amylopectin, and the commonest structural polysaccharide is cellulose, although some groups also secrete xylan and mannan. The most widespread sterols are cholesterol, brassicasterol, sitosterol and their close relatives. The known secondary metabolites of the green algae are rather limited in structural range and are mostly confined to terpenoids of relatively common skeleton, and a range of aromatics. Halogenation is uncommon. Nitrogenous compounds found in green algae tend to be low-molecular-weight amines related to the amino acids or peptides and modified peptides. There are few more highly elaborated alkaloids except for purines.

5.4.6.2 Rhodophytes (Red Algae) (ZF)

The red algae are characterised by a unique and complex reproductive cycle involving three alternating generations. The great majority of the 4000 species known are marine, sometimes inhabiting deep water. They may be mono- or multicellular with a complete absence of flagellae. The chloroplasts have a double membrane similar to those of cyanobacteria and presumably arose by endosymbiosis with these organisms. There is no general agreement about the subclassification of red algae. At the highest level, a division into two unequal subclasses is usually recognised. The Bangiophyceae, considered the more primitive group, is the smaller and consists of either unicellular or very simple multicellular organisms. The larger subgroup is the Florideophyceae, comprising the better-known more highly differentiated macroscopic plants.

An important biochemical similarity between the red algae and the cyanobacteria is the presence of the phycobilins, Phycocyanobilin (blue-green) and Phycoerythrobilin (red); the latter is responsible for the red colour of the tissues.

The secondary metabolites of the red algae are characterised by a high proportion of halogenated terpenoids, especially linear C_{15} compounds, and aromatics.

Nitrogenous natural products are relatively scarce in the majority of red algae, and mostly limited to widely distributed small molecules and cyclic peptides.

5.4.6.3 Phaeophytes (Brown Algae) (ZH1000)

About 1500 species of brown algae are known, almost exclusively marine. The term Phaeophyta is to be preferred, since modern studies have shown that they are only very distantly related to the other algae and the term 'brown alga' is therefore a misnomer, although it remains in widespread use. Together with the diatoms and the chrysophytes, they constitute the Stramenopiles. Whereas the other two subgroups are entirely monocellular, the vast majority of brown algae are multicellular and macroscopic, sometimes attaining very large size. Most species inhabit cold and temperate, often rough, seas, and are sessile, demonstrating a well-defined differentiation into a foot (holdfast), stem (stipe) and frond, and growing in surface or relatively shallow waters. Two superorders are recognised, based on life-cycle criteria. The Fucales do not show generational alternation, producing haploid gametes, which reproduce the diploid stage (cf. higher plants and animals). The other brown algae show generational alternation between a haploid gametophyte and a diploid sporophyte (cf. ferns). In some families, the two forms are both macroscopic and may be indistinguishable to the naked eye; in others, the gametophyte is microscopic.

The photosynthetic pigments of all organisms of the Stramenopiles are chlorophylls c1 and c2. (characteristic absence of chlorophyll b). Both the structural and the storage carbohydrates of the phaeophytes differ from those present in other classes of algae.

Brown algae contain a wide range of terpenoids, phenolics and meroterpenoids, but few halogenated compounds. There may be a high content of sulphated phenolics. There are also a few simple alkaloids. The reproductive cycle of brown algae is mediated by a range of hydrocarbon pheromones secreted by the female gametes which are small, non-halogenated aliphatic molecules, mostly C_{11}, of a type not found in terrestrial plants.

The tissues of brown algae uniquely concentrate inorganic arsenic and some of this is converted via biomethylation to dimethylarsinate and thence into a range of ribosides.

5.4.7 FUNGI (ZG)

Fungi are now considered part of the Eukaryote kingdom, and are characterised by the lack of a photosynthetic mechanism and by a mode of life which is saprophytic, parasitic or symbiotic. Another major biochemical difference from algae lies in their cell wall structure usually based on chitin rather than cellulose. Fungi are found throughout a wide range of environments.

The lower fungi or Chytridiomycotes (ZG1000) are taxonomically difficult to classify, resembling the higher fungi in their lack of photosynthesis, but resembling the monocellular algae in having a cellulosic cell wall structure and a flagellate stage, sometimes with alternation of generations. Chemical studies are limited.

The majority of fungi fall into the category of higher fungi or Eumycetes having typical fungal biochemistry, and which can be subdivided into the four main classes of Zygomycetes (ZG2000), Ascomycetes (ZG3000), Basidiomycetes (ZG4000) and Deuteromycetes (ZG5000). These groups are distinguished by their method of spore formation. The Deuteromycetes are an ill-defined group roughly corresponding with the term fungi imperfecti (the terms Mycelia sterilia and Hyphomycetes are also found according to various schemes). These are fungi in which no reproduction is observable and which are therefore extremely difficult to identify. Spore formation can be induced in some of them under laboratory conditions, work which shows that they are a loose collection of unrelated fungi rather than a cladistic group, and can lead to reclassification into one of the other classes. This can lead to taxonomic duplication, with the organism allotted a new name based on the reproductive form (which should take priority) while still retaining its old name.

Plants, including algae, harbour a wide variety of endophytic fungal species; for example, 116 different fungal strains were cultivated from a single specimen of *Fucus serratus*. It is not, in general, known whether any particular relationship should be considered as symbiotic, benign or pathogenic. Fungal mycelia are also found in animal tissues.

Most fungal secondary metabolites are based on a polyketide biogenesis, but some terpenoids are found. The alkaloids obtained from fungi are dominated by diketopiperazines/indoles (Bugni et al., 2004; Jensen et al., 2002; Kirk et al., 2001; Moore, 2005).

5.4.8 Bacillariophytes, Chrysophytes and Haptophytes (ZH5000)

These groups of unicellular organisms (including diatoms and golden algae) belong to the 'brown' major biochemical line and together with the brown algae (mostly multicellular) constitute the Stramenopiles. The term 'algae' formerly applied to some of these groups is now considered on biochemical and submicrostructural studies to be a misnomer (cf. brown algae). Although these organism types are linked together in the classification scheme used here, this is tentative as they may be rather unrelated. They have been relatively little studied chemically.

The range of terpenoids isolated is very narrow and limited so far to simple phytanes. It is noteworthy that in two studied diatom species, the biosynthesis of these (in the chloroplasts) is by a non-mevalonate pathway, while the steroids, produced in the cytoplasm, are mevalonate-derived.

Nitrogenous compounds are similarly few in number (Moore, 2005).

5.4.9 Dinoflagellates (ZH7000)

These monocellular organisms are economically important as the causative agents of toxic red tides. Biochemical and other studies have shown clearly that they are more closely related to the ciliates and to certain other groups than they are to other flagellate organisms. Only just over half are photosynthetic; some are carnivorous. Their main anatomical characteristic is the possession of two flagellae, one equatorial and one longitudinal. Most are unicellular but some are filamentous. They participate in a range of symbiotic associations, especially with corals and with molluscs.

The known toxins of dinoflagellates fall into two main groups, though the exact type of toxin produced is genus-specific. There are numerous polyketide-derived toxins which dinoflagellates biosynthesise by a totally different pathway from that in other organisms. The other main class is composed of nitrogenous guanidinoid toxins of which Saxitoxin is the prototype (Moore, 2005).

5.4.10 Liverworts (Marchantiophyta) (ZI0001)

These are a subgroup of the nonvascular higher plants, which were traditionally grouped together with the mosses and hornworts as the bryophytes (plants lacking vascular tissue and reproducing by spores), but which are now considered distinct. They are exclusively terrestrial plants of small size, mostly growing in humid locations. About 9000 species probably exist. They are distinguished from the mosses, which are often similar in appearance, by their unicellular rhizoid. Like the other bryophytes but unlike higher plants the dominant plant (gametophyte) is haploid; the diploid sporophyte is very short-lived. They have no economic importance but natural product isolations have been quite extensive, with at least 2000 so far identified. These include a range of terpenoids (including some unusual ones such as Hodgsonox), flavonoids and simple aromatics. Perhaps unexpectedly, a range of lignans such as Bazzania acid has been isolated (Takeda et al., 1990).

5.4.11   Mosses (ZJ0001)

These are small non-flowering nonvascular terrestrial plants, which like the liverworts have a dominant haploid gametophyte phase. They do not have fully developed roots but are anchored by threadlike rhizoids. The sporophyte stage is longer-lived than in the liverworts. Mosses may be mono- or dioicous.

Chemical studies are less numerous than for the liverworts. Natural products which have been characterised include terpenoids and flavonoids, and also some lipid-derived compounds including icosanoids, e.g. Dicranenones. Nitrogenous compounds are rare. Nitrogenous maytansinoids isolated from mosses seem to have arisen historically by gene transfer from microorganisms or by archaic symbiosis.

Some mosses emit allomones, which attract microarthropods such as mites and springtails in order to effect fertilisation, although chemical studies of the components have not yet been carried out (Cassady et al., 2004; Goffinet et al., 2004; Takeda et al., 1990).

5.4.12   Hornworts (ZJ5000)

This is a small group (about 100 species), again formerly classified under the Bryophyta but now considered distinct from the mosses and liverworts. They are relatively small terrestrial plants growing in damp habitats and like the liverworts and mosses have a haploid gametophyte as the main plant body.

Chemical studies have been limited in number. Apart from common flavonoids and some lignans, the most notable isolate has been the unusual alkaloid Anthocerodiazonin from cultured *Anthoceros agrestis*. However, many hornworts develop internal cavities which become invaded by photosynthetic cyanobacteria, and it seems likely that this may be a cyanobacterial metabolite (Takeda et al., 1990).

5.4.13   Lichens (ZK0001)

Lichens are organisms arising from symbiosis between a fungus (usually an ascomycete) and an alga or cyanobacterium (or both); they may be considered as self-contained ecosystems. They are widespread small terrestrial organisms, which can grow on almost any surface in a wide range of environments, including extreme ones. About 20,000 species are known. By convention they are named according to the fungal component of the symbiosis. They can be divided morphologically into different types (fruticose, crustose, etc.), which differ in the internal tissue organisation as well as exterior appearance.

Lichens are well studied chemically. Their metabolites include a range of aromatics including polycyclics and the characteristic depsidones and depsides, also polycyclics such as a wide range of naphthoquinones and anthraquinones. They have a variety of colours other than green, which are caused by these and other pigments such as the yellow usnic acid. Terpenoids are mostly confined to the triterpenoids such as hopanes. The range of alkaloids is limited. Lichens are used in perfumery and dyeing and also appear in traditional Chinese medicines.

5.4.14   Fern Allies (ZL)

This fairly small group is subject to varying taxonomic views but essentially consists of five families containing a limited number of surviving genera, including the horsetails (Equisetaceae, one surviving genus, *Equisetum*), club mosses (Lycopodiaceae, four genera), spikemosses (Selaginellaceae, one surviving genus *Selaginella*), quillworts (Isoetaceae, two genera) and Psilotaceae (two genera, now thought to be more closely related to the true ferns). They are vascular plants that reproduce by spores and are structurally similar to the earliest evolving vascular plants. The horsetails were dominant in the paleozoic era before the evolution of flowering plants. The plant body consists mainly of stems and the leaves are minute.

Natural products characterised from *Equisetum* are mostly simple flavonoids and low-MW aliphatic and aromatic compounds. Few alkaloids have been characterised; Nicotine occurs in both horsetails and lycopodiums. However, *Lycopodium* is a rich source of alkaloids with nearly 300 identified, mostly based on polyalicyclic skeletons derived from lysine. There are a number of still-undetermined structure. Related alkaloids have also been obtained from the related *Huperzia* spp.

Horsetails are used in some regions as food and in folk medicine but are weakly toxic.

5.4.15 FERNS (ZM)

Ferns are a group of about 12,000 species of vascular plants having stems, leaves and roots but distinguished from flowering plants in reproducing by spore formation. Whereas in the flowering plants the haploid phase is dependent on the saprophyte, in ferns it is a smaller free-living organism consisting of a prothallus resembling a liverwort in structure. The sporophyte is much larger consisting of green frond leaves attached to stems anchored to roots resembling those of flowering plants. The classification of ferns is difficult owing to often similar morphology and the occurrence of cryptospecies (morphologically indistinguishable species which are genetically distinct and do not interbreed).

Ferns are widespread and some are used as food plants; however, some contain carcinogenic toxins, most notably bracken *Pteridium aquilinum*, which is a cause of stomach cancer in Japan. The toxins responsible are aromatic terpenoids. Ferns contain a wide range of terpenoids, flavonoids and other types of nonnitrogenous natural products but only a small number of different alkaloids. Notable is an extensive range of phloroglucinol-derived metabolites such as the Albaspidins (Soeder, 1985).

5.4.16 GYMNOSPERMS (ZN)

These comprise the conifers together with the smaller cycad and ginkgo groups. Together with the angiosperms, these form the spermatophytes or seed-bearing plants. In the gymnosperms, the seeds are naked (not enclosed in an ovary) and develop on the leaf, stalk or scale surface, which in the case of conifers is modified to form the cone. Conifers are mostly evergreen trees, which are widespread in temperate and subarctic habitats, whereas the cycads are tropical and subtropical. There are more than 1000 species of gymnosperm in 14–16 recognised families.

Conifers produce a wide range of terpenoids, the harvesting of which (e.g. turpentine, camphor) is of major economic importance. Another economically important medicinal group is the taxoids (diterpenoids) from family Taxaceae which include the anticancer drug taxol (paclitaxel) (an alkaloid with pendant aromatic amine group) and its relatives. The range of sterols is limited to a small number of characterised cholestanes and relatives. There are also complex alkaloids of the harringtonine type, but 'traditional' indole alkaloids widely present in the angiosperms are absent.

5.4.17 ANGIOSPERMS (DICOTS) (ZQ); ANGIOSPERMS (MONOCOTS) (ZR)

The angiosperms are the flowering plants which reproduce by specialised floral organs, having ovules enclosed in an ovary and male organs which are reduced in size, producing light pollen aiding insect or wind pollination. The seeds are produced within fruits. They are widespread worldwide in a very wide range of habitats and are the dominant plants in the majority of ecosystems. They may be herbs, shrubs or trees, all with distinctive and complex root systems and well-developed tissue differentiation in both the above-ground and root tissues. The photosynthetic leaves are covered in a waxy cuticle. Many species are diploid or polyploid.

The traditional division of angiosperms into dicots (Dicotyledonous) and monocots (Monocotyledonous) is now considered an oversimplification although it remains in widespread use. Cladistic research shows that the monocots (approximately 98 families) are a monophyletic group but the dicots (approximately 327 families) are not. The total number of species extant is up to 400,000.

The angiosperms are of profound economic importance and the range of natural products produced by different members is immense. Angiosperms dominate all areas of plant economic importance except timber and paper production. However, as well as much research into the metabolites (including toxins, etc.) produced by the economically important species such as food plants, since the nineteenth century there has been equally significant parallel research into natural products produced by unusual, especially tropical, families, for example the Apocynaceae and Loganiaceae, sources of many alkaloids.

The number of different natural product in angiosperms reported in the *Dictionary of Natural Products* in 2013 was more than 110,000 (cf. gymnosperms 4000).

5.4.18   PORIFERA (SPONGES) (ZS)

The sponges are the most primitive of the multicellular organisms, providing an evolutionary bridge between the monocellular eukaryotes and the rest of the animal kingdom. They are multicellular organisms lacking all organ differentiation (including gonads) and some can reconstitute themselves after passing through a sieve. They are mostly marine, with a few freshwater species.

The taxonomy of sponges is complex. Three main subdivisions have been generally recognised, depending on the nature of the skeletons that they secrete; calcareous (ZS1000), siliceous or askeletal. The largest group is the demosponges, about 95% of known species, in which the skeleton is of spongine, a proteinaceous polymer similar to keratin. The Hexactinellida sponges, characterised by silica spicules of sixfold symmetry, are found only at great depth and have been little studied chemically.

In DNP a simple classification into four groups is used, which divides sponges into Calcareous sponges (ZS1000), Homoscleromorphous demosponges (ZS3000), Tetractinomorphous demosponges (ZS4000) and Ceractinomorphous demosponges (ZS5000).

Sponges participate in a wide range of symbiotic/commensal relationships, and a large number of the isolations of natural products earlier reported from them are in fact owing to the presence of cyanophytes in particular. As a result, the range of natural products reported from sponges and sponge aggregates covers the whole range of known types of marine natural product. Demosponges are the most prolific of all marine organisms in terms of the secondary metabolites that have been isolated from (but not necessarily produced by) them (Fattorusso et al., 1997; Faulkner et al., 1994; Hooper, 2002; Kuniyoshi et al., 2001; Moore, 2006; Watanabe et al., 1997).

5.4.19   CNIDARIA (ZT)

This class of organisms (medusa, sea anemones, hydroids and corals) represent the first major development in animal body-plan over the undifferentiated sponges, showing cellular differentiation into cells with different functions, but in general no well-defined organs. The term Cnidarian replaces the older 'Coelenterate'. The phylum is typified by a carnivorous lifestyle, the presence of specialised stinging cells (cnidocysts) used in the capture of prey and defensively, and a digestive system consisting of a sac with only one opening. They have a basically radial body plan, which may be modified either in the direction of a fixed polyp with a central gastric cavity (hydras) or a free-swimming medusa form (jellyfish) in which the gastric cavity is underneath. Reproduction is sexual, producing a free-swimming larval planula followed by a polyp form, although in some species only one of these is formed. About 10,000 species are documented, marine and freshwater, classified into two subphyla.

The first subphylum (Anthozoa) comprises the sea anemones, gorgonians, crinoids and corals, which have no free-floating phase, and a skeletal structure consisting either of secreted calcareous minerals or of proteinaceous material (gorgonine, analogous to the spongine found in the sponges). The anthozoa are divided into two groups depending on their symmetry; eightfold in the octocorals (alcyonians or soft corals and gorgonians) (ZT1000) or sixfold or a multiple of sixfold (hexacorals, including the sea anemones and hard corals) (ZT2000). The former subphylum is the most studied group of the cnidarians chemically.

In general, relatively few nitrogenous secondary metabolites have been isolated. The proportion of halogenated metabolites is also relatively low. Some sea anemones owe their colour to carotenoids. Octocorals are rich in prostanoids, steroids, terpenoids (but only sesqui- and diterpenes) and aromatics. The hexacorals contain a range of polyunsaturated long-chain acids. In general, the hexacorals and octacorals contain a wide range of terpenoids and steroids but with wide variations in the types isolated from the different groups. Alkaloids have also been isolated.

The other subphylum of cnidarians (Medusozoa) comprises the Cubozoa (box jellyfish, ZT5000), Hydrozoa (hydras, ZT6000) and Scyphozoa (true jellyfish, ZT7000). Chemical studies have been mostly confined to their venoms, which are peptides, although some smaller molecules have been characterised (Anderluh et al., 2002; Faith et al., 2001).

5.4.20 PLATYHELMINTHS (ZU1000)

These are flukes, tapeworms and flatworms characterised by a bilateral body plan and the complete absence of a digestive cavity. About 18,000 species are known. They may be terrestrial, freshwater or marine and many belong to orders which are exclusively parasitic (e.g. flukes). The majority of marine species belong to the class of planarians (turbellarians). These are mobile, carnivorous animals having no physical means of defence and relying entirely on substances absorbed or modified from their diet or produced by symbiotic organisms, as chemical antifeedants.

5.4.21 ANNELIDS (ZU3000)

These are the segmented worms, having an alimentary canal. They include the polychaetes, oligochaetes (earthworms; mostly terrestrial), hirudineans (leeches; mostly freshwater), echiurians and Vestimentifera. They include hydrothermal vent-living species. Natural products of interest from them include aromatics, pigments and toxins.

5.4.22 OTHER VERMIFORM GROUPS (ZU5000)

The ribbon-worms or nemerteans have yielded the powerful nicotinic receptor agonist, Anabaseine, used as a venom by the worm, together with several related oligopyridines. Toxins of the tetrodotoxin series are also found in the tissues and also some peptide toxins, which have only been investigated fully for one species, *Cerebratulus lacteus* (Neurotoxin B-IV). The unsegmented phoronidian worms Phoronopsis have yielded antibacterial bromophenols like those obtained from the annelids.

5.4.23 BRYOZOA (ZU6000)

These colonial organisms are entirely aquatic and mostly marine. They are distinguished by their unique form of gastric cavity, which is surrounded by tentacles forming an organ called the lophophore. The colonies are produced by budding and therefore consist of genetically identical individuals, each of which is surrounded by a bilayered exoskeleton, the inner layer calcareous (not always continuous) and the outer layer chitinous. They are suspension feeders, feeding on plankton and bacteria, and are found at all depths. Bryozoans have so far been less studied than the number of known species (5700) would justify given the range of interesting natural products already isolated from them. This is probably a consequence of the difficulty of harvesting them.

Their defence chemicals comprise numerous alkaloids and the Bryostatins, an important series of polyether polyketide toxins with anticancer properties, some of which have also been found in other marine organisms. The ultimate source of these may be *Candidatus* bacteria (Kem, 2001; Kerr et al., 2001; Prinsep, 2001).

5.4.24 MOLLUSCS (ZV)

This is a diverse and widely distributed phylum. The body plan is basically nonsegmented and bilateral, although in some molluscs (the gastropods), it is often modified by torsion into a spiral surrounded by a shell. There are well-developed organs inside a more or less thickened outer layer, the mantle, which secretes a shell formed of calcareous matter and protein. This shell may be external, as in the gastropods, internal as in squids, or may be totally lacking (octopuses and nudi-branchs). There is generally a muscular foot and a cephalic region, which may be highly developed into tentacles and other organs, as in the cephalopods. The alimentary canal is well developed and furnished with a rasping radula used in feeding. Different classes of molluscs show variation in this general body plan, for example the bivalves have a hinged shell, no cephalic region and no radula, and some of them also lack the foot.

They are present in marine, freshwater and terrestrial environments, ranging in size from micro-scopic to very large and show a wide range of dietary behaviour (carnivores, herbivores, filter feed-ers and detritus feeders). They also undergo a wide range of symbiotic relationships. In particular, in some molluscs, the mantle incorporates symbiotic algae providing toxic antipredator substances.

The phylum is usually divided taxonomically into seven unequal classes, but of these, four (including Chitons, ZV1000) are numerically limited and have been studied chemically little or not at all. The most important classes both in terms of number of species, economic importance and chemical studies are the gastropods (ZV2000, ZV3000, ZV4000), the bivalves (ZV6000) and the cephalopods (ZV8000).

The bivalves with their well-developed physical protective mechanism of the double shell, appear to have less need for chemical defence mechanisms and their secondary metabolites are less profuse. They have mostly been studied in terms of their economically important shellfish toxins, which are in fact microbial/dinoflagellate products. The cephalopods too have been rather little studied; their most characteristic metabolites are Adenochromines. The most studied organisms chemically have been various types of gastropod, which have little or no physical defence and rely almost entirely on chemical defence against predators.

The numerous gastropods are sometimes further divided into three subclasses, the Prosobranchia (ZV2000), the Opisthobranchia (ZV3000), and the Pulmonata (ZV4000) (Cimino et al., 1999, 2006). (This division is not recognised by the *Catalogue of Life*, but since it is a convenient subdivi-sion of a large group of natural product-producing organisms, it is followed in *Dictionary of Natural Products*.)

5.4.25 ECHINODERMS (ZW)

These organisms are characterised by a radially symmetrical body plan (which is acquired in the adult stage, the larvae being bilaterally symmetrical) and a unique system of respiration through water-filled tube feet which also provide locomotion. There is a calcareous endoskeleton. This is the largest phylum of exclusively marine animals, with about 7000 species known. They may be herbivores, suspensivores, detritivores, carnivores or necrophages. There are many examples of commensalism and parasitism between echinoderms and other organisms.

Although various classification schemes for echinoderms differ in detail, five main groups are generally recognised. The most primitive are the Crinoids (ZW1000) in which the mouth and anus are on the same surface. They have a planktonic larval form followed by an adult form, which may be sessile (sea lilies) or mobile (feather stars). The other four groups, in which the mouth and anus are on opposing faces, are the starfish (Asteriodea) (ZW2000) sea urchins (Echinoidea) (ZW3000), sea cucumbers (Holothuroidea) (ZW4000) and Ophiurioids (ZW5000).

The presence of steroidal saponins of different types in the starfish and in the sea urchins is unique in the animal kingdom, and serves to delineate them from the other echinoderms and from each other.

Throughout the echinoderms, halogenation is rare. There are also few alkaloids. The few exceptions to this generalisation are very probably derived from organisms such as dinoflagellates present in the food chain. In the starfish, the range of glycosides and of steroid sulfates is extensive. The holothurians contain many steroids and specialised triterpenoid glycosides of the holostane type (Moore, 2006; Stonik et al., 1999).

5.4.26 ARTHROPODS (ZX)

This is the largest and most successful group of land invertebrates. The principal groups are subdivided as indicated below (the scheme is that used in *Dictionary of Natural Products*).

Functionally, the most important types of natural product studied from arthropods are various kinds of toxin, and a wide range of pheromones, especially from economically destructive species such as various moths and beetles. As with marine organisms, many isolated products may not be biosynthesised *de novo* but may be from metabolic transformation of food components.

5.4.26.1 Arachnids (zx0500)

These are the spiders, scorpions and relatives.

Spider and scorpion venoms are a rich source of nitrogenous compounds, some peptides and others polyamine alkaloids, many of them powerful neurotoxins having significance in neurological research. Branched long-chain aliphatic compounds have been characterised from spider silk.

5.4.26.2 Acarids (zx0600)

Mites.

5.4.26.3 Beetles (Coleoptera) (zx1500)

These are the most numerous of all animal groups; the number of species that exist is probably in the region of a million. They are mostly characterised by biting mouth parts and a hard exoskeleton formed partly by the foremost pair of wings being hardened into wing covers (elytra), which may be fused to give an exclusively ground-dwelling or freshwater habit. There are some pest species, particularly among the weevils, and in many species it is the larvae which are pests of stored food, etc.

About 500 natural products have been described from beetles including pheromones (e.g. Brivicomin) and a range of alkaloids from defensive secretions. But given the vast number of extant species this must represent only a minute proportion of the possible metabolites.

5.4.26.4 Diptera (zx2000)

These are the two-winged flies. The most studied species is the genetically important *Drosophila melanogaster*, where the range of secondary metabolites includes its eye pigments and a range of aliphatic pheromone components.

5.4.26.5 Lepidoptera (zx2500)

These are the butterflies and moths. Natural product studies have mostly been concerned with the pheromones of economically important species (pests but also silkworms) but also butterfly wing pigments (pteridines) and other nitrogenous compounds.

5.4.26.6 Hymenoptera (zx3000)

These are bees, wasps, ants and related groups. A wide variety of venom constituents have been isolated which may be peptides or alkaloidal (piperidines, pyrrolidines and pyrazines). A variety of long-chain aliphatics have been characterised from trail and swarming pheromones of ants and bees and from royal jelly.

5.4.26.7   Hemiptera (ZX3500)

The hemipterans are the true bugs characterised by sucking mouth parts and feeding mostly on plant sap. They include the economically destructive greenfly (*Aphis*). The order includes one genus (*Halobates*) of bugs which are the only truly marine insects. Some unusual terpenoids have been isolated as well as a range of condensed aromatic pigments. As with other insects, aggregation and other pheromones contain straight-chain aliphatics.

5.4.26.8   Other Insects (ZX4000)

Among the most important under this heading are the pest cockroaches (*Periplaneta* and *Blatta* spp.) and locusts (*Locusta* spp.). These insects undergo partial metamorphosis and their moulting and juvenile hormones have been much studied.

5.4.26.9   Crustacea (ZX8000)

The crustaceans are the only arthropods that occur to any extent in the sea. In the past the crustaceans have been considered a separate phylum, but they are now often considered a major subphylum of the arthropods having in common the hard chitinous exoskeleton and the body divided into head, thorax and abdomen. About 60,000 species are known, some of them (ostracods, e.g. *Cypridina*; and copepods, e.g. *Calanus*) very small planktonic organisms. The best-known large species are the decapods (crabs and lobsters).

Chemical studies on crustaceans have been fragmentary and are mostly confined to their carotenoid pigments. The moulting hormones of crustaceans are terpenoid (Methylfarnesate) and steroidal (Ecdysone/Crustecdysone), the latter being common to insects also. The simple pyridine Homarine and the arsenical Arsenobetaine are widespread in nature but were first isolated from crustaceans. Chitin, derived industrially from crab shells, is an important industrial material.

5.4.27   HEMICHORDATES (ZY1000)

This is a numerically limited class of animal (about 100 species recorded), consisting of two surviving types of organism; the acorn worms (Enteropneusts) inhabiting temperate and tropical waters, and the pterobranchs, colonial animals inhabiting chitinous tube galleries and found in polar waters. There is differentiation of the body into three well-defined zones, and they have some but not all of the morphological characters that define the chordates. Two types of natural product have been identified from them. The first is a range of toxic cyclohexanes (e.g. Bromoxone), halogenated phenols (e.g. 2,4-dibromophenol) and halogenated indoles (Glossobalol) isolated from *Ptychodera*, *Balanoglossus* and *Glossobalanus* spp. These are used by the worms as defence chemicals and are of environmental significance. Of greater biochemical interest are the highly cytotoxic disteroidal metabolites, the Cephalostatins from *Cephalodiscus gilchristi*. Owing to the difficulty of culturing hemichordate species, there is no information currently available concerning their possible distribution elsewhere in the phylum or on their biosynthesis.

5.4.28   PROTOCHORDATES (ZY5000)

These simplest chordate animals are generally divided into two unequal groups; the urochordates (tunicates) and the cephalochordates, which are free-swimming bilaterally symmetrical animals (*Amphioxus*). Protochordates are the most developed of the invertebrates and have a notochord, which is the evolutionary precursor of the spinal column characteristic of the vertebrates. They are exclusively marine. The larger group of urochordates is divided into three classes. Of these, two, including the free-floating salps, have been little investigated. Most chemical studies have been on the third group, the sessile ascidians or sea squirts. These are filter feeders, often harbouring commensal cyanobacteria and other organisms, which may be the true source of some of the reported natural products. Their chemistry is dominated by the presence of an extraordinary range of mostly biologically potent nitrogen compounds. The ascidians also accumulate vanadium to a very high concentration in specialised cells, as well as other metals. Some have highly acidic tissues (down to pH 1).

Nonnitrogenous secondary metabolites are few in number. There is an extensive range of modified peptides and depsipeptides. Several series of these are macrocyclic thiazoles and oxazoles. Ascidians also contain a wide range of sulfur compounds (Davidson, 1993).

5.4.29 Fish (ZZ1000)

The fish represent the most numerous of marine vertebrates. They are well studied taxonomically and extensively documented in the online database *Fishbase* (a contributor database to the *Catalogue of Life*). They can be divided into cartilaginous fishes (e.g. sharks) and the larger category of bony fishes, considered to be the more highly evolved. However, the common term 'fish' is polyphyletic, i.e. by strict taxonomic criteria it is now known that different groups of fish are evolutionarily separate.

Their commercially important lipids have been extensively studied. The cartilaginous fishes contain a wide range of polyhydroxylated nitrogenous and nonnitrogenous sterols. The colour of different fish is due to the presence of various carotenoids and xanthophylls. The sexual development of many species of fish is determined by the presence of steroids in the water. The fish products that have received the most chemical attention, apart from the lipids, are the toxins produced by various species. These may be steroidal (e.g. Pavoninins), peptide (Grammistins) or alkaloid-like, such as the much-studied Tetrodotoxin and its relatives which are, however, metabolites of commensal *Pseudomonas* bacteria or dinoflagellates, and are also found in other marine organisms and even terrestrial ones. Some fish secrete peptide venoms in specialised spines.

5.4.30 Amphibians (ZZ2000)

These consist of the frogs, salamanders and some other groups. The majority of extant species are frogs and toads. Natural product research has focussed almost entirely on their venoms, produced by skin glands. In the case of toads (*Bufo* spp.), these consist of a wide range of bufanolide steroids and low-molecular-weight alkaloids. The more recently studied highly poisonous secretions of South American frogs, e.g. *Dendrobates* and *Mantella*, contain a large number of different alicyclic alkaloids such as quinolizidines.

5.4.31 Reptiles (ZZ3000)

Natural product research among reptiles mostly deals with snake venoms, which are exclusively peptide or glycopeptide. The bile acids and cloacal gland secretions of some species have also been investigated.

5.4.32 Birds (ZZ4000)

The wing pigments of various species of bird have been investigated and there has been some research on their bile acids. The toxins of some frog species, e.g. batrachotoxinins, have also been found in some bird species, presumably by a dietary route.

GLOSSARY OF TAXONOMIC AND RELATED BIOLOGICAL TERMS

The following glossary contains taxonomic and general biological terms that are likely to be encountered in the natural products/biochemical literature. Cross-references to other terms in the glossary are italicised.

Aberration: A (typically uncommon) variant on the typical form of a species. Aberrations are treated as infrasubspecific and therefore have no status in the codes.

Abiotic: The non-living parts of the environment.

Acaricide: Substance killing acarids.

Acarid: Class of arthropods including mites and ticks.

Accessions collection: Material held in museum collections that has not yet been accurately sorted and incorporated into the main collection.

Acrosome: Part of a spermatozoon.

Actinomycetes: Large class of mostly anaerobic Gram-positive bacteria that produce a wide range of antibiotics.

Aff. or Sp. aff. (abbreviation for Species Affinis): A taxon that is similar to but apparently not identical to a known one. Subsequent research may assign it to, e.g. a new species or a subspecies of a known species. More or less interchangeable with *cf.*

Agamospecies: Any plant which obligately or habitually reproduces by means of *agamospermy* and which, as a result, forms genetically isolated microspecies.

Agamospermy: Any of many mechanisms by which a plant sets seed without fusion of *gametes.*

Agonomycete: Type of *imperfect fungus* that fails to produce either fruit bodies or thalli.

Allele: One of a number of alternative forms of the same gene at the same chromosomal position.

Allelochemical: Substance produced by an organism which influences other organisms, e.g. their growth or reproduction.

Allelopathy: Production by an organism of *allelochemicals* which influences other organisms, e.g. their growth or reproduction.

Allomone: An *allelochemical*, which benefits the sender but not the receiver, e.g. a plant-produced insect defence substance.

Allopatry: Pertaining to taxa or populations that occupy geographically separated areas.

Allopolyploidy: *Polyploidy* that results from the combination of a chromosome set from each of two different individuals or more usually taxa (cf. *autopolyploidy*).

Allotype: A member of a type series of a species representing the opposite sex to that of the *holotype*, considered by the original author to belong to the same species.

Allozyme: Any of one or more variants of an enzyme coded by different *alleles* at the same gene locus.

Amphidiploid: An organism of *polyploid* hybrid origin which now behaves as a *diploid.*

Anabolism: Metabolic steps building larger molecules from smaller ones.

Analogy: Superficial similarity between two character or character states that does not reflect common evolutionary origin; cf. *homology.*

Anamorph: Asexual stage of an *ascomycete* or *basidiomycete* fungus.

Anamorphic fungi: Another name for *imperfect fungi.*

Aneuploidy: State when chromosome number varies from normal (2n) by some amount other than an integer multiple of a *haploid* chromosome set (cf. *euploidy*).

Angiosperms: Flowering plants, a major group of the higher plants.

Anther: The part of the stamen in *angiosperm* flowers where pollen is produced.

Anticodon: Sequence of three nucleotides in transfer RNA that passes on the coding for an amino acid from the codon.

Antigen: Any substance producing an immune response.

Apomorphy: An innovated character present in a *clade*; any novel feature found in a species and its descendents.

Archaea (archaebacteria): Primitive group of bacteria.

Archaeon: Member of the *archaea*.

Ascidians: Sea squirts. See above under ZY5000.

Ascomycetes (Ascomycota): Major group of the fungi. See above under ZG.

Autapomorphy: An *apomorphy* that is unique to a particular group.

Authority: Taxonomic term for the author of a *taxon* cited after its scientific name.

Autonym: In the botanical code, a computer-generated name.

Autopolyploidy: An individual or strain that has more than two genetically identical or nearly identical sets of chromosomes derived from the same ancestral species.

Autotroph: Organism that produces its own nutrients, by photosynthesis or chemosynthesis.

Axenic: Pure, uncontaminated, in reference to living cultures.

Bacteriophage: Virus that replicates within bacterial cells.

Basidiomycetes (Basidiomycota): A major group of fungi. See above under ZG.

Benthic: Ecological region at the bottom of a sea or lake.

Bifidobacteria: Bacteria of genus *Bifidobacterium*.

Binomen: The combination of a generic name and a trivial (species) name which comprises the full scientific (Latin) name of an organism.

Biota: The total collection of organisms present in a given location.

Biovar (biotype): In bacterial taxonomy, an infrasubspecific rank of no official standing denoting a strain distinguishable on the basis of biochemical or physiological properties.

Bryophytes: Mosses, liverworts and hornworts, the most primitive group of land plants. See above under ZJ.

Bryozoans: A group of marine invertebrates. See above under ZU.

Cambium: A tissue of vascular plants giving rise to other types of cell by division and specialisation.

Carpophore: Stalk supporting the *pistil* in some flowering plants; fruiting body of some fungi.

Catabolism: Metabolic processes breaking down large molecules into small ones.

Centrotype: The most typical strain of a bacterial species in culture. This has no special standing in bacterial nomenclature.

Cephalopod: A class of molluscs: squids, octopuses and relatives. See above under ZX8000.

cf.: See *aff.*

Character: Any physical structure (macroscopic, microscopic or molecular) or behavioural system that can have more than one form, the variation which potentially provides *phylogenetic* information.

Chemotaxis: Movement of an organism in response to a chemical stimulus.

Chemotaxonomy: The process of obtaining and applying taxonomic information about organisms by studying their metabolic pathways and small-molecule metabolites. Generally taken to exclude protein sequence data and other properties of macromolecules.

Chimera: Organism composed of genetically distinct cells.

Chloroplast: Photosynthetic cellular body in plants.

Chordates: Animals processing a notochord at some point in their development. Including but not limited to vertebrates.

Cilia: Slender protuberances from *eucaryotic* cells, e.g. in protozoa responsible for cell motility.

Clade: Any supposedly monophyletic group of taxa in a *phylogenic* hypothesis.

Cladistics: The process of defining evolutionary relationships between taxa using evidence from extant taxa.

Cladogram: A *dendrogram* (tree diagram) specifically depicting a *phylogenetic* hypothesis.

Class: A formal classificatory group between *phylum* and *order* (also stem of subclass and superclass).

Clone: A group of genetically identical organisms resulting from nonsexual cell division processes.

Cnidarians: Phylum consisting of corals, jellyfish and their relatives. See above under ZT.

Codes: Any of the *ICBN, ICNB, ICNCP* or *ICZN*, which formally regulate the application of names to their respective groups of organisms through their rules and additionally provided recommendations and principles. At present the *ICTF* only has a recommendatory role.

Codon: Sequence of three nucleotides coding for a single amino acid.

Coelenterate: Older term encompassing the *cnidaria* and the *ctenophores*.

Commensal: Two or more organisms living together to mutual benefit.

Congeneric: Of two or more species believed to belong to the same genus; hence congeners.

Congruence: Degree of similarity between two phylogenetic or classificatory systems, two identical classifications being said to show perfect (100%) congruence.

Conidia: Nonmotile fungal spores.

Consensus tree: A tree displaying as resolved only those features that are the same in all (or in some cases, the majority) of a set of trees suggesting different *phylogenies* for the same set of taxa.

Conserved name (Nomen conservandum): A name that is invalid according to strict interpretation of the appropriate code rules, but which is validated by the plenary powers of the appropriate commission.

Consortium: An association of organisms, especially in bacteriology.

Copepods: A group of planktonic crustaceans.

Cortex: In botany, the outer portion of the root or stem.

Cosmid: A type of *plasmid*.

Cotyledon: The seed leaf or leaves of *angiosperms* (see *dicotyledons, monocotyledons*).

Cotype: Additional specimen from the same collection as the *type specimen*.

Crinoids: Sea lilies. Marine animals in *echinoderm* phylum.

Cryptic character: Character that is difficult to observe perhaps because of small size or hidden location.

Cryptic species: Pairs or groups of biological species that are difficult to define or distinguish from the other members on the basis of external morphological characters.

Ctenophores: Phylum of marine animals (comb jellies).

Cultivar: In horticulture and agriculture equivalent to a plant (or fungus) variety.

Cuticle: Waxy outer covering of plants or fungi produced by the epidermis.

Cutin: Waxy polymer component of the cuticle.

cv.: Abbreviation for cultivar.

Cytokines: Signalling proteins secreted by cells.

Cytoplasm: In *prokaryotes* the whole of the cell content enclosed by the membrane. In *eukaryotes* the whole cell content except for the nucleus.

Cytotoxin: Any substance toxic to cells.

Decapods: Crabs and lobsters. See above under ZX8000.

Deme: A small, local subunit of a species' total population.

Dendrogram: Tree diagram as used in cladistic analysis.

Designation: The act by a revising taxonomist, whereby a single specimen from a series of *paratypes, syntypes* or *cotypes* is selected as the *lectotype*, or selection as a type species of a genus of one of a number which the original author did not specify as a type species.

Detritivore: Organism feeding on detritus.

Dichotomy: A branch point (node) tree or a decision point in a key where two new branches arise from the stem.

Dicotyledons (dicots): Major division of *angiosperms* having two *cotyledons*.

Dinoflagellates: A large group of flagellate monocellular organisms. See above under ZH7000.

Diploid: Chromosome set of 2 in the cell, as found in the majority of adult *eucaryote* cells. The ploidy number is normally assumed to be 2 unless otherwise stated.

Diptera: Two-winged flies.

Discomycetes: A subdivision of fungi. See above under VG.

Division: In zoology, a classificatory group between *class* and *order* or cohort. In botany, a classificatory group between *kingdom* and *class* equivalent to *phylum* in zoological classifications.

Domain: Informal term for a hierarchic level above kingdom to emphasize the fundamental difference between pro- and eukaryotes.

Ecdysis: The characteristic form of moulting shown by many arthropods and other animals.

Echinoderms: Phylum of (marine) animals including starfish, sea lilies and their relatives.

Echinoids: An *echinoderm* group, sea urchins.

Ecotype: A distinct form of an organism that develops under a given set of environmental conditions.

Ectoprocta: Alternative name for *bryozoans*.

Elasmobranchs: A subclass of cartilaginous fishes; sharks and relatives.

Endemic: Of a taxon (usually a species) restricted to an area in which it originated.

Endophyte: Organism, e.g. fungus, living within the tissues of a plant.

Endoplasmic reticulum: Network of membranes present in the cytoplasm of most *eukaryotic* cells.

Endosperm: Tissue inside a plant seed which nourishes the embryo.

Enterococci: Type of gram-positive bacteria.

Enterotoxin: External toxin produced by bacteria.

Eosinophil: Type of white blood cell.

Epicotyl: The upward-growing embryonic shoot in higher plants.

Epigeal: Organism (usually plant) growth or activity above the soil surface.

Epithelial cells: Cells lining the surface of organs or cavities.

Epithet: A botanical or a bacteriological name following the generic (or subgeneric) name. More or less equivalent to species name but also includes subspecies.

Epitope: The part of an antigen to which an antibody binds.

Erythrocytes: Red blood cells.

Eubacteria: True bacteria (distinction from *archaea*). See above under ZB.

Eucaryote: Organism whose cells show differentiation into separate nucleus and other organelles.

Euglenids (euglenoids): Free-living unicellular flagellate organisms.

Eukaryote: An organism (or cell) in which the genetic material is contained in a nucleus enclosed within a membrane (distinction from *prokaryotes*).

Eumycotes: True fungi.

Euploid, euploidy: State when chromosome number is changed from the normal haploid number.

Eutrophication: Enhanced growth of an ecosystem in response to increased nutrition, e.g. algal blooms.

False character: A character whose distribution of states among taxa suggests a *phylogeny* which must have shown convergence, parallelism or reversal (cf. *true character*).

Family: A formal classificatory group between *order* and *tribe* (also stem of subfamily and superfamily).

Flavedo: Outer peel of a citrus fruit.

Form: Lowest supplementary taxonomic rank in botanical nomenclature. Not accepted for taxonomic purposes by the *ICZN*.

Gamete: Cell produced during sexual reproduction (*meiosis*) containing a single set of chromosomes (cf. *haploid*).

Gametophyte: A *haploid* sexual form of an organism (plant, e.g. fern) showing alternating generations. Produces gametes by *mitosis*.

Gastropods: Large animal group, a subdivision of molluscs. See above under ZV.

Genetic distance, genetic similarity: Any of several measures of the degree of divergence (or lack of it) between populations calculated using evidence of *allele* frequencies.

Genetic drift: Random mutations and changes in gene frequency within a population that over time causes its average genetic composition to depart further away from its starting conditions.

Genospecies: A group of bacterial taxa (*species*) that are capable of exchanging genetic material.

Genotype: The genetic composition of an organism; often used in connection with one or a few specified gene loci.

Genus: A formal taxonomic grouping that comprises one or more species that are (usually) believed to be closely related.

Genus-group name: Generic and subgeneric name.

Golgi bodies: Organelles formed in most eukaryotic cells in which protein biosynthesis occurs.

Gymnosperms: Major division of the higher plants, comprising conifers and their relatives. See above under ZN.

Halophile: Organism capable of living in high salt concentrations.

Halophyte: Plant growing in highly saline conditions.

Hapanotype: A special concept of type used in connection with the *protozoa* comprising several apparently related individuals at different stages in their developmental cycle mounted on one or more slides, etc.

Haploid: In *eukaryotic* organisms in which reproduction depends upon the combining and subsequent halving of chromosome number, the stage displaying the smaller (halved) chromosome number.

Hapten: A small molecule that elicits an immune response when attached to a larger carrier molecule.

Helminths: Parasitic flatworms. See above under ZU 1000.

Hermatypic: Reef-building organisms (stony corals).

Heterologous: Relating to the cross-reaction of one component derived from one species with a complementary component derived from a different species. For example, in DNA hybridisation between taxa or interaction of an antigen from one taxon with an antibody raised against antigens from another taxon.

Heterozygosity: The possession in a *diploid* (or higher ploid) organism of two (or more) forms of a gene at a given locus.

Holomorph: The whole body of an *ascomycete* or *basidiomycete* fungus, comprising *anamorph* and *teleomorph*.

Holothurians: Sea cucumbers, a subdivision group of *echinoderms*.

Holotype: The specimen on which the concept of a species or subspecies is based. In most instances the holotype must be single specimen identified as such by the original author. In the case of protozoa the concept may be extended to cover a number of individuals (believed to be conspecific) on a microscope slide although in these cases they should more properly be referred to as a type slide. See also *hapanotype*.

Homeotype: Same as *metatype*.

Homology: Similarities between structures or other characters in two or more taxa that are the result of heritance from a common ancestor.

Homonym: Either of two or more identical scientific names that could cause a conflict of interpretation in taxonomy. In practice, homonymy applies either to species names that are placed within a single genus or genus group names within different domains of the international regulatory bodies dealing with nomenclature.

Hybrid name (nomen hybridum): A scientific name derived from two or more languages (usually Latin and Greek). Not recommended but not prohibited.

Hydrozoans: A subgroup of the *cnidaria* consisting of hydras and their relatives.

Hyphae: The filaments which form the main growth of most fungi. Also formed by *actinomycetes*.

Hypocotyl: The downward-growing embryonic shoot in higher plants.

Ichnofossil: A fossil of a construction of an organism rather than of the organism itself, for example animal tracks, burrows, nests, etc.

Ichnotaxon: A taxon based upon an *ichnofossil*.

Ichthyotoxin: Substance toxic to fish.

ICN: International Code of Nomenclature for algae, fungi and plants. Before 2011 called ICBN.

ICNB: International Code of Nomenclature of Bacteria.

ICNCP: International Code of Nomenclature of Cultivated Plants.

ICTF: International Committee for the Taxonomy of Fungi. A joint commission of the mycology division of *IUMS* and the *IMA*. Makes recommendations on fungal taxonomy and nomenclature.

ICZN: International Code of Zoological Nomenclature.

IMA: International Mycological Association.

Imperfect fungi (fungi imperfecti): Fungi for which a sexual stage has not been observed making them difficult to classify.

Inflorescence: A group of flowers arranged on a stem, see for example *raceme, umbel, spadix*.

Informative character: Any character whose distribution of states among the taxa under consideration could potentially provide information about the *phylogenetic* relationships of these taxa. Excludes invariant characters and characters displayed by only a single taxon within the group.

Inotropic agent: A substance which alters the strength of muscular contraction.

Instar: Stage between moults on the way to maturity shown by insects or other arthropods showing incomplete metamorphosis.

Introgression: Spread of genetic information from one gene pool or species to another through hybridisation and subsequent backcrossing. Extremely important in the evolution of plants but probably considerably rarer in animals.

Intron: A non-coding nucleotide sequence which is removed from messenger DNA prior to translation.

Invalid name: A name for a taxon that violates *ICZN* rules.

Isoform: Any of several different forms of the same protein having minor structural differences but functionally essentially identical (cf. *isozyme*).

Isotype: In botanical nomenclature, a duplicate of a *holotype* which was collected at the same time and place as the holotype.

Isozyme (isoenzyme): Form of an enzyme differing in amino acid sequence but carrying out essentially the same function.

IUMS: International Union of Microbiological Societies.

Kairomone: An *allelochemical* which benefits the receiver but not the sender.

Karyotype: The chromosomal complement of a cell or organism.

Key: A series or set of questions requiring decisions about taxa aimed at leading to identification.

Kingdom: In most classificatory systems, the highest taxonomic category, normally comprising the Animalia, Plantae, Fungi, Protista and Monera (Prokaryota, bacteria). Some modern phylogenetic classifications subdivide the Monera into three major groups (cf. *archaea*).

Kleptotype: Unofficial term for a specimen or part of a specimen removed without consent to another collection.

Lectin: Carbohydrate-binding protein involved in cell recognition.

Lectotype: An individual specimen selected from the *type series* of a previously described species so as to fix the identity of that species.

Legitimate name: In bacterial and botanical nomenclature, a name that is validly published and must be taken into account when deciding patterns of priority, synonymy and homonymy.

Lepidoptera: Butterflies, moths and their relatives.

Lophophore: Organ present in *bryozoans*.

Lumper: Unofficial term for a taxonomist who tends to include groups with only slight differences within the same taxon (cf. *splitter*).

Meiosis: The process during sexual reproduction which produces the gametes, reducing the cell chromosome count to *haploid*.

Meristem: Plant tissue consisting of undifferentiated cells capable of growth and differentiation.

Meristic character: A character whose states are determined by the number of times a component structure is repeated, for example the number of segments in an insect antenna.

Metaspecies, metataxon: A species or other taxon, respectively, which while only defined by *symplesiomorphies* is not known to be *paraphyletic*.

Metatype: A specimen compared with a *holotype* and believed by a competent taxonomist (frequently the original author) to belong to the same taxon.

Microsome: Vesicle body formed from the *endoplasmic reticulum* when cells are broken up. Not present in living cells.

Mitochondria: Organelles present in most eukaryotic cells that are responsible for catabolism and energy production.

Mitosis: Cell duplication by fission to produce daughters with the same number of chromosomes (ploidy count). As distinct from *meiosis*.

Mitosporic fungi: Another name for *imperfect fungi*.

Monocotyledons (monocots): Major division of *angiosperms* having a single *cotyledon*.

Monophyletic: A taxon or group of taxa all members of which have a common ancestor and which includes all the descendents of that ancestor.

Monothetic: Taxonomic classification based on a single character.

Monotypic: Of a taxon which contains only one member of a subordinate taxon; for example a genus containing only one species or an order containing only one family.

Morphocharacter: Short for morphological character. Generally refers to all forms of physical features from gross morphology to ultrastructure although typically excludes chromosomal features and cell biochemistry.

Morphorvar or morphotype: In bacterial taxonomy, an infrasubspecific rank of no official standing denoting a strain distinguishable on morphological grounds.

Muscarinic receptors: Type of acetylcholine receptors present in certain neurons.

Mycelium: Collective term for the *hyphae* of a fungus.

Mycobacteria: Family of *actinobacteria* including the tuberculosis bacterium.

Myelin: Lipoproteinoid material forming a sheath around neuronal cells.

Myosins: Family of proteins playing a role in muscle contraction.

Myxobacteria: Group of bacteria notable for swarming (gliding) behaviour.

Myxomycetes: Slime moulds.

Naked name (nomen nudum): Scientific name that refers to an organism which does not exist or has not been validly described.

Necrophage: Organism feeding on dead tissues.

Neotype: A specimen selected to represent and fix a previously described species whose original type material is irretrievably lost or destroyed.

New name (nomen novum): A name proposed to replace an existing name that is invalid, e.g. due to *homonymy*.

Nothomorph: A distinct variant of an interspecific hybrid usually resulting from separate hybridisation events.

Nothotaxon: A taxon resulting from hybridisation of two or more taxa of the same rank. Term used principally in botanical nomenclature.

Notochordates: Subgroup of the protochordates. See above under ZY5000.

Objective synonym: Synonym where two or more named taxa (genus or species) have the same type (species or type specimens, respectively) and, therefore, about which there can be no difference of opinion (cf. *subjective synonym*).

Operculum: A covering, e.g. in gastropod anatomy.

Order: A taxonomic category above *family* and below *class*.

Organelle: Structure within *eukaryotic* cytoplasm (e.g. *mitochondria*).

Original description: Generally the first publication that names a new taxon and attempts to characterise it such that it can be distinguished from other existing taxa (also see *protologue*).

Original designation: The designation of the *type species* of a *genus* by an unambiguous statement in the original description.

Orthoptera: Grasshoppers, crickets and their relatives.

Osteoblast: Type of bone cell responsible for building bone tissue.

Osteoclast: Type of bone cell responsible for reabsorption of bone tissue.

Panmixis: Breeding in which all members of a population or species are equally likely to cross-fertilise any other.

Paralectotype (also but less often **lectoparatype**): In zoological nomenclature, the remaining specimens of a type series (i.e. *paratypes*) after a *lectotype* has been designated. Not recognised in botanical nomenclature where only one type specimen is acknowledged.

Paraphyletic: Describes a group defined by possession of a uniquely derived character inherited from a common ancestor, but which may not include all descendents of the common ancestor.

Paratype: In zoology, specimens of the type series other than the *holotype*.

Parenchyma: In higher plants, the bulk of the nonwoody tissues consisting of undifferentiated cells. In animal organs, the bulk of the functional tissue.

Parsimony: *Cladistic* principle that the most likely *phylogenetic* explanation must be the one requiring the least number of evolutionary steps.

Pathovar (pathotype): In bacterial taxonomy, an infrasubspecific rank of no official standing denoting a strain distinguishable on the basis of its pathogenicity spectrum.

Patristic: Relating to similarity due to having a common ancestor.

Patronym: A scientific name based on and in honour/recognition of a named person.

Pedicel: Flower stalk.

Pelagic: Zone of sea or lake not in the surface layer or the bottom (*benthic*) layer.

Peroxisome: Type of *organelle* in eucaryotic cells responsible for lipid and other metabolism.

Petiole: The stalk attaching leaf to stem in flowering plants.

Phage: Synonym for *bacteriophage*.

Phagovar (phagotype): In bacterial taxonomy an infrasubspecific rank of no official standing denoting a strain distinguishable on the basis of its susceptibility to certain phages.

Pheromone: An *allomone* which triggers an intraspecific response (e.g. clustering behaviour, sexual attraction).

Phloem: Specialised tissue present in vascular plants, responsible for water and nutrient transport.

Phylogenetic systematics: Another name for *cladism*.

Phylogeny: The evolutionary history of a group of taxa.

Phylogram: A *dendrogram* indicating a hypothesised evolutionary history which additionally indicates by means of branch length, the degree of evolutionary change believed to have occurred along each lineage.

Phytohaemagglutinins: Plant *lectins*.

Pistil: Female reproductive part of an *angiosperm* flower, consisting of the ovary and conjoined parts.

Plankton: Free-floating microscopic marine organisms, taxonomically diverse.

Planula: The free-living larval form of *cnidarians*.

Plasmid: A (usually cyclic) small DNA fragment separate from the main chromosome(s) in *prokaryotes* and widely employed in genetic manipulations (see also *cosmid*).

Plenary powers: Powers taken by the international commissions to make nomenclatural decisions varying the rules of their respective codes if they deem it to be sufficiently important.

Pleomorphic: Of asomycete and basidiomycete fungi in which a species is represented by separate asexual and sexual spore-producing morphs (*anamorphs* and *teleomorphs*, respectively). The *ICBN* allows independent names for anamorphs until the teleomorph is known.

Plesiomorphy: The ancestral character state for a character in a group of organisms (cf. *apomorphy*).

Ploidy: The number of sets of chromosomes in a cell. See also *haploid, diploid, polyploidy*.

Polyclade: A multiple-entry identification key now normally accessed by computer program.

Polymorphism: (1) The occurrence in one species of more than one discrete phenotypic state (either genetically based or otherwise); (2) the occurrence of two or more states of a character among the members of a taxon.

Polyphyletic: A group which does not include the common ancestor of all of its members (cf. *paraphyletic*). For example, fish, which do not all have a common ancestor.

Polyploidy: Chromosome set greater than two in a cell. Common in plants, less common in animals. See also *aneuploidy, euploidy*.

Polythetic: (1) Of a taxon whose membership is defined not by possession of a single characteristic, but by having a defined minimum number of a set of attributes; (2) A *key* in which each lead of a couplet may make use of more than one character (cf. *monothetic*).

Polytomy: A branch point in a tree at which three or more branches arise from the ancestral line.

Porifera: Sponges (zoological phylum).

Primary homonym: Either of two identical species names that were both originally raised for (usually but not necessarily different) taxa in the same genus; only one can be a valid name, the other needing replacement, cf. *secondary homonym*.

Priority: A principle establishing that the correct name of a taxon, if more than one name has been given, is the one which was published on the earliest date. Priority only extends back as far as a defined date and may be overruled for the sake of stability (cf. *conserved name*).

Prokaryote: Organism which lacks a membrane-bound nucleus and complex *organelles*.

Propagule: General term for biological entity used by an organism for propagation, e.g. seed, spore, cutting (in horticulture).

Protologue: Collectively all information pertaining to a taxon or its *type specimen*(s) given in the original description.

Protozoa: Unicellular eukaryotic organisms. A diverse group.

Pseudogene: A gene which is no longer functional due to some change in its control region. Its selective neutrality makes it particularly attractive for *phylogeny* reconstruction.

Pseudomonad: A type of bacterium.

Psychrophile: (Micro)organism tolerant to cold (arctic) conditions.

Pteropods: A group of gastropods (sea slugs and relatives).

Punctuated equilibrium: A model of Darwinian evolution in which changes result from periods of rapid evolution separated by periods of relative constancy.

Raceme: A type of *inflorescence* where the flowers are carried on short stalks from a main stem.

Radula: Rasping organ present in molluscs.

Regnum: Latin for *kingdom* most often employed in botanical nomenclature.

Rejected name: Any name for an organism other than its *valid name*.

Rejected work: A published work in which any nomenclatural acts are either unavailable under the relevant code or which have been rejected (and therefore have no standing in nomenclature) by a specific decision of the relevant commission.

Replacement name: A scientific name that has been found to be a junior *homonym*.

Retroviruses: A group of viruses carrying their genetic information as RNA and replicating by reverse transcription.

Rhizome: A subterranean stem.

Rhizophore: Downward-growing leafless shoot found in some club mosses.

Ribosome: Large cellular organelle involved with protein synthesis.

Salp: A free-floating planktonic *tunicate*.

Sarcoplasm: The cytoplasm of muscle fibre cells.

Schizotype: An implied *lectotype*.

Scientific name: For species, it is *binomen*, for a subspecies, it is *trinomen*.

Secondary homonym: A homonym which arises when a taxon with a certain *specific name* is moved from one genus to another, thus creating a duplication with a pre-existing scientific name.

Section: An informal taxonomic rank between *genus* and *series* used principally in botanical and bacteriological nomenclature.

Selectin: Type of glycoprotein involved in cell adhesion.

Series: An informal supplementary taxonomic rank between *section* and *species* used principally in botanical and bacteriological nomenclature.

Serovar (serotype): In bacterial taxonomy an infrasubspecific rank of no official standing denoting a strain distinguishable on the basis of its antigenic properties.

Seta: Stiff hair-like structure or bristle.

Sibling species: Either of two very closely related species which can only be distinguished by minute or *cryptic characters* but are genetically isolated from one another. By implication, these are *sister groups*.

Siderophore: Iron-chelating compound produced by microorganisms or plants.

Sister group: Either of a pair of taxa or groups whose closest common ancestor is not shared by any other group.

Soredium: Asexually produced *propagule* produced by some lichens comprising both a fungal and an algal component.

sp.: Abbreviation for species (singular). Used when the true identity is unknown, e.g. *Pieris* sp.

Spadix: Type of inflorescence in which the flowers are borne directly on a fleshy stem.

Species: A taxonomic rank below *genus* and above the subspecific ranks of subspecies, variety, etc. It is subject to various definitions.

Species group name: Specific plus subspecific names.

Specific name: In zoology, the combination of a generic and a trivial name although commonly now applied to just the trivial part of a binomen (cf. *epithet*). In botany, the term specific epithet should be used.

Sphaeroplast: Spherical body formed from a cell after removal of the cell wall.

Splitter: Informal term for a taxonomist who tends to divide into separate taxa on the basis of small differences (cf. *lumper*).

Sporophyte: The asexual (spore-producing) form of an organism (e.g. fern) showing alternating generations. *Diploid*, producing spores by *meiosis*.

spp.: Abbreviation for species (plural). Used to indicate several species usually of the same genus.

Squamous cell: Type of flattened epithelial cell found in the skin and other locations.

ssp.: Abbreviation for subspecies (singular).

Staphylococci: Spherical gram-positive bacteria responsible for many infections.

Stomata: Pores on the leaf undersides of higher plants through which transpiration and gas exchange occurs.

Strain: In bacteriology, a pure culture of the descendents of a single isolation.

Streptococci: A type of rod-shaped gram-positive bacteria responsible for infections.

Stromata: The connective tissue of an organ.

Subjective synonym: Synonymy resulting from two or more taxa with different names which are, in fact, the same (cf. *objective synonym*).

Suspensivore: (Aquatic) organism feeding on suspended material.

Symbiont: An organism living in symbiotic relation with another.

Sympatry: The occurrence together at the same locality or in overlapping areas of two populations.

Symplesiomorphy: *Plesiomorphous* character states shared by a group of taxa due to shared ancestry.

Synapomorphy: An *apomorphous* character shared by two or more taxa and thus indicating common ancestry for the members of this group.

Synomone: An *allomone* which benefits both the sender and receiver.

Synonym: Each of a set of different generic or specific names that can be applied to a single taxon having been described on more than one separate occasion (usually but not always by different authors) (see also *subjective synonym, objective synonym*). Unlike common grammatical uses, taxonomic synonyms do not have equal status but are subordinate to the scientific name.

Synoptic collection: A collection of correctly identified representatives of all taxa from a given area used as a reference collection to help confirm subsequent identifications.

Syntype: Any member of a type series for which no *holotype* or *lectotype* has been designated.

Tautonymy: The identical (or sometimes virtually identical) spelling of a generic, specific or sub-specific name.

Taxon: Any definable taxonomic unit whether described or not, e.g. subspecies, species, tribe, genus, family, etc.

Teleomorph: Sexually reproductive phase of an *ascomycete* or *basidiomycete* fungus, typically a fruiting body.

Teleosts: Bony fishes.

Thermophile: (Micro)organism tolerant to hot conditions.

Tribe: A formal classificatory group between family and genus (also stem of subtribe and supertribe).

Trichomes: General term for hairs and hair-like structures.

Trichomonads: An order of unicellular organisms, mostly parasites or endosymbionts.

Trinomen: A scientific name comprising a genus, species and subspecies name. In zoology, the abbreviation ssp. may be omitted, e.g. in *Blaps japanensis yunnanensis*.

True character: A character whose state distribution among taxa correctly reflects the evolutionary history of the group, i.e. all taxa possessing a given *apomorphous* character states comprise a *monophyletic* group and all those lacking it do not belong to the group, cf. *false character*.

Tunicates: A phylum of simple chordate animals. See above under ZY5000.

Type culture: A living pure culture derived from a single isolation of a *prokaryote* being the effective type material upon which the prokaryote species is based. Incorrectly used for Protista, Fungi or other multicellular organisms for which there should be a permanently preserved *holotype*.

Type series: All specimens of a species upon which the original author based the original description but excluding any that the author specifically mentions as variants or as doubtful members of the species.

Type slide: In zoological nomenclature, a special form of type specimen consisting of a series of directly related individuals of a protozoan mounted as a microscopic preparation and used to define the species concept. No one individual in such a type series can be regarded as either a *holotype* or *lectotype*.

Type species: The species designated as the type of a genus-group name.

Type specimen: The specimen or any member of a series of specimens (*type series*) on which the original description and hence the concept of the species is fixed.

Umbel: A type of *inflorescence* in which the pedicels spread from a common point, characteristic of the Umbelliferae.

Uncertain name (nomen dubium): A name that cannot be associated with certainty with a known taxon because the original description is inadequate and the type specimen is lost or in poor condition.

Uninformative character: A character whose variation among a set of taxa provides no information about their phylogenetic relationships, e.g. *autapomorphies*.

Unrooted tree: A network connecting a set of taxa which has not been connected to the common ancestor.

Urochordates: A subdivision of the protochordates. See above under ZY5000.

Valid name: The correct scientific name for a taxon (cf. *rejected name, invalid name*).

Variety: A supplementary taxonomic category between subspecies and *form* in botanical nomenclature. Not accepted for taxonomical purposes by the *ICZN*.

Vascular plants: The most highly developed group of plants, defined as having a vascular system. Including but not limited to *angiosperms* and *gymnosperms*.

Vicariance: The presence of closely related taxa in different geographic areas as a result of the formation of natural barriers dividing an existing population, as opposed to through jump dispersal.

Voucher specimen: A specimen (or its remains) deposited in a permanent collection to provide identity evidence should there be a subsequent need.

x: Denotes a natural hybrid, e.g. *Cattleya* x *hardyana* is a natural hybrid of *Cattleya dowiana* and *Cattleya warscewiczii*.

Xylem: Specialised tissue present in vascular plants, responsible for support (e.g. wood) and nutrient transport.

Zygote: The initial cell produced by fusion of two gametes in sexually reproducing organisms.

REFERENCES

Anderluh, G. et al., *Toxicon*, 2002, **40**, 111–124 (rev, anemone toxins).

Blunt, J.W. et al., *Dictionary of Marine Natural Products*, Chapman & Hall/CRC, Boca Raton, FL, 2008. (The introduction gives a more detailed overview of the taxonomy of marine organisms.)

Brummitt, R.K., *Vascular Plant Families and Genera*, Royal Botanic Gardens, Kew, London, 1992.

Bugni, T.S. et al., *Nat. Prod. Rep.*, 2004, **21**, 143–163 (rev).

Burja, A.M. et al., *Tetrahedron*, 2001, **57**, 9347–9377 (rev).

Cassady, J.M. et al., *Chem. Pharm. Bull.*, 2004, **52**, 1–26 (maytansinoids in mosses).

Cimino, G. (ed.) et al., *Progress in Molecular and Subcellular Biology*, Vol. 43, Springer, Berlin, Germany, 2006.

Cimino, G. et al., *Curr. Org. Chem.*, 1999, **3**, 327–372 (rev, opisthobranchs).

Cimino, G. et al., *Marine Chemical Ecology*, eds. McClintock, J.B. et al., CRC Press, Boca Raton, FL, 2001.

Davidson, B.S., *Chem. Rev.*, 1993, **93**, 1771–1791 (rev).

Erskens, R.H.J., *Nat. Prod. Rep.*, 2011, **28**, 11–14.

Faith, F.M.Y. et al., *Rec. Adv. Mar. Biotechnol.*, 2001, **6**, 85–100.

Fattorusso, E. et al., *Progr. Chem. Org. Nat. Prod.*, 1997, **72**, 215–301 (rev, sponge glycolipids).

Faulkner, D.J. et al., *Pure Appl. Chem.*, 1994, **66**, 1983–1990 (rev).

Garrity, G.M. et al., *Taxonomic Outline of the Prokaryotes, Bergey's Manual of Systematic Bacteriology*, 2nd edn., version 5.0, Springer, New York, May 2004. http://141.150.157.80/bergeysoutline/outline/bergeysoutline_5_2004.pdf

Gerwick, W.H. et al., *Alkaloids: Chem. Biol. Perspect.*, 2001, **57**, 75–184 (rev).

Goffinet, R. et al., *Monogr. Syst. Bot.*, 2004, **98**, 205–239 (Missouri Botanical Garden Press) (Chemosystematics of bryophytes).

Hooper, J.N.A., *Systema Porifera: A Guide to the Classification of Sponges*, Kluwer/Plenum, New York, 2002.

Jensen, P.R. et al., *Br. J. Pharmacol.*, 2002, **137**, 293 (rev, natural products from marine fungi).

Kem, W.R., *Rec. Adv. Mar. Biotechnol.*, 2001, **6**, 187 (revs).

Kerr, R.G. et al., *Rec. Adv. Mar. Biotechnol.*, 2001, **6**, 149–164.

Kirk, P.M. et al. (eds.), *Dictionary of the Fungi*, 9th edn., CABI Publishing, Wallingford, U.K., 2001.

König, G.M. et al., *ChemBioChem*, 2006, **7**, 229–238 (rev, natural products from associated microbes).

Kornprobst, J.M. (ed.), *Encyclopedia of Marine Natural Products*, Wiley-Blackwell, Hoboken, NJ, 2010.

Kuniyoshi, M. et al., *Rec. Adv. Mar. Biotechnol.*, 2001, **6**, 29–84 (rev).

Lepage, S.P. et al., *International Code of Nomenclature of Bacteria*, American Society for Microbiology, Washington, DC, 1992.

Moore, B.S., *Nat. Prod. Rep.*, 2005, **22**, 580–593 (rev).

Moore, B.S., *Nat. Prod. Rep.*, 2006, **23**, 615–629 (rev, biosynth).

Piel, J., *Nat. Prod. Rep.*, 2004, **21**, 519–538 (rev, symbiotic bacteria).

Prinsep, M.R., *Rec. Adv. Mar. Biotechnol.*, 2001, **6**, 165–186.

Soeder, R.W., *Bot. Rev.*, 1985, **51**, 442–536 (rev, natural products).

Stonik, V.A. et al., *J. Nat. Toxins*, 1999, **8**, 235–238 (rev, holothuroid toxins).

Takeda, R., Hasegawa, J. and Shinozaki, M., *Bryophytes: Their Chemistry and Chemical Taxonomy*, eds. Zinsmeister, H.D. and Mues, R., Clarendon Press, Oxford, U.K., 1990.

Venkateswarlu, Y., *Rec. Adv. Mar. Biotechnol.*, 2001, **6**, 101–143 (revs, corals).

Watanabe, Y. et al., *Sponge Sciences: Multidisciplinary Perspectives*, Springer, Tokyo, Japan, 1997.

6 Natural Product Skeletons
Occurrence and Classification of Natural Products

This chapter gives an overview of the known types of natural product skeletons with their numbering, or where there are skeletal variations within the group, an illustration is given of a representative example compound. Groups are listed in the order of the type of compound classification scheme of the *Dictionary of Natural Products* (DNP) and the codes given with each heading refer to that scheme.

The occurrence numbers taken from the 2013 release of DNP give an indication of the number of natural products so far reported for each type. However, this is only an approximate indication since

1. There are many other structure codes in DNP for the simpler types of natural products. Nearly all of these are self-explanatory (e.g. VA1950, monocarboxylic alcohols) and are not included in the charts.
2. Natural products of a hybrid type (e.g. flavonolignans) are frequently given more than one type of compound code and, therefore, appear in the chart statistics twice.

The numbering schemes shown are generally those appearing in DNP. Where no numbering is given for a particular skeleton, this may mean that there is more than one scheme in the literature, in which case the individual DNP entries need to be consulted for more information on numbering schemes. For some groups of more recently characterised natural products, no scheme exists in the literature and DNP has not yet published the standardised numbering (Table 6.1).

TABLE 6.1
Natural Product Parent Skeletons

Aliphatic-related skeletons (VA)

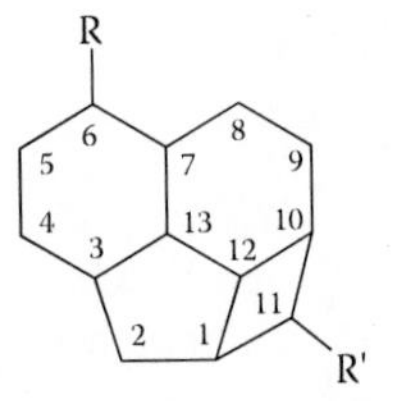

Endiandric acids (53)
VA2640

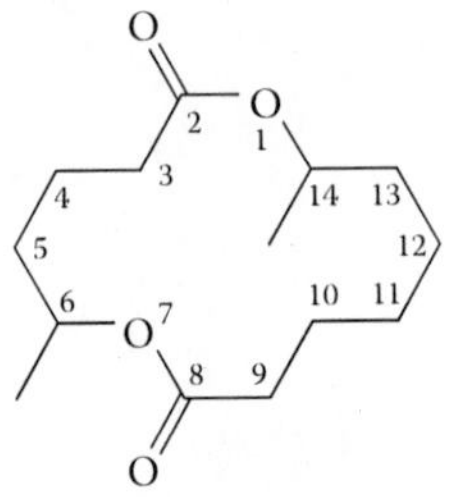

Colletodiols (14)
VA6000

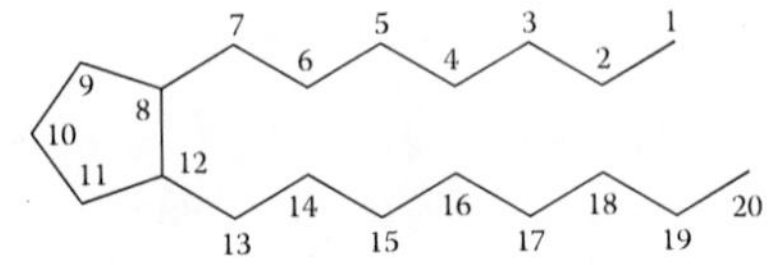

Prostaglandins group (332)
VA6100-VA6130

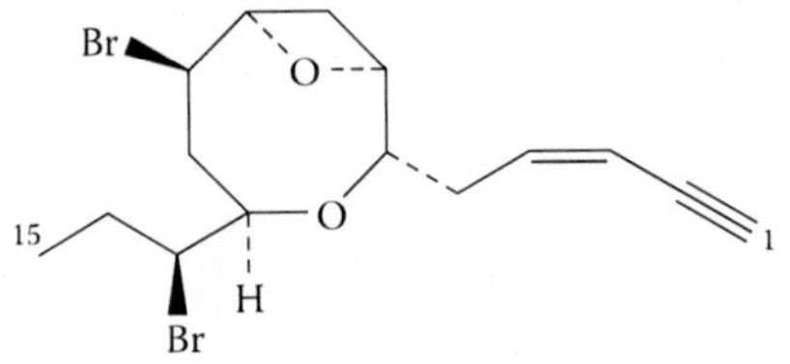

e.g. Laureatin
Marine acetogenins (217)
Linear C_{15} compds.
with various epoxidations
VA8500

Polyketides (VC)

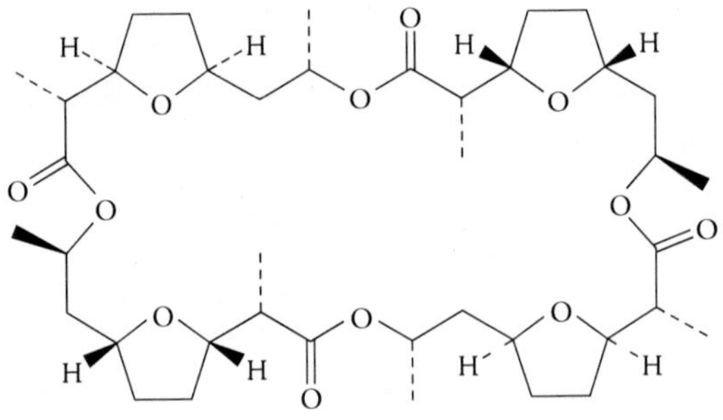

e.g. Nonactin
Nactins (51)
VC0060

(Continued)

TABLE 6.1 (*Continued*)
Natural Product Parent Skeletons

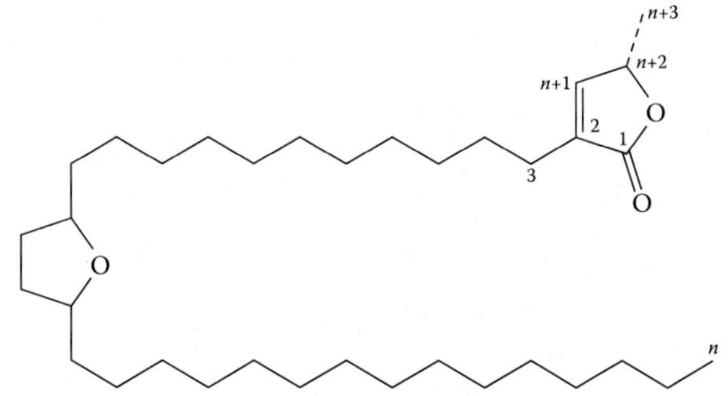

Acetogenins (542)
Various chain lengths and epoxidation
patterns
VC0080-VC0087

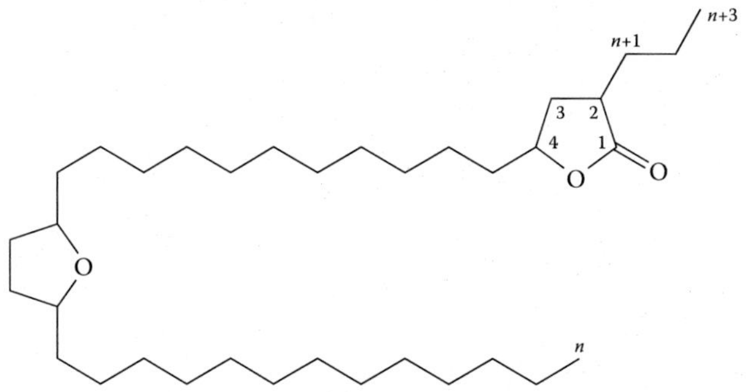

Isoacetogenins (45)
Various chain lengths and
epoxidations
VC0088

(*Continued*)

TABLE 6.1 (*Continued*)
Natural Product Parent Skeletons

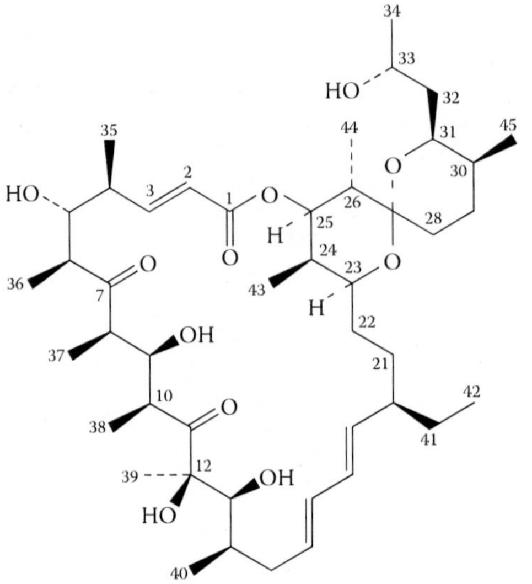

e.g. Oligomycin A
Oligomycins/Cytovaricins (58)
VC0120

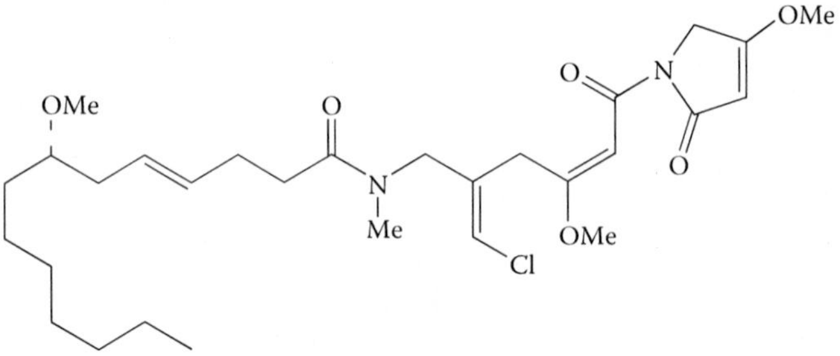

e.g. Malyngamide A
Malyngamides (72)
VC0180

(*Continued*)

TABLE 6.1 (*Continued*)
Natural Product Parent Skeletons

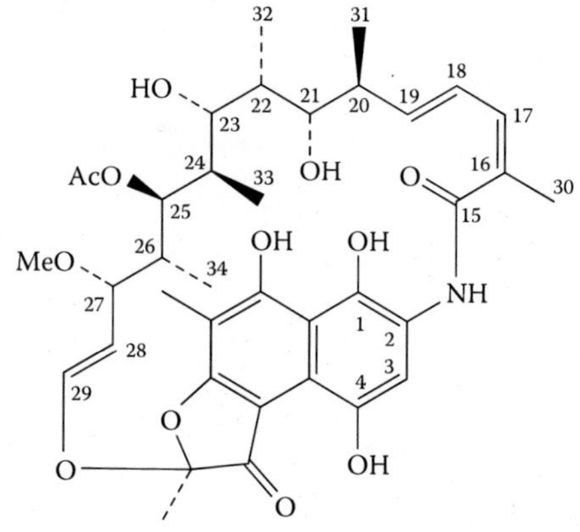

e.g. Rifamycin
Ansamycins (233)
VC0200

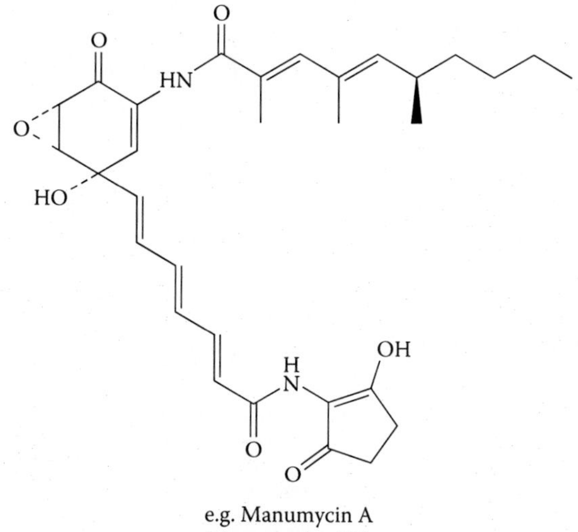

e.g. Manumycin A
Manumycins (32)
VC0280

(*Continued*)

TABLE 6.1 (*Continued*)
Natural Product Parent Skeletons

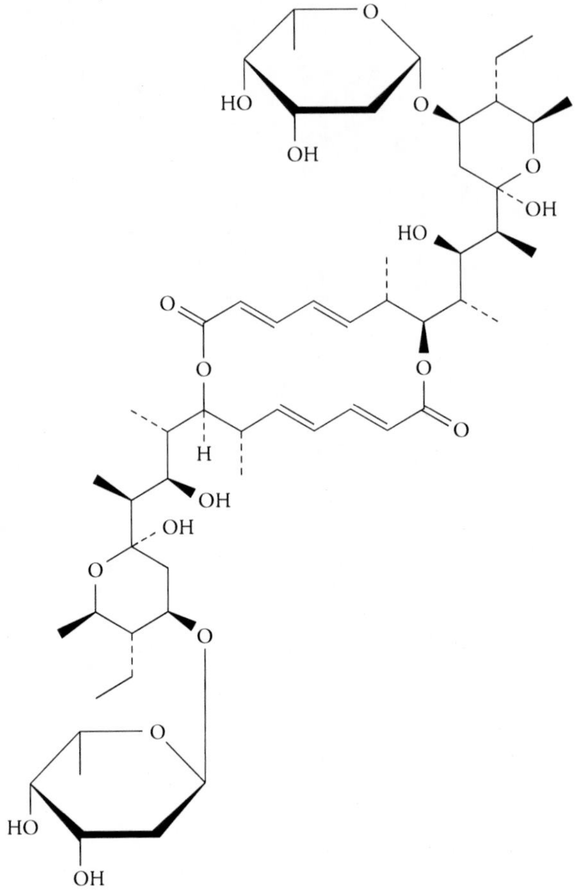

e.g. Elaiophylin
Efomycins (13)
VC0295

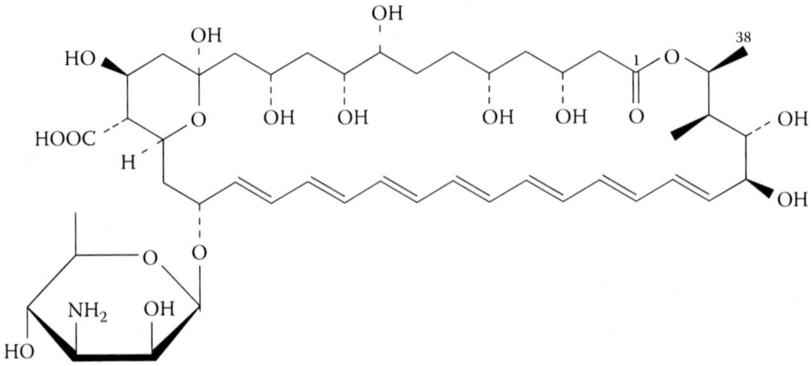

e.g. Amphotericin B
Macrolactone polyketides (416)
VC0300

(*Continued*)

TABLE 6.1 (*Continued*)

Natural Product Parent Skeletons

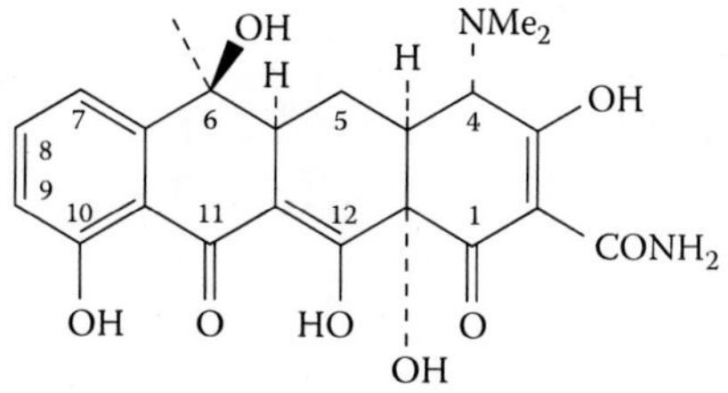

e.g. Tetracycline
Linear tetracyclines (111)
VC0400

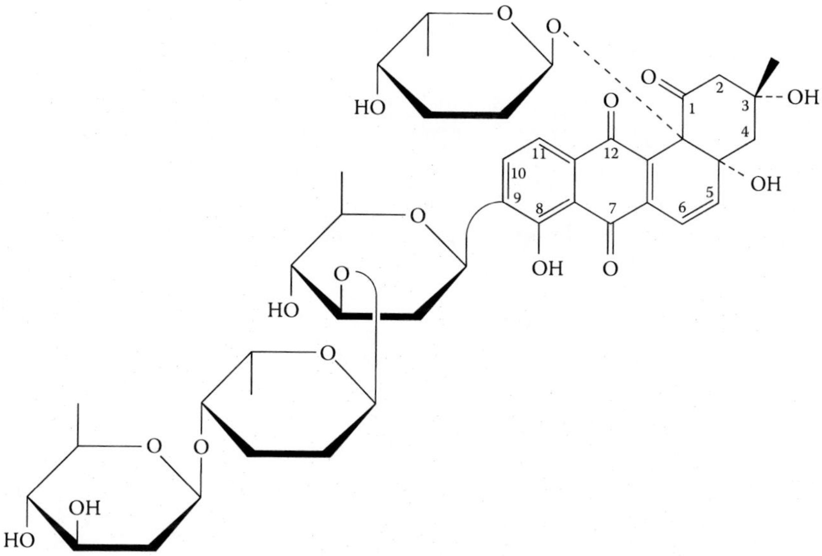

e.g. Urdamycin A
Angucyclines (293)
VC0450

(*Continued*)

TABLE 6.1 (*Continued*)
Natural Product Parent Skeletons

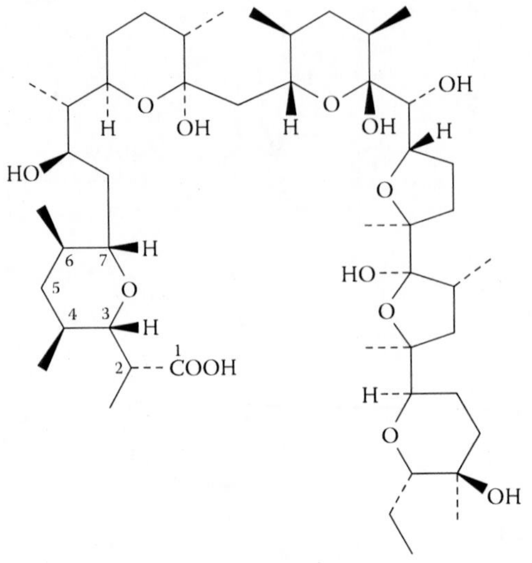

e.g. Alborixin
Monensin/Nigericin polyethers (115)
VC0460

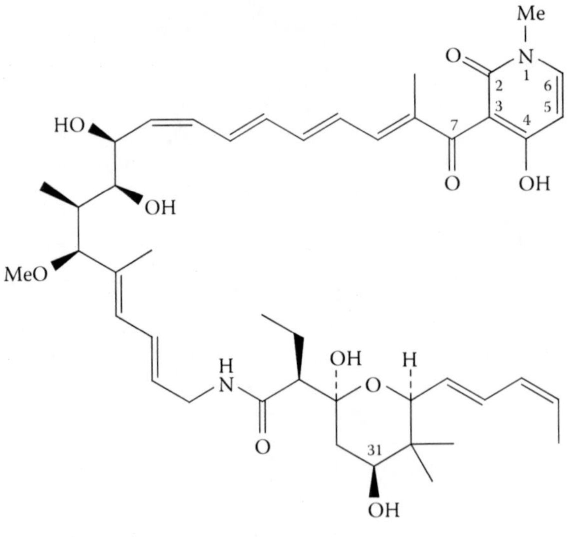

e.g. Factumycin
Elfamycins (50)
VC0520

(*Continued*)

TABLE 6.1 (*Continued*)
Natural Product Parent Skeletons

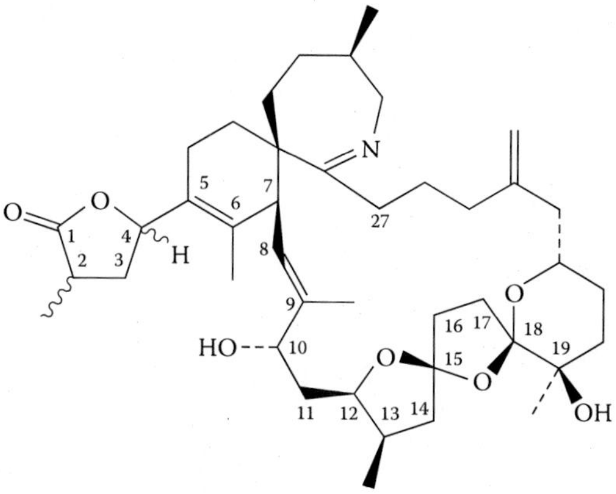

e.g. Spirolide B
Spirolides and pinnatoxins (30)
VC0550

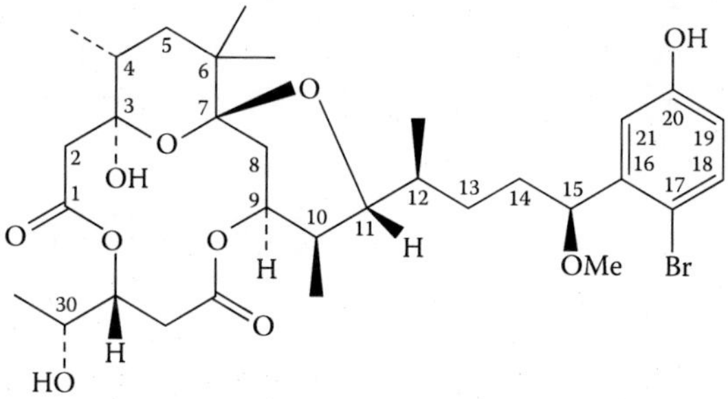

e.g. Aplysiatoxin
Aplysiatoxins (17)
VC0570

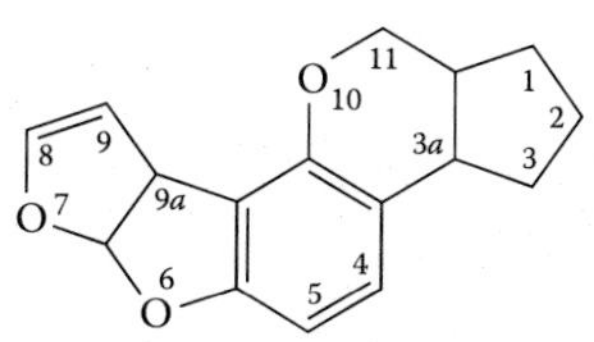

Aflatoxins (52)
VC0600

(Continued)

TABLE 6.1 (*Continued*)
Natural Product Parent Skeletons

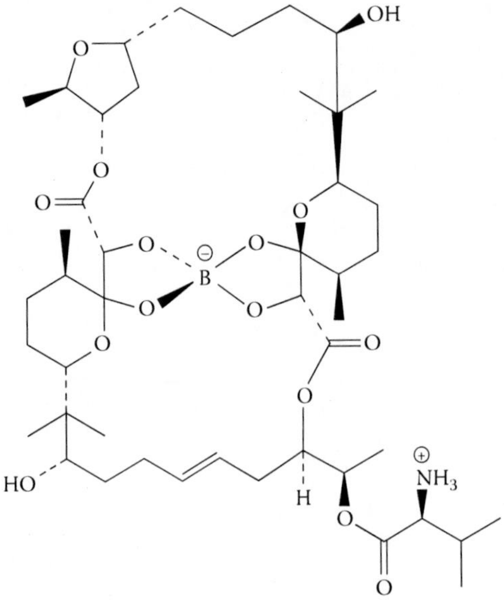

e.g. Boromycin
Boromycins (13)
VC1050

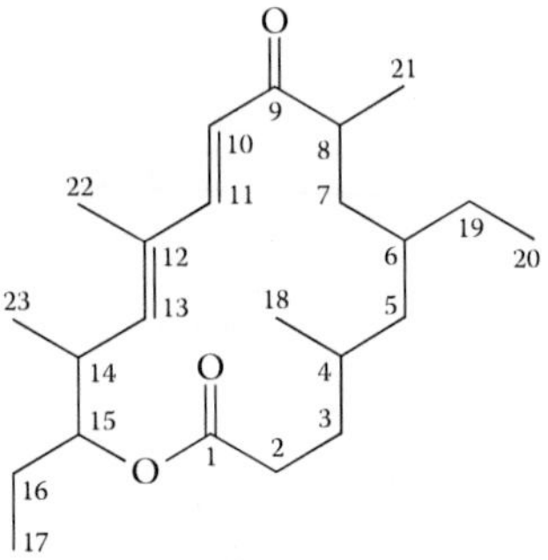

Tylosins (199)
VC1100

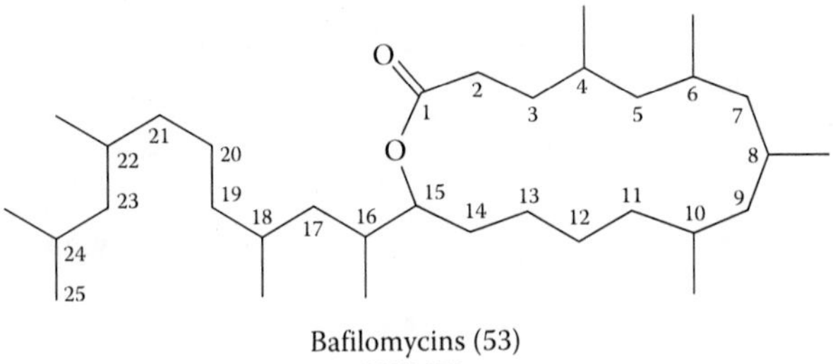

Bafilomycins (53)
VC1150

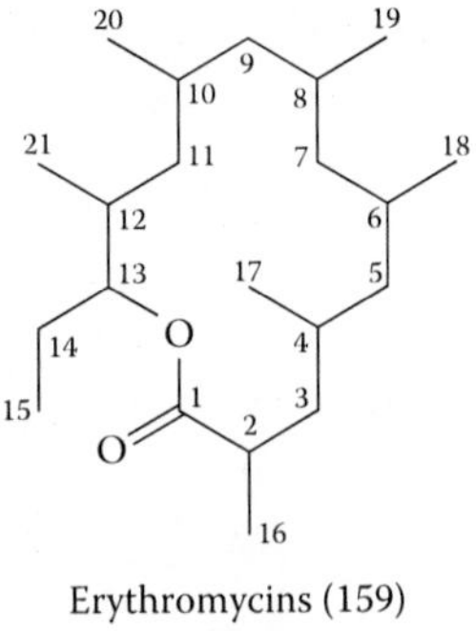

Erythromycins (159)
VC1200

(*Continued*)

TABLE 6.1 (*Continued*)
Natural Product Parent Skeletons

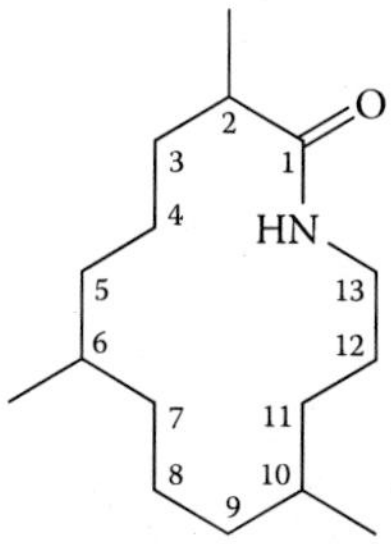

Fluvirucins (18)
VC1210

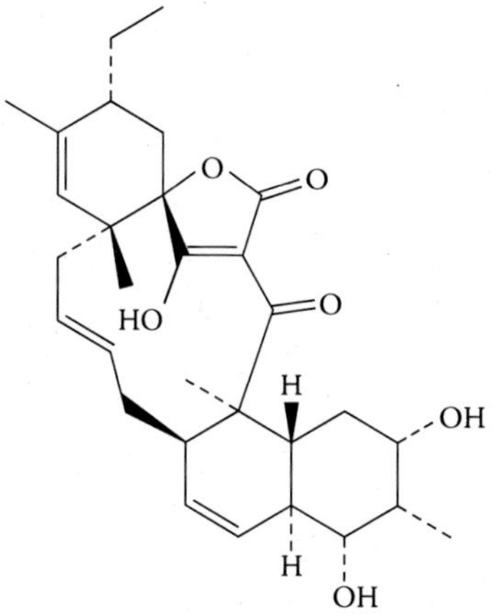

e.g. Nomimicin
Chlorothricins (83)
VC1300

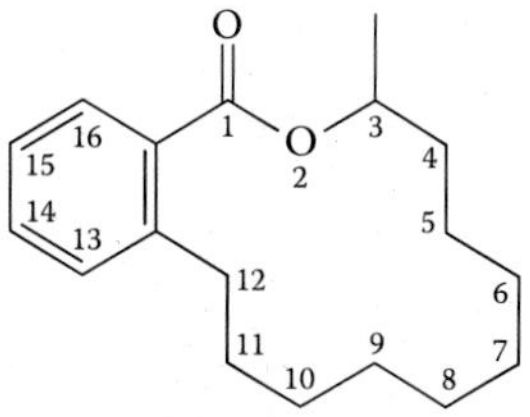

Zearalenones (154)
VC1350

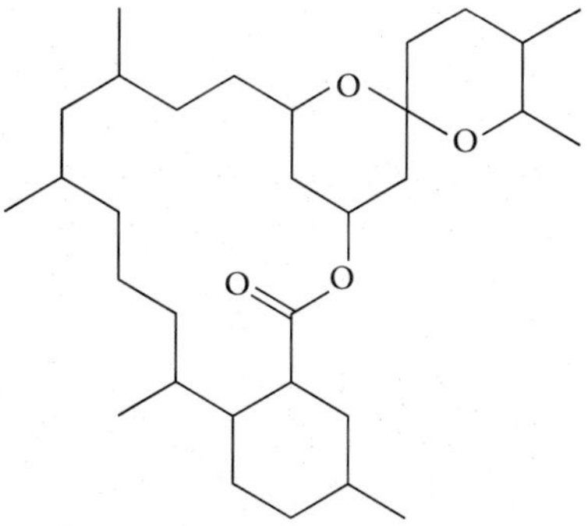

Avermectins/Milbemycins (169)
VC1400

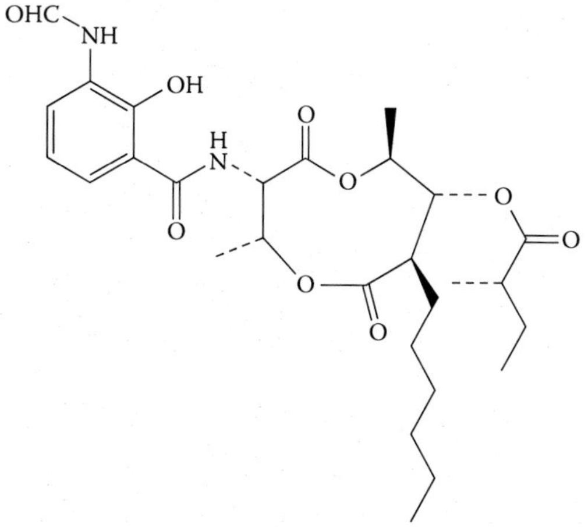

e.g. Antimycin A_{1a}
Antimycins (72)
VC1450

(*Continued*)

TABLE 6.1 (*Continued*)
Natural Product Parent Skeletons

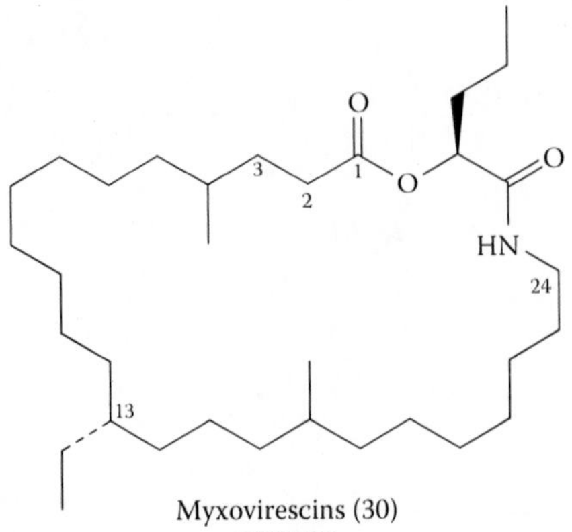

Myxovirescins (30)
VC1500

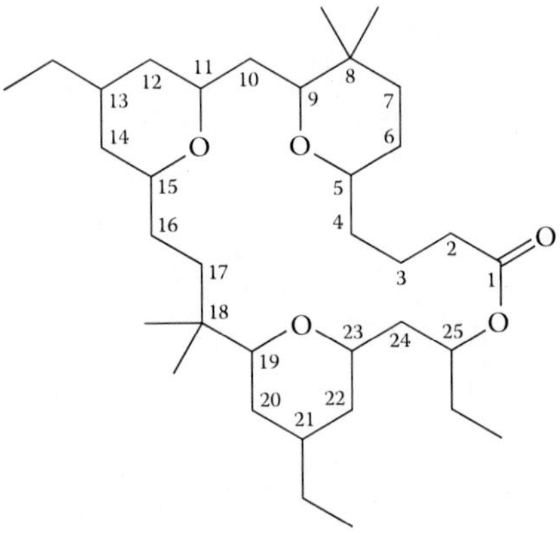

Bryostatins (32)
VC1520

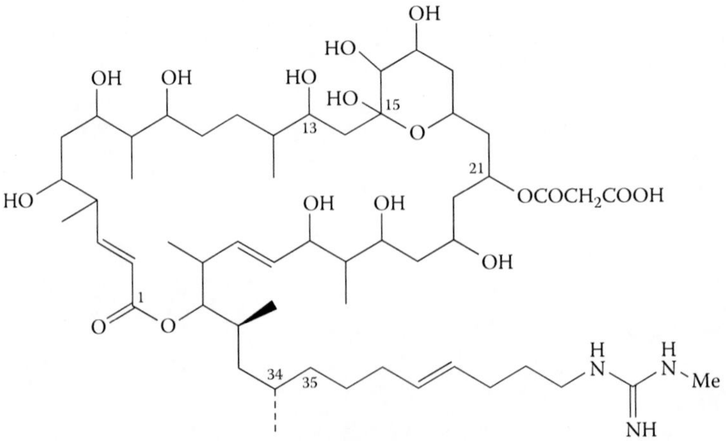

e.g. Copiamycin
Copiamycins (58)
VC1600

(*Continued*)

TABLE 6.1 (*Continued*)
Natural Product Parent Skeletons

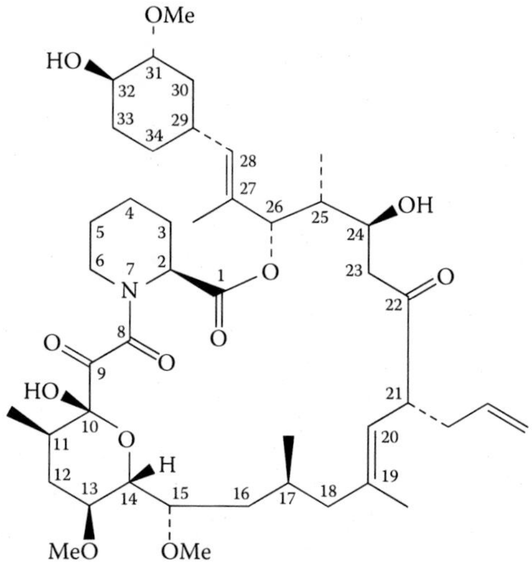

e.g. Fujimycin
Fujimycins (43)
VC1700

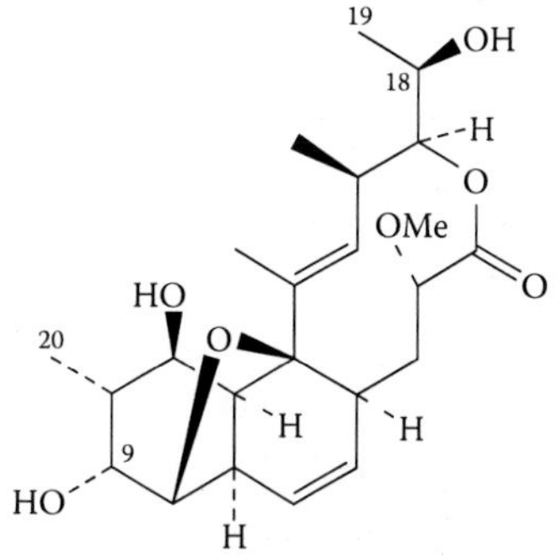

e.g. Nodusmicin
Nodusmicins (11)
VC1750

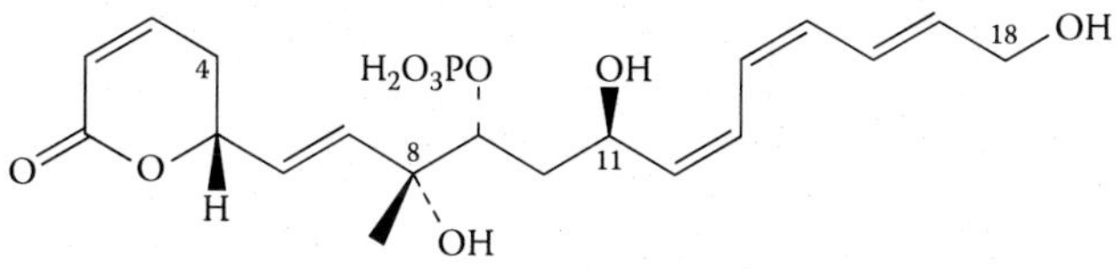

e.g. Fostriecin
Phoslactomycins/Phosphazomycins (21)
VC1900

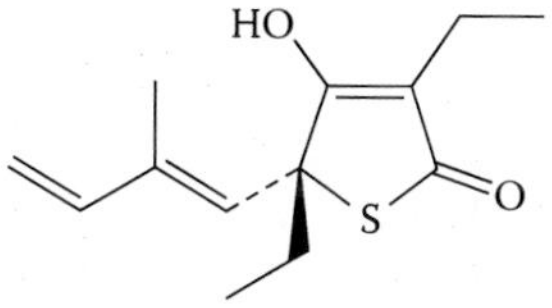

e.g. Thiotetromycin
Thiotetronic acids (5)
VC1950

Carbohydrates (VE)

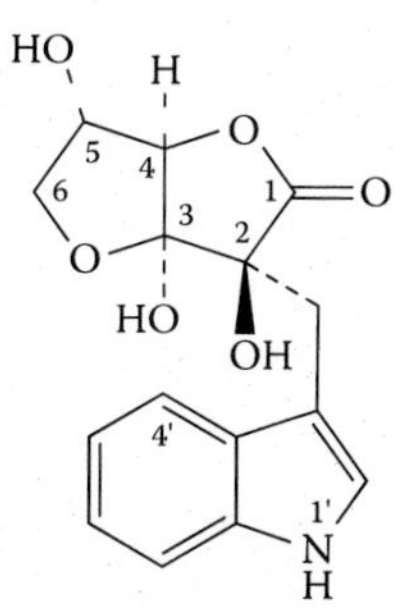

e.g. Ascorbigen
Ascorbic acid and
ascorbigens (46)
VE8250

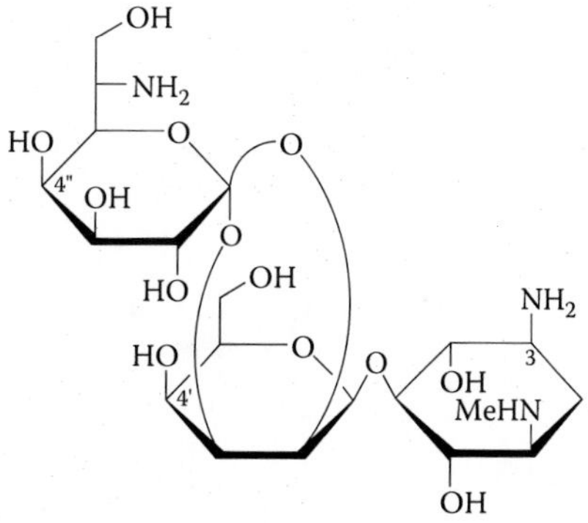

e.g. Destomycin A
Orthosomycins (58)
VE9350

(*Continued*)

TABLE 6.1 (*Continued*)
Natural Product Parent Skeletons

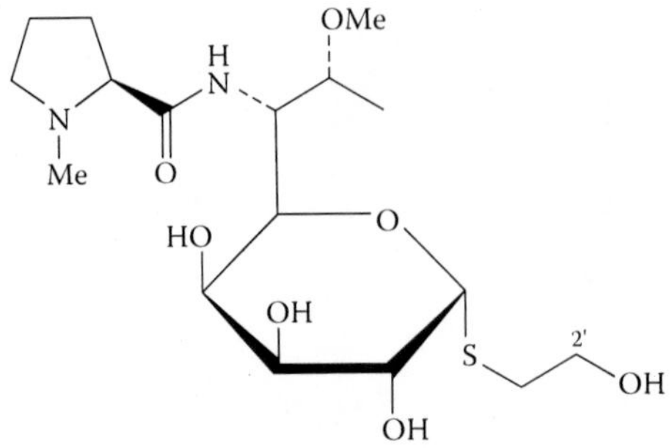

e.g. Desalicetin
Lincomycins (21)
VE9550

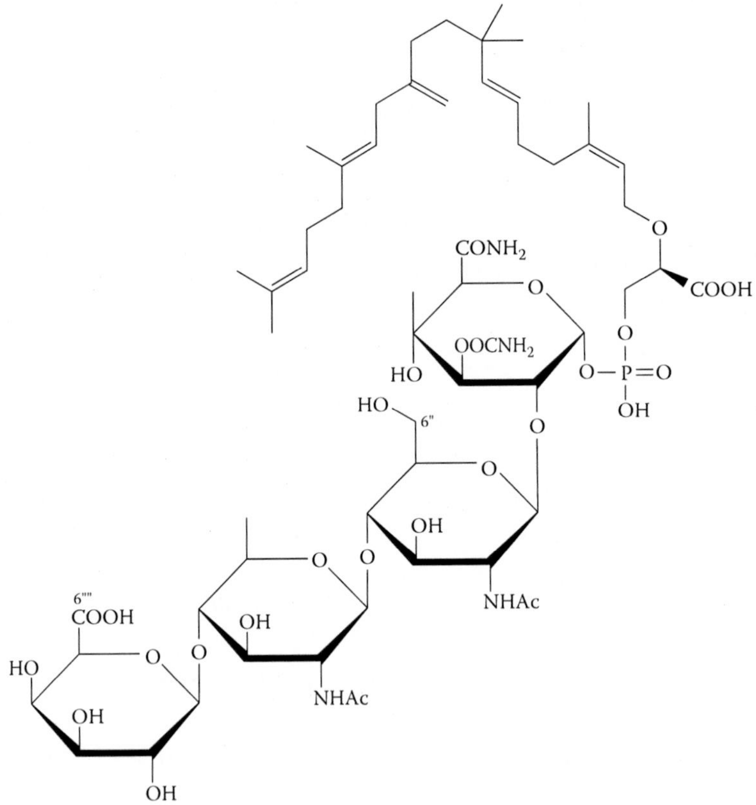

e.g. Nosokomycin C
Moenomycins (25)
VE9560

(*Continued*)

TABLE 6.1 (*Continued*)
Natural Product Parent Skeletons

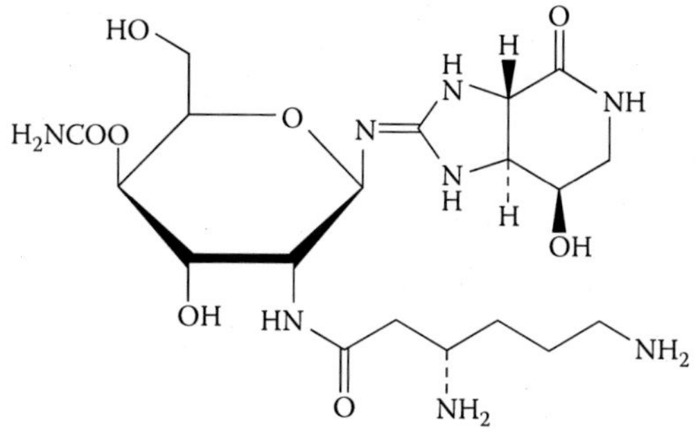

e.g. Streptothricin F
Streptothricins (58)
VE9570

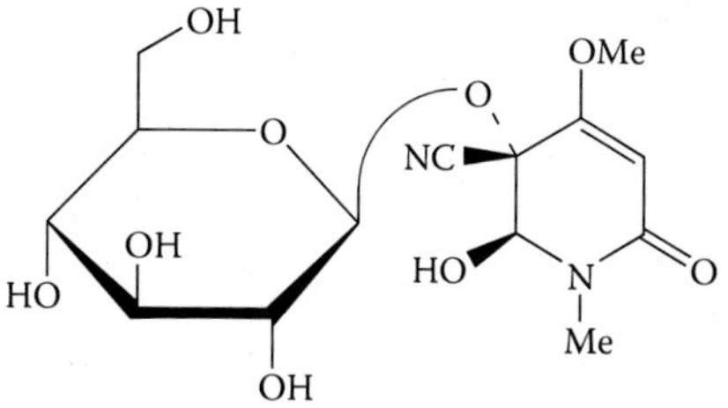

e.g. Acalyphin
Cyanogenic glycosides (150)
VE9700

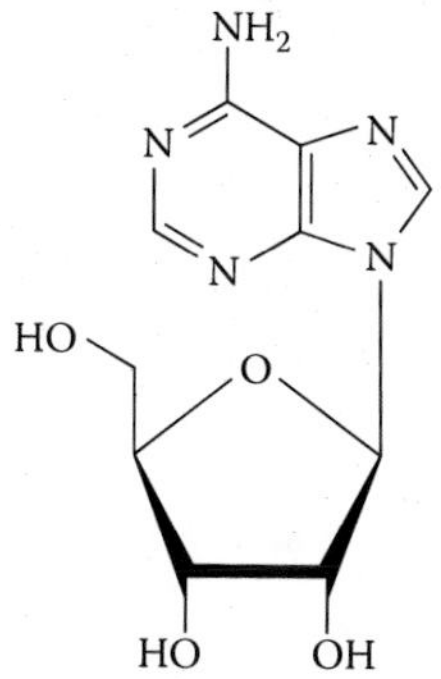

e.g. Adenosine
Nucleosides (681)
VE9900

Oxygen heterocycles (VF)

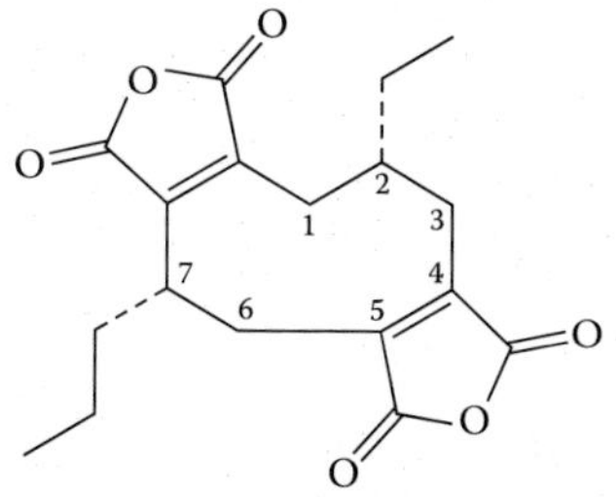

e.g. Byssochlamic acid
Nonadrides (28)
VF5200

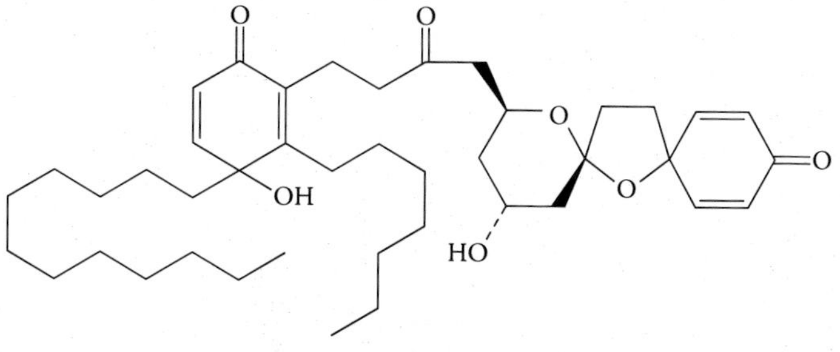

e.g. Aculeatin C
Spiroketals (344)
VF8000

(*Continued*)

TABLE 6.1 (*Continued*)
Natural Product Parent Skeletons

Simple aromatic natural products (VG)

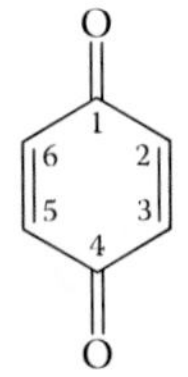

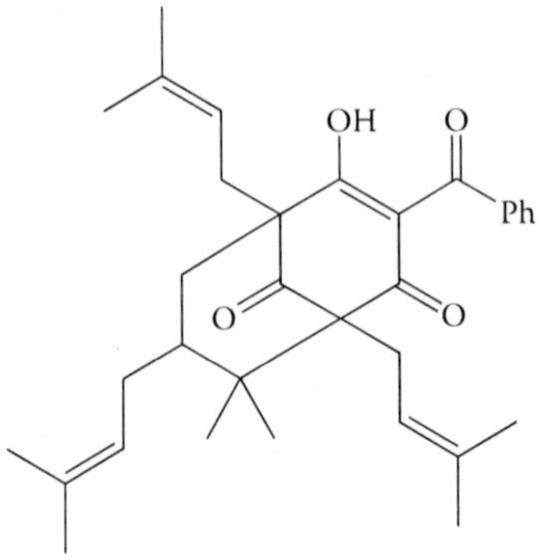

e.g. Clusianone
Acylphloroglucinols (757)
VG0460

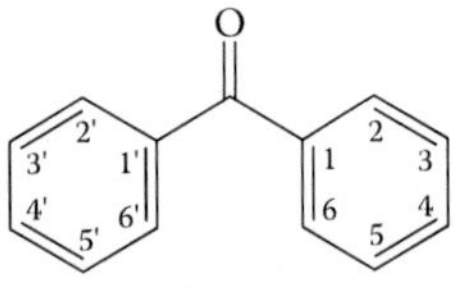

Benzoquinones (432)
VG0300-VG0390

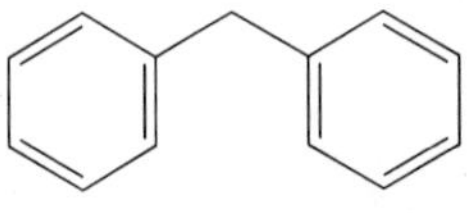

Diphenylmethanes (113)
VG0450

Benzophenones (216)
VG0500-VG0508

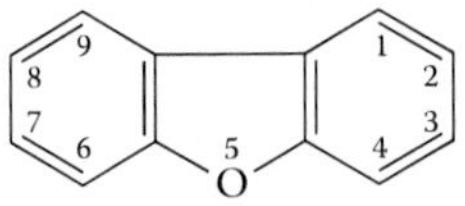

Dibenzofurans (220)
VG0520

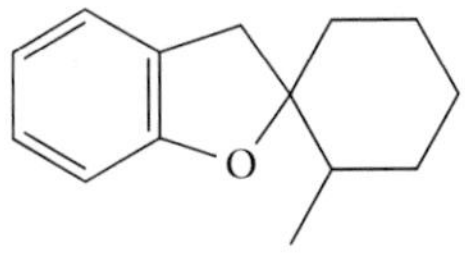

Griseofulvins (30)
VG0530

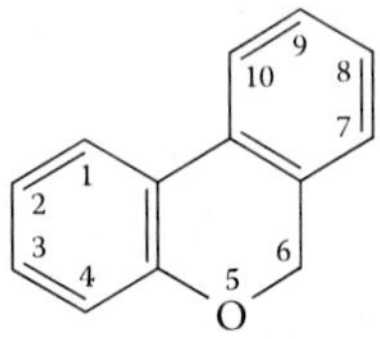

Dibenzo[*b,d*]pyrans (81)
VG0535

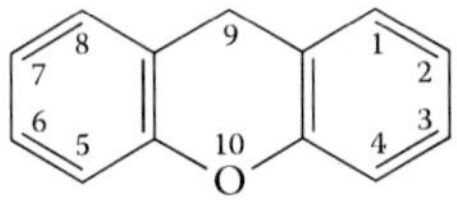

Dibenzo[*b,e*]pyrans (45)
VG0540

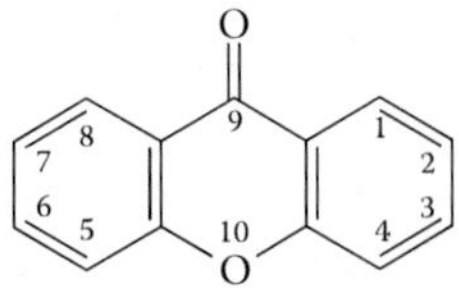

Xanthones (1720)
VG0550-VG0558

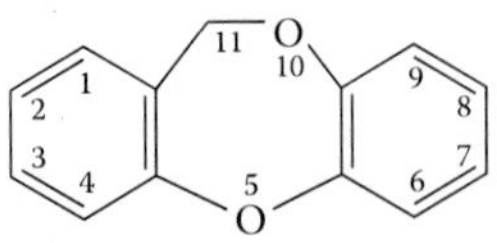

Depsidones (244)
VG0600-VG0610

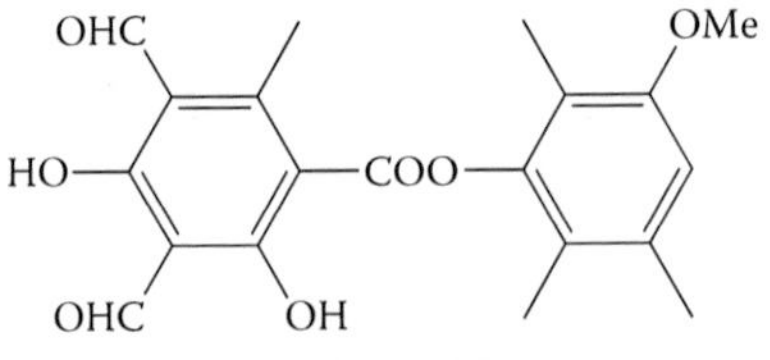

e.g. Nephroarctin
Depsides (307)
VG0620-VG0660

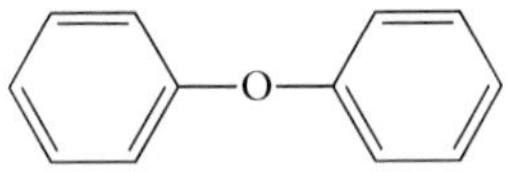

Diphenyl ethers (406)
VG1000

(*Continued*)

TABLE 6.1 (*Continued*)
Natural Product Parent Skeletons

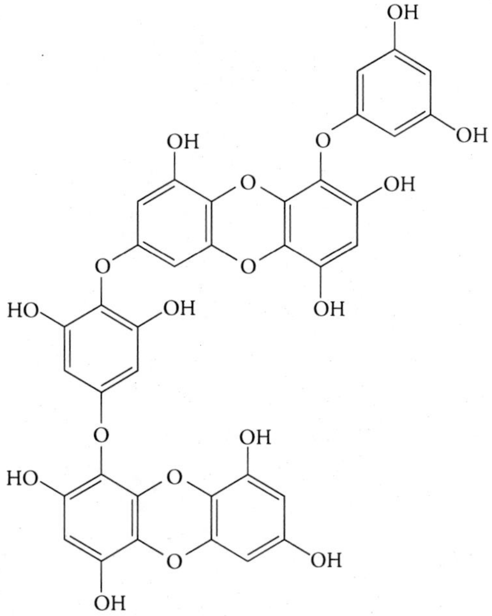

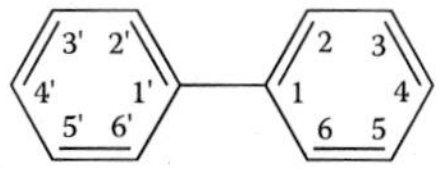

e.g. Dieckol
Phlorotannins (201)
VG2500

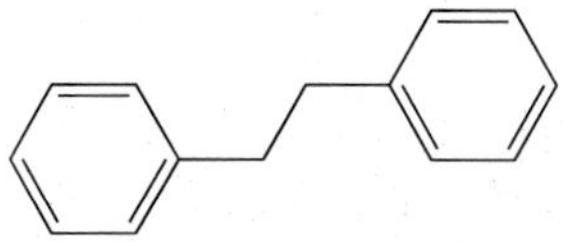

Simple biphenyls (384)
VG2000

Bibenzyls (523)
VG3000

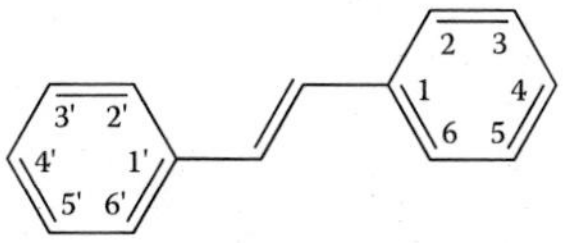

Stilbenes (367)
VG4000

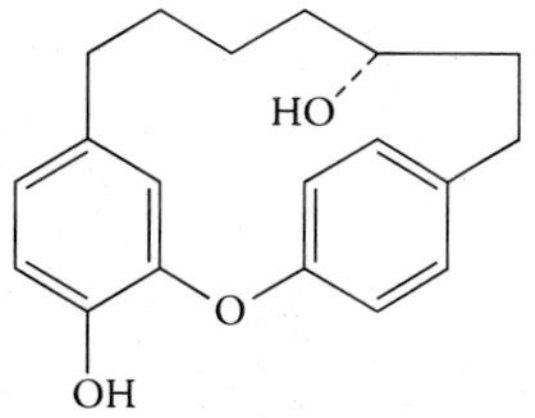

e.g. Acerogenin A
Diarylalkyls (617)
VG7000

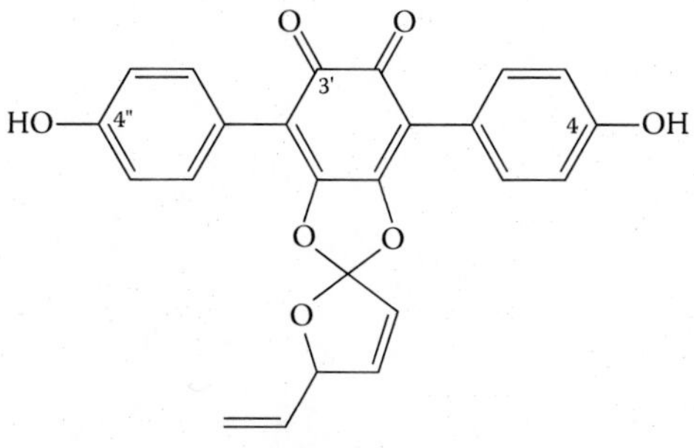

e.g. Spiromentin A
Terphenyls (177)
VG7500

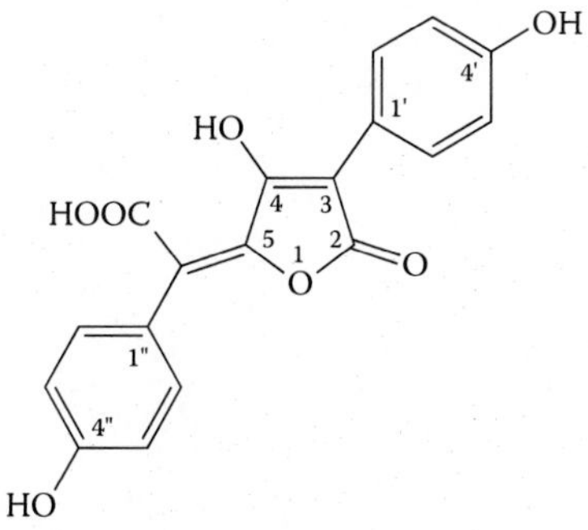

e.g. Atromentic acid
Pulvinones (121)
VG7600

(*Continued*)

TABLE 6.1 (*Continued*)
Natural Product Parent Skeletons

e.g. Bisvertinoquinol
Sorbicillin oligomers (59)
VG8000

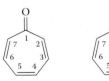

Tropone   Tropolone
Tropones and tropolones (158)
VG9800

Benzofuranoids (VH)

Benzofurans (620)
VH1000

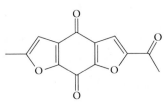

e.g. Conicaquinone
Benzodifurans (19)
VH2000

Isobenzofurans (318)
VH3000

e.g. Angeolide
Angeolides (25)
VH3200

e.g. Rocagloic acid
Flavaglines (130)
VH3500

(*Continued*)

TABLE 6.1 (*Continued*)
Natural Product Parent Skeletons

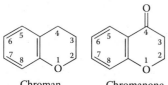

e.g. Papulacandin B
Papulacandins (21)
VH4000

Benzopyranoids (VI)

Chroman
3,4-Dihydro-
2*H*-1-benzopyran

Chromanone
2,3-Dihydro-
4*H*-1-benzopyran-
4-one

Chromone
4*H*-1-Benzopyran-4-one

2-Chromene
β-Chromene
4*H*-1-Benzopyran

3-Chromene
α-Chromene
2*H*-1-Benzopyran

1-Benzopyrans (1433)
VI0030

e.g. Khellin
Furo-1-benzopyrans (80)
VI0050

e.g. Octandrenolone
Pyrano-1-benzopyrans (155)
VI0070

2*H*-1-Benzopyran-2-ones
Coumarins (2624)
VI0100–VI7000

(*Continued*)

TABLE 6.1 (*Continued*)
Natural Product Parent Skeletons

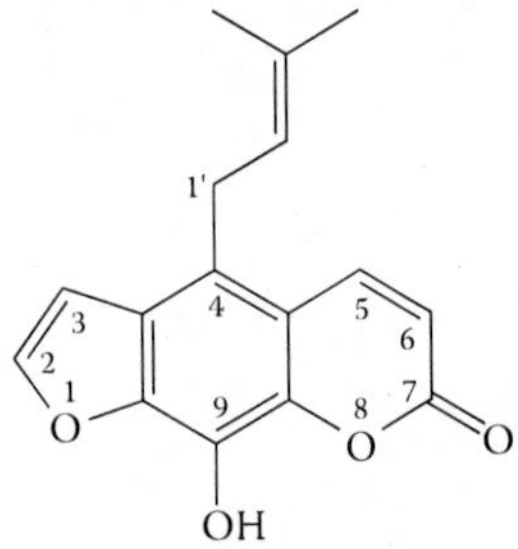

e.g. Alloimperatorin
Furanocoumarins (600)
VI8000-VI8500

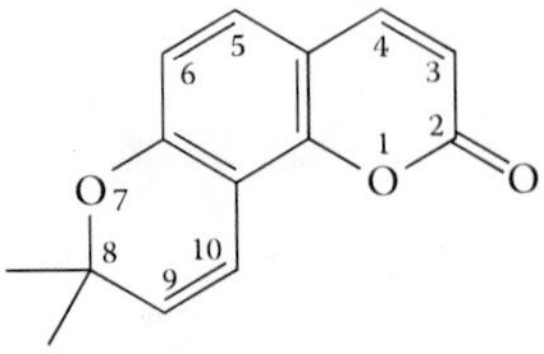

e.g. Seselin
Pyranocoumarins (515)
VI9000-VI9500

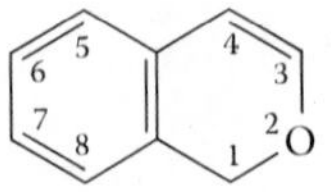

Isochromene
Isobenzopyran
1*H*-2-Benzopyran
2-Benzopyrans (160)
VI9600

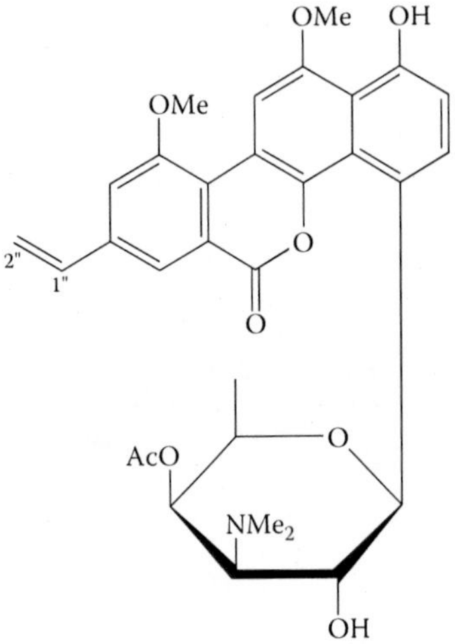

e.g. Ravidomycin
Benzo[*d*]naphtho[1,2-*b*]pyrans (55)
VI9610

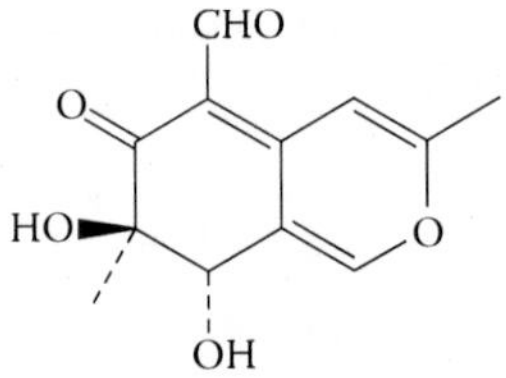

e.g. Austdiol
Azaphilones (341)
VI9620

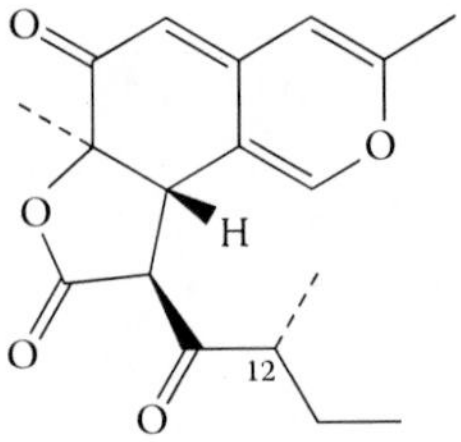

e.g. Monochaetin
Furo-2-benzopyrans (53)
VI9650

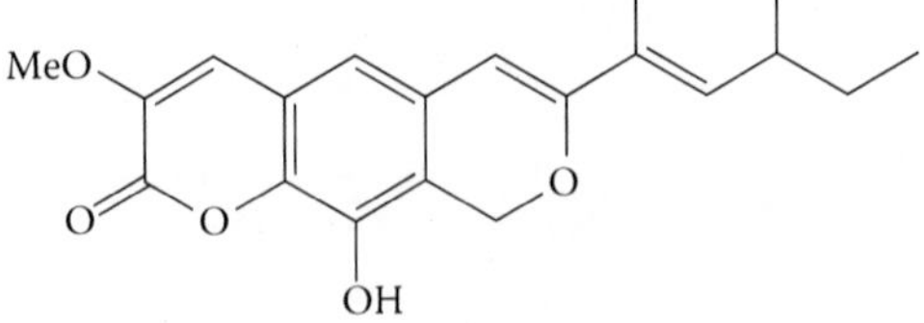

e.g. Leptosphaerolide
Pyrano-2-benzopyrans (60)
VI9670

(Continued)

TABLE 6.1 (*Continued*)

Natural Product Parent Skeletons

e.g. γ-Rubromycin
Rubromycins (22)
VI9680

1*H*-2-Benzopyran-1-one
Isocoumarins (495)
VI9700

2-Benzothiopyrans (5)
VI9800

Flavonoids (VK)

Anthocyanidins (756)
VK0010-VK0090

Pyranoanthocyanidins (49)
VK0095

Flavans (208)
VK1000

Flavan-3-ols (442)
VK1100

Leucoanthocyanidins (98)
VK1200

Flavan-4-ols (57)
VK1250

(*Continued*)

TABLE 6.1 (*Continued*)
Natural Product Parent Skeletons

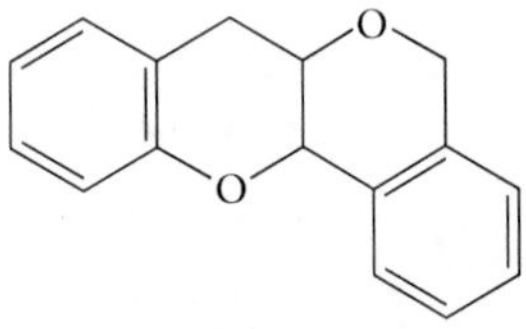

Peltogynoids (67)
VK1300

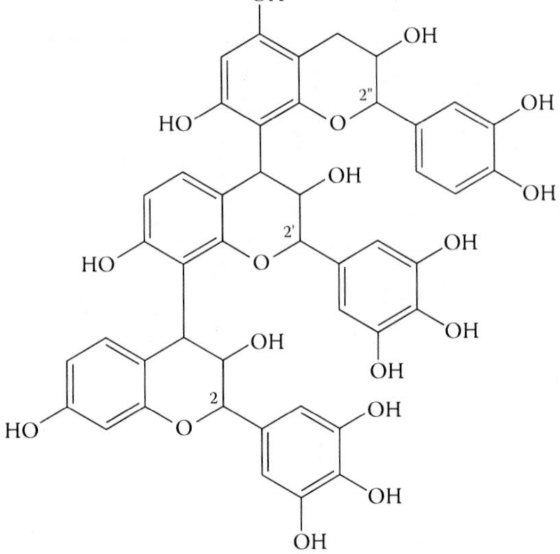

e.g. Dolabriproanthocyanidin
Proanthocyanidins (530)
VK1500

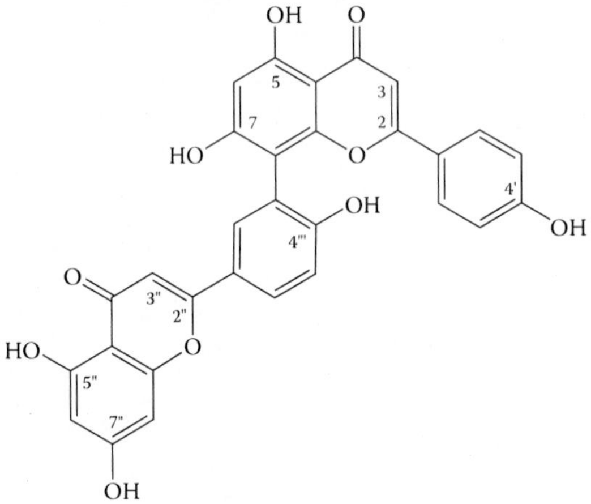

e.g. Amentoflavone
Biflavonoids and oligoflavonoids (767)
VK2000

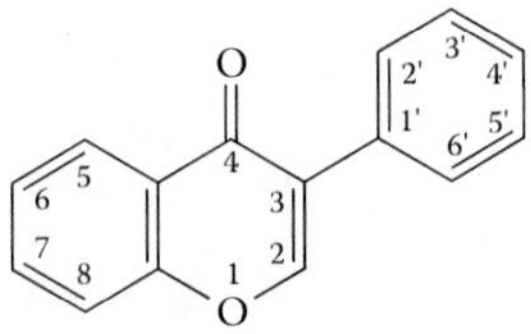

Isoflavones (1155)
VK3000-VK3090

(*Continued*)

TABLE 6.1 (*Continued*)
Natural Product Parent Skeletons

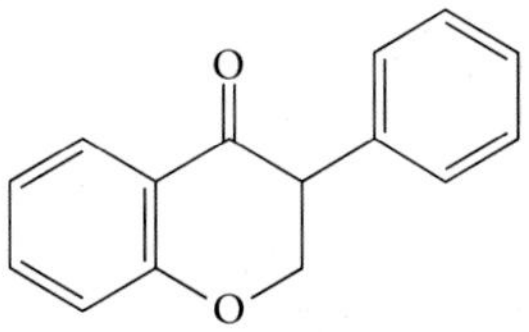

Isoflavanones (253)
VK3100

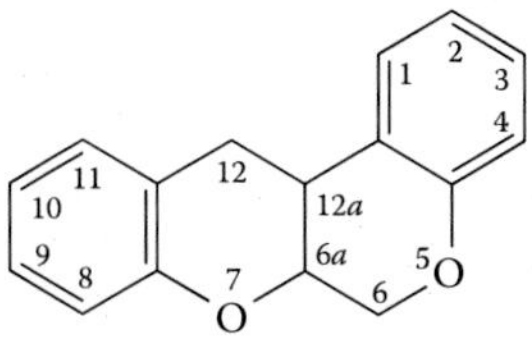

Rotenoids (181)
VK3200-VK3300

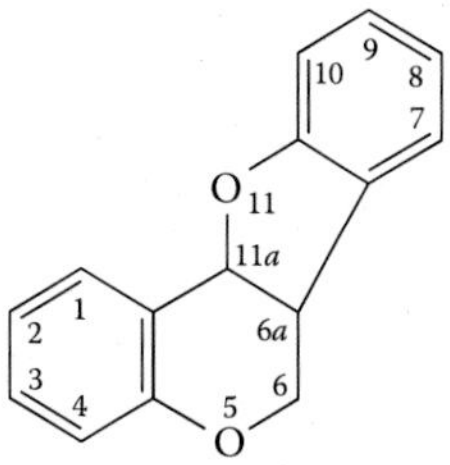

Pterocarpans (382)
VK3400-VK3550

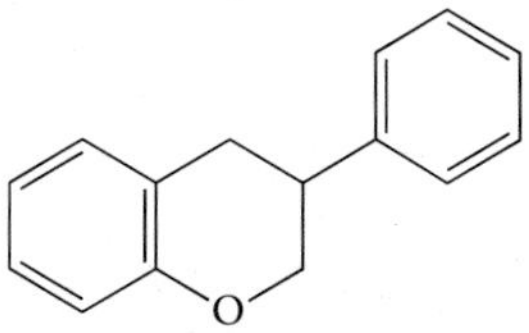

Isoflavans (212)
VK3600-VK3650

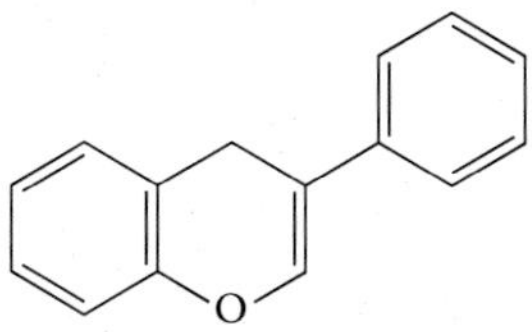

Isoflav-2-enes (3)
VK3680

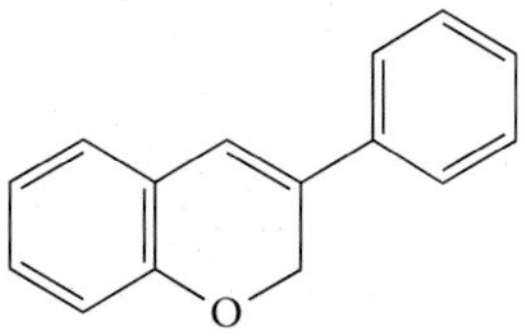

Isoflav-3-enes (33)
VK3700

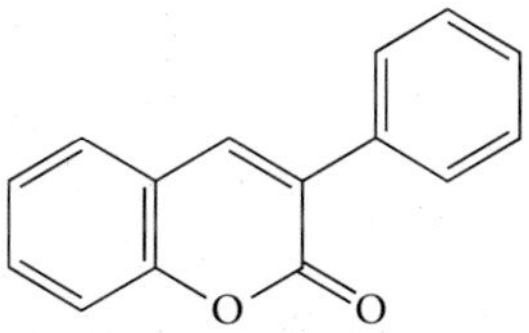

3-Arylcoumarins (58)
VK3720

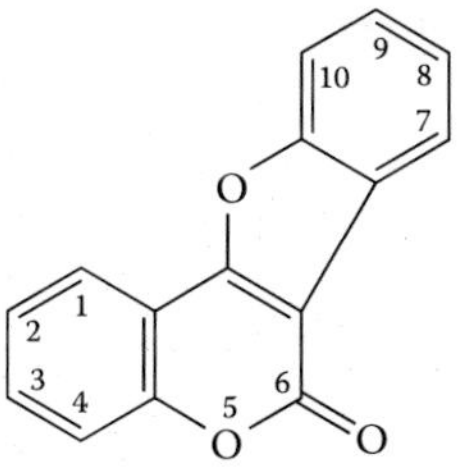

Coumestans (99)
VK3750

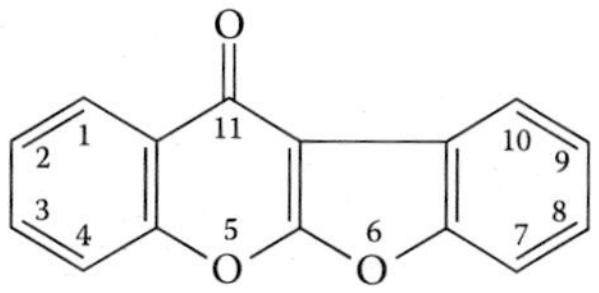

Coumaronochromenes (74)
VK3770

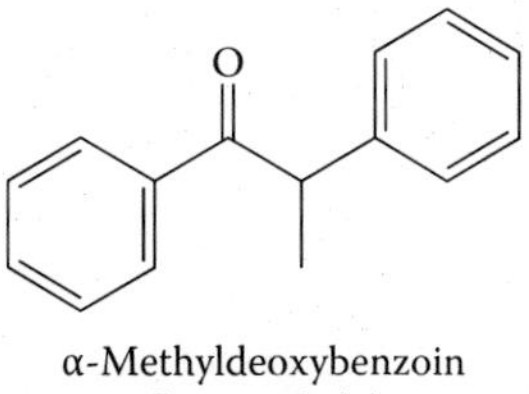

α-Methyldeoxybenzoin
flavonoids (8)
VK3800

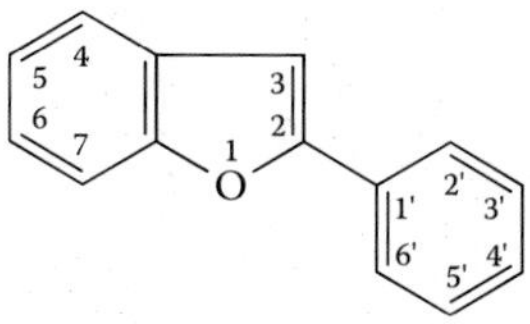

2-Arylbenzofurans (339)
VK3820

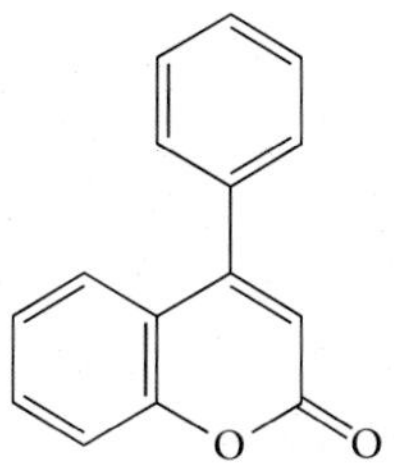

Neoflavonoids (386)
VK4000

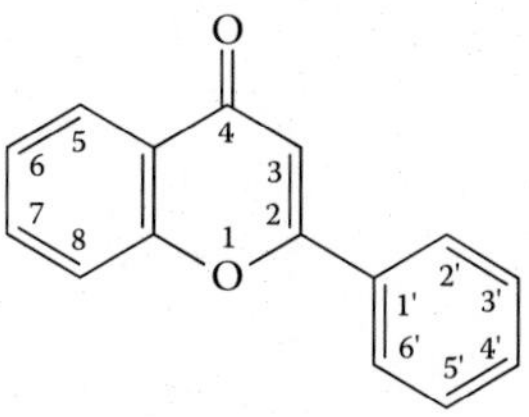

Flavones (2685)
VK5010-VK5090

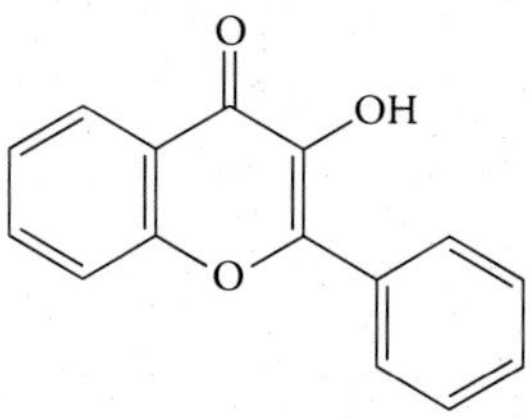

Flavonols (3307)
VK5210-VK5290

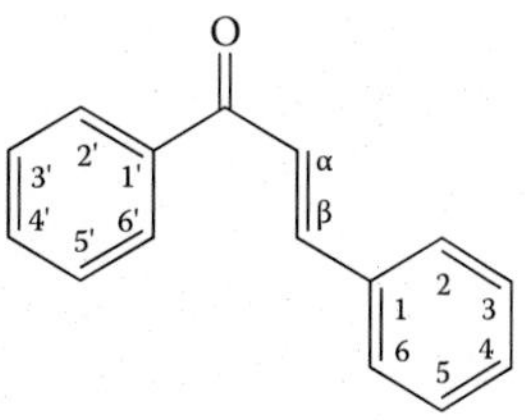

Chalcones (685)
VK6010-VK6095

(*Continued*)

TABLE 6.1 (*Continued*)
Natural Product Parent Skeletons

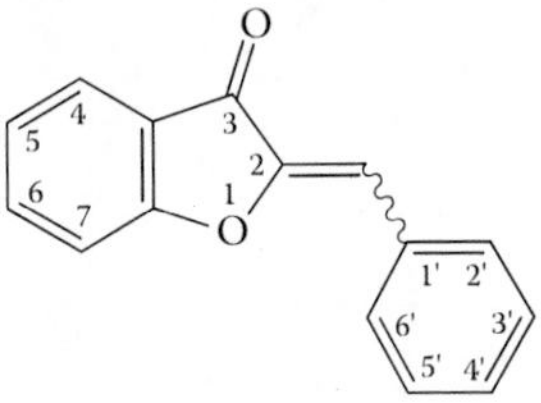

Aurones (113)
VK6100

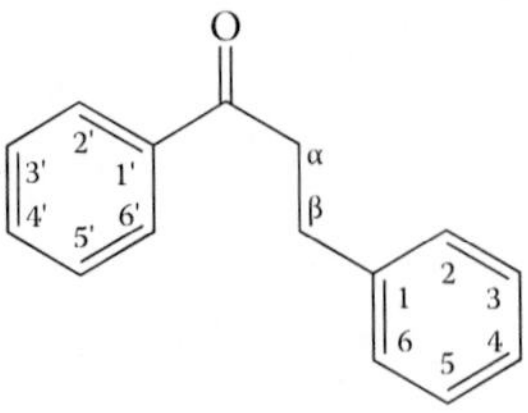

Dihydrochalcones (425)
VK6200

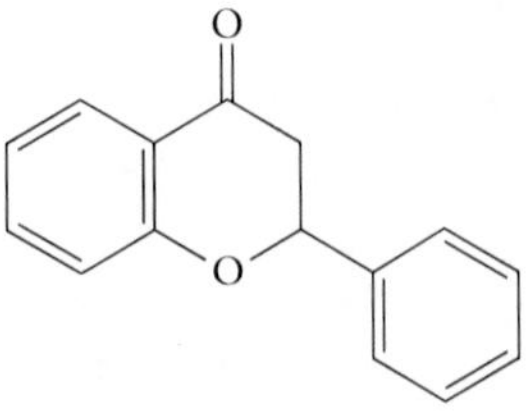

Flavanones (1499)
VK6300–VK6390

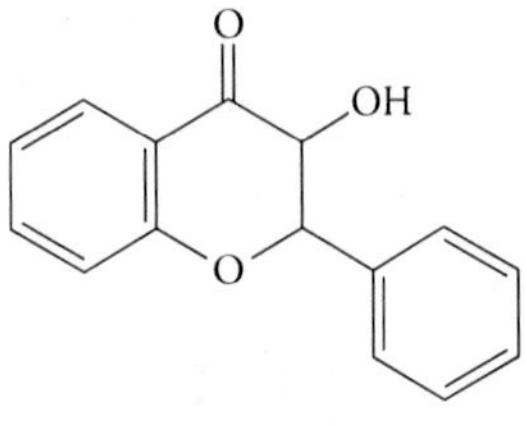

Dihydroflavonols (454)
VK6410–VK6490

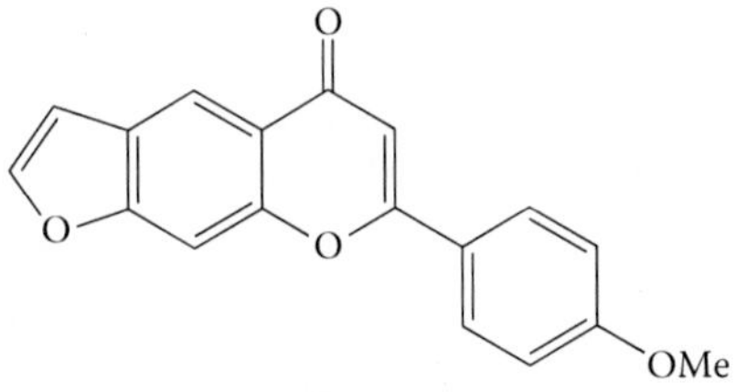

e.g. Glabone
Furanoflavonoids (132)
VK6500

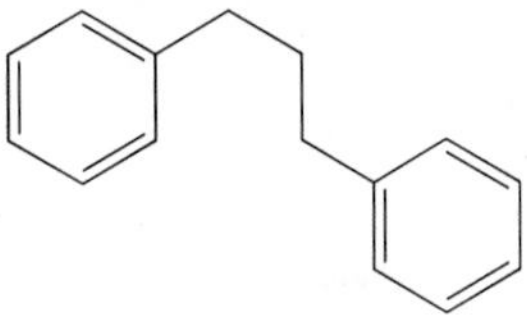

1,3-Diarylpropanes (130)
VK6600

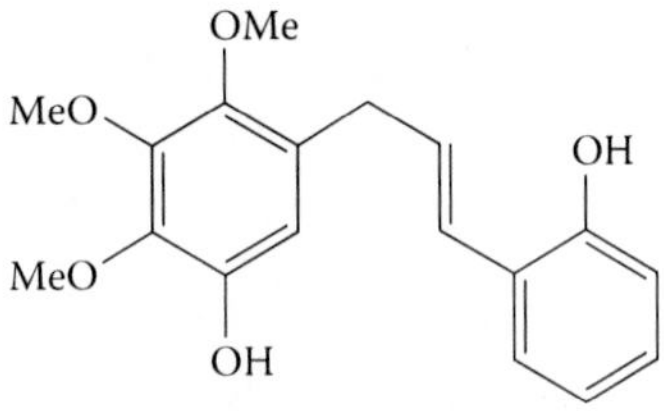

e.g. Petrostyrene
Cinnamylphenols (63)
VK6700

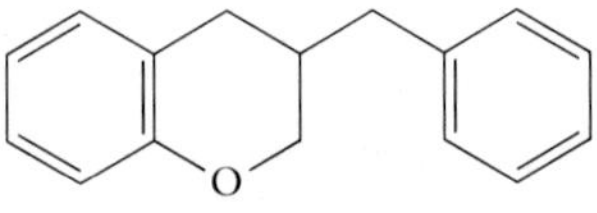

Homoisoflavonoids (225)
VK6800

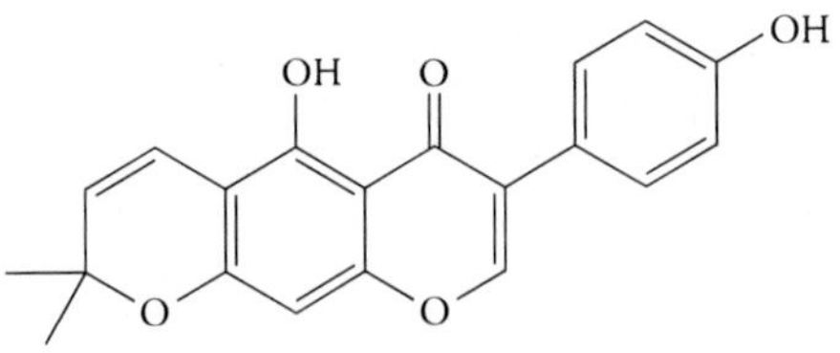

e.g. Alpinumisoflavone
Cyclised C-prenylated
flavonoids (1457)
VK8300

(*Continued*)

TABLE 6.1 (*Continued*)
Natural Product Parent Skeletons

Tannins (VM)

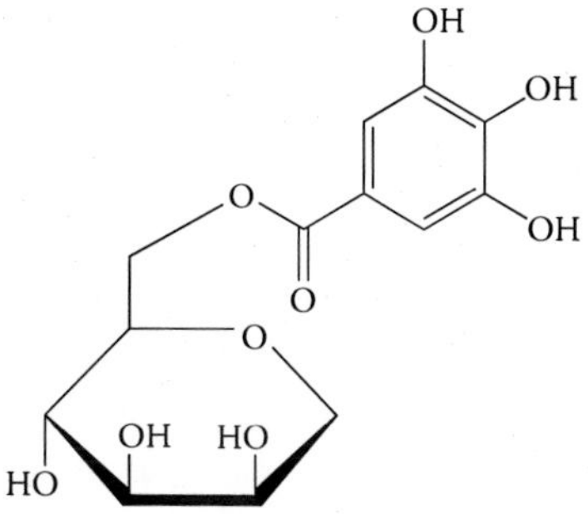

e.g. Pycnalin
Simple gallate ester tannins (622)
VM6000

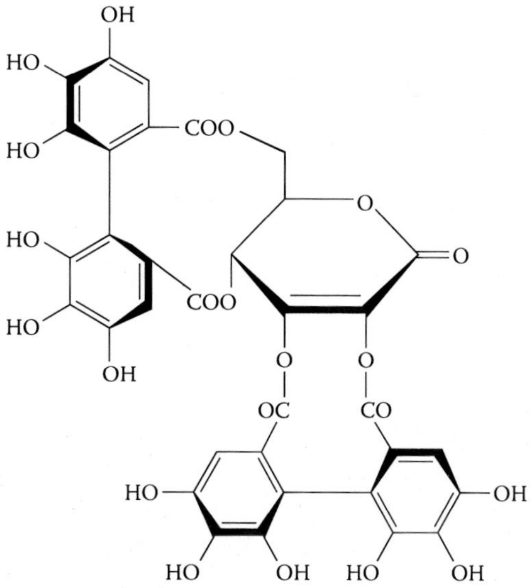

e.g. Emblicanin B
Hexahydroxydiphenoyl ester tannins (503)
VM6100

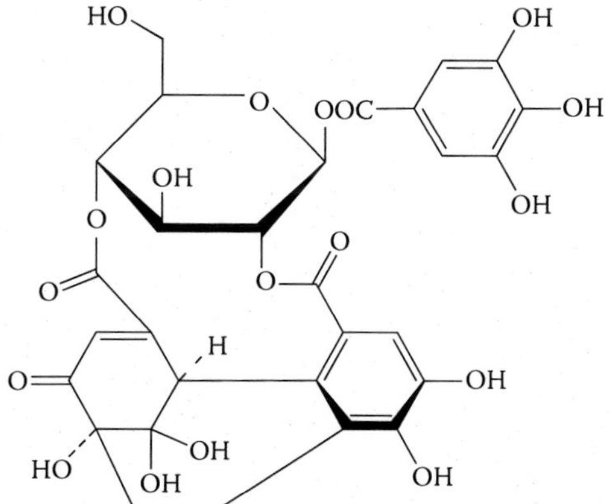

e.g. Furosin
Dehydrohexahydroxydiphenoyl
ester tannins (56)
VM6200

(*Continued*)

TABLE 6.1 (*Continued*)
Natural Product Parent Skeletons

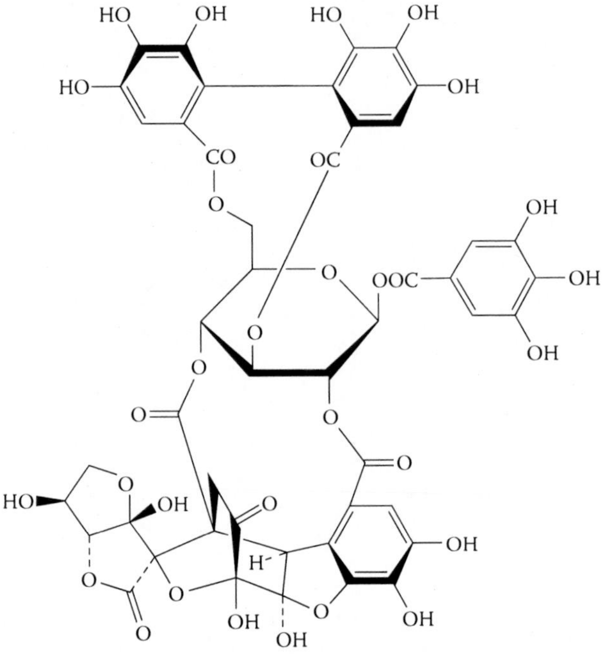

e.g. Elaeocarpusin
Elaeocarpusinoyl ester tannins (8)
VM6300

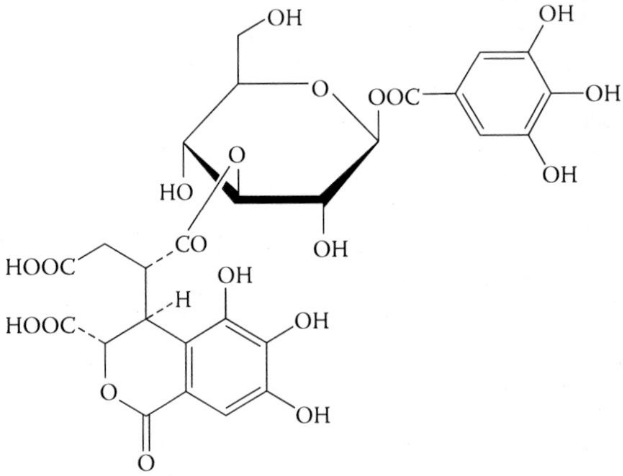

e.g. Phyllanemblinin D
Chebuloyl ester tannins (13)
VM6500

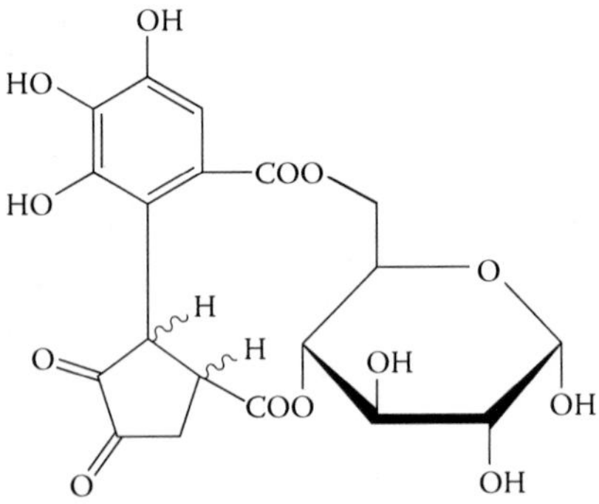

e.g. Algarobin
Brevifoloyl ester tannins (19)
VM6600

(*Continued*)

TABLE 6.1 (*Continued*)
Natural Product Parent Skeletons

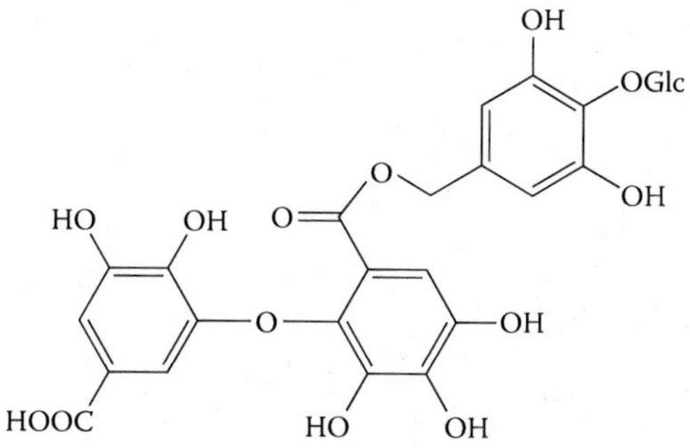

e.g. Chesnatin
Dehydrodigalloyl ester tannins (62)
VM6700

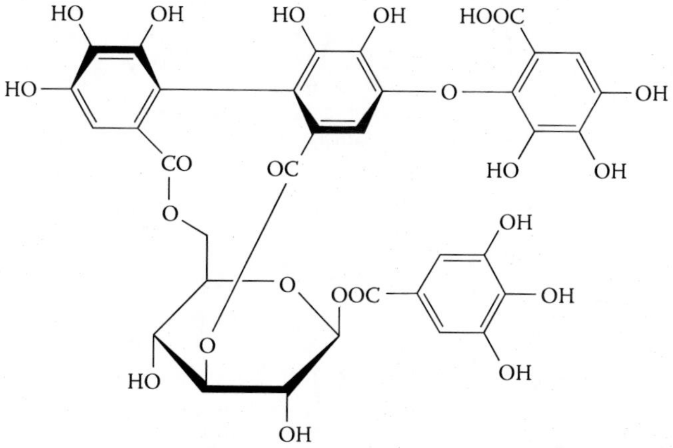

e.g. Mallotinic acid
Valoneoyl ester tannins (172)
VM6800-VM6850

(*Continued*)

TABLE 6.1 (*Continued*)
Natural Product Parent Skeletons

e.g. Sanguiin H7
Sanguisorbyl ester tannins (22)
VM6900

e.g. Grandinin
Flavogallonoyl ester tannins (22)
VM7000–VM7050

(*Continued*)

TABLE 6.1 (*Continued*)
Natural Product Parent Skeletons

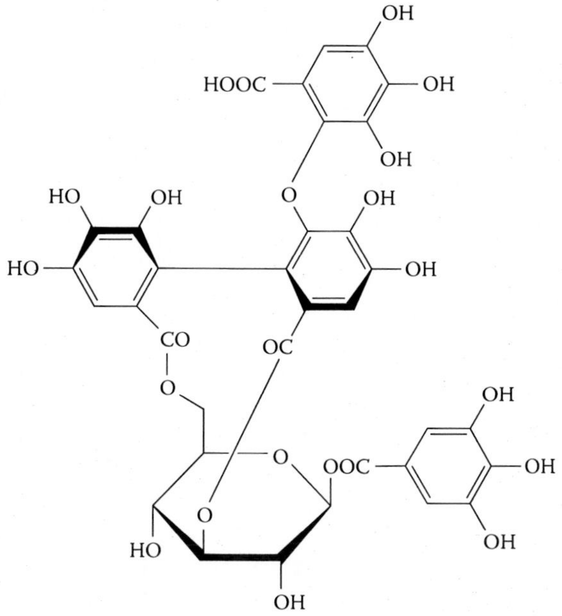

e.g. Macaranin B
Macaranoyl ester tannins (6)
VM7200

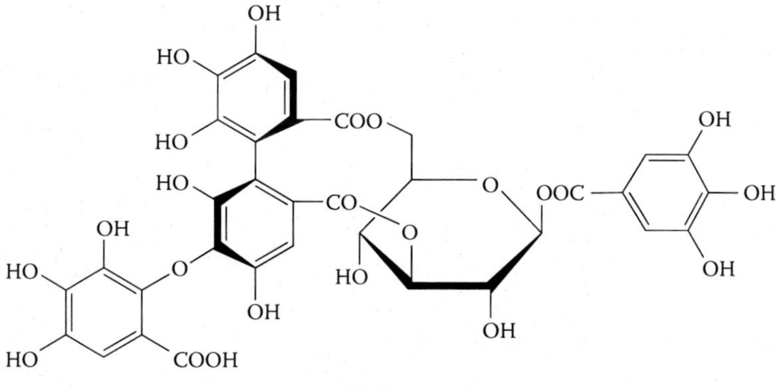

e.g. Mallorepanin
Tergalloyl ester tannins (14)
VM7300-VM7350

(*Continued*)

TABLE 6.1 (*Continued*)
Natural Product Parent Skeletons

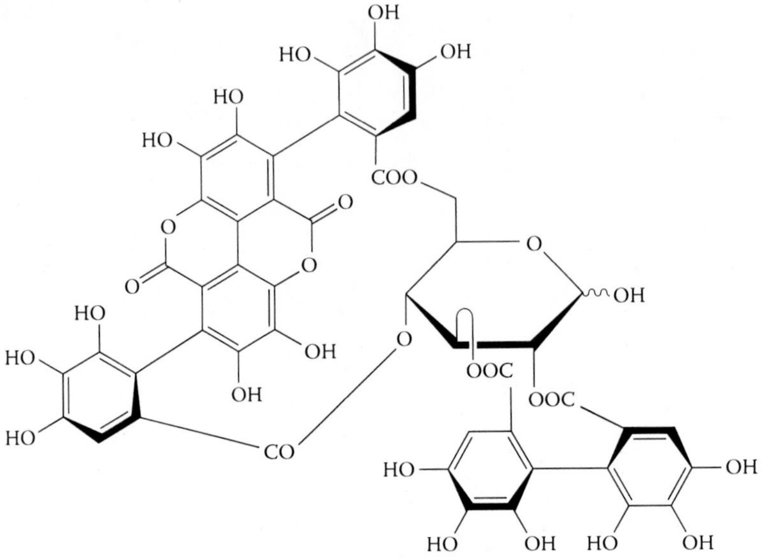

e.g. Punicalagin
Gallagyl ester tannins (10)
VM7600

Lignans (VO)

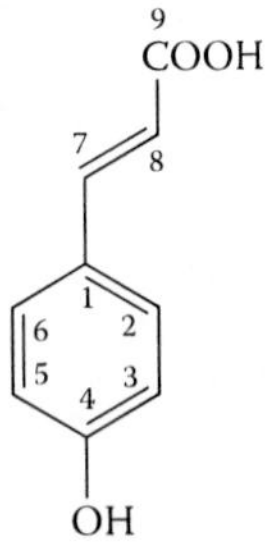

e.g. *p*-Coumaric acid
Lignan monomers (603)
VO0020

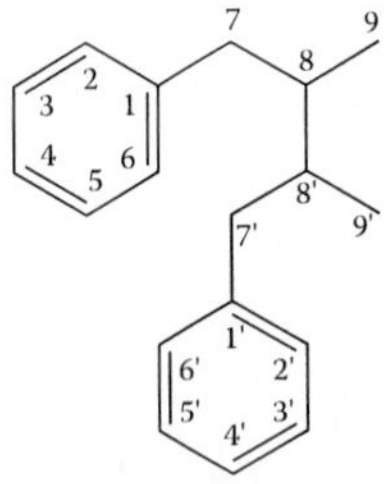

Dibenzylbutane
lignans (231)
VO0050-VO0100

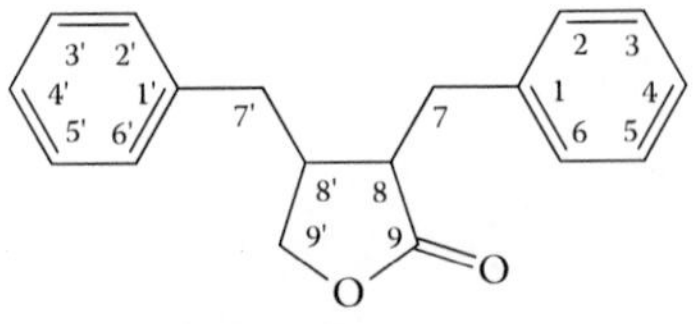

Dibenzylbutyrolactone
lignans (280)
VO0150-VO0200

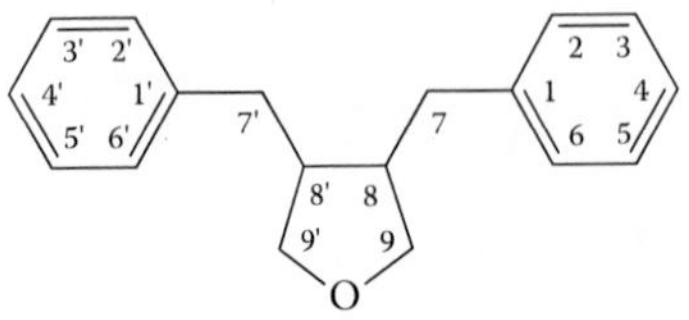

9,9'-Epoxytetrahydrofuranoid
lignans (29)
VO0250

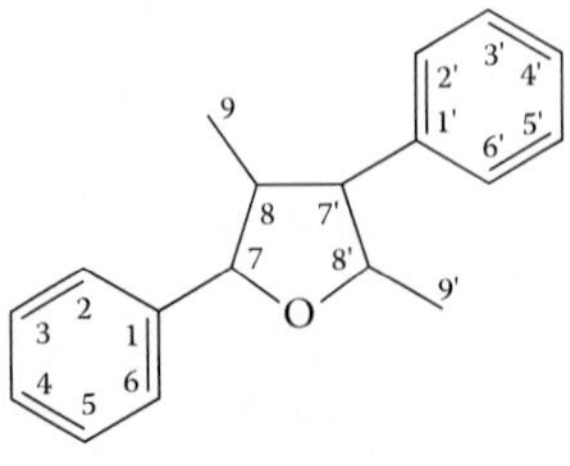

7,8'-Epoxytetrahydrofuranoid
lignans (4)
VO0280

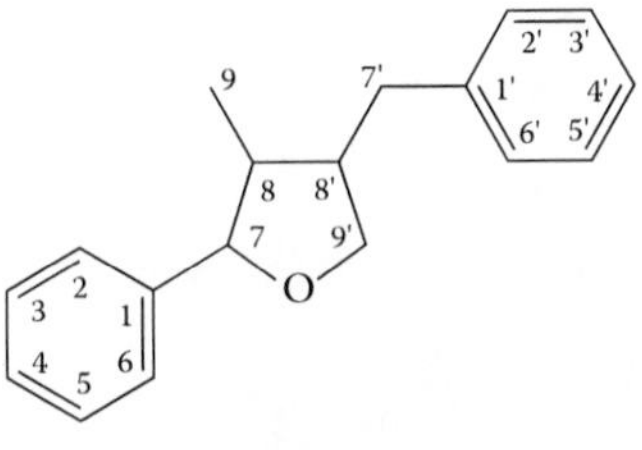

7,9'-Epoxytetrahydrofuranoid
lignans (234)
VO0300

(*Continued*)

TABLE 6.1 (*Continued*)
Natural Product Parent Skeletons

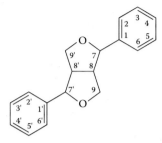

7,7'-Epoxytetrahydrofuranoid
lignans (175)
VO0350

Furofuranoid lignans (356)
VO0400-VO0470

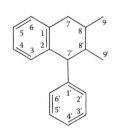

2,7'-Cyclolignans (413)
VO0500-VO0550, VO0640

2,7'-Cyclo-9,9'-
epoxylignans (22)
VO0600

2,7'-Cyclolignan-
9,9'-olides (64)
VO0620

2,7'-Cyclolignan-
9',9-olides (211)
VO0630

7,8'-Cyclolignans (71)
VO0670

Dibenzocyclooctadiene
lignans (419)
VO0750

e.g. Samin
Norlignans
VO0800

e.g. Rayalinol
Homolignans
VO0850

(*Continued*)

TABLE 6.1 (*Continued*)
Natural Product Parent Skeletons

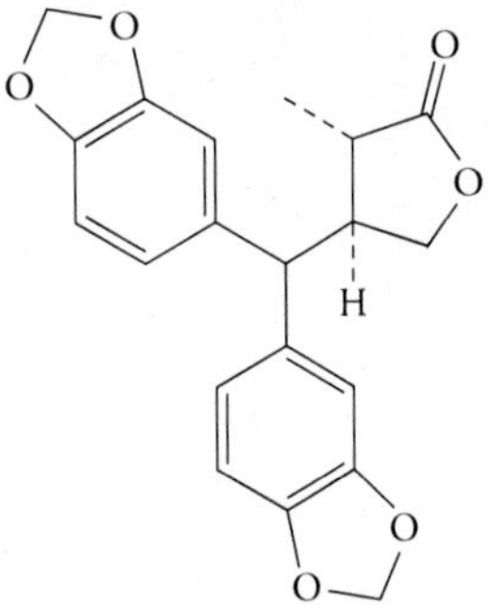

e.g. Peperomin D
Secolignans
VO0900

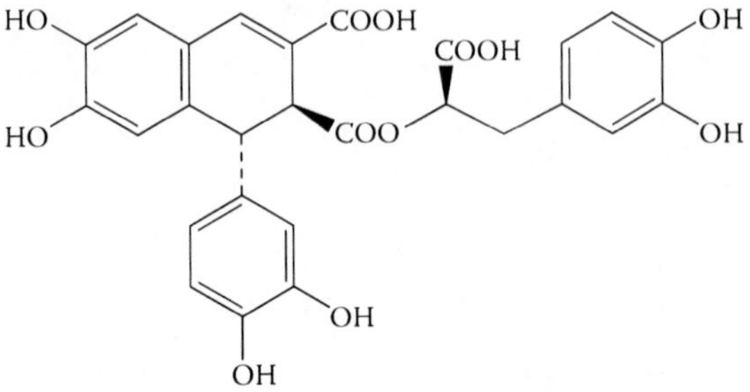

e.g. Salvianolic acid R
Sesquilignans
VO1000

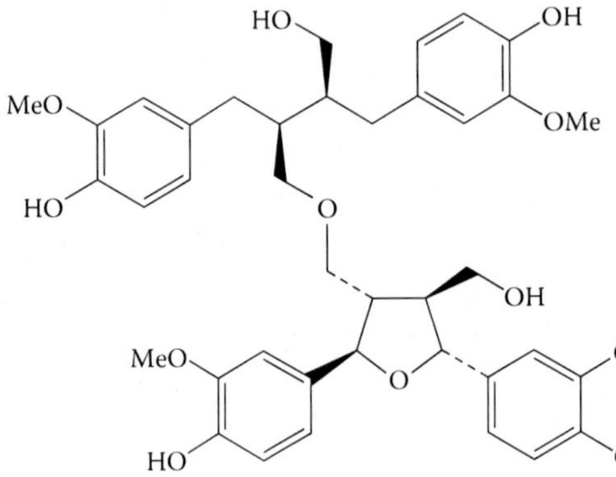

e.g. Larixsin
Bilignans
VO1200

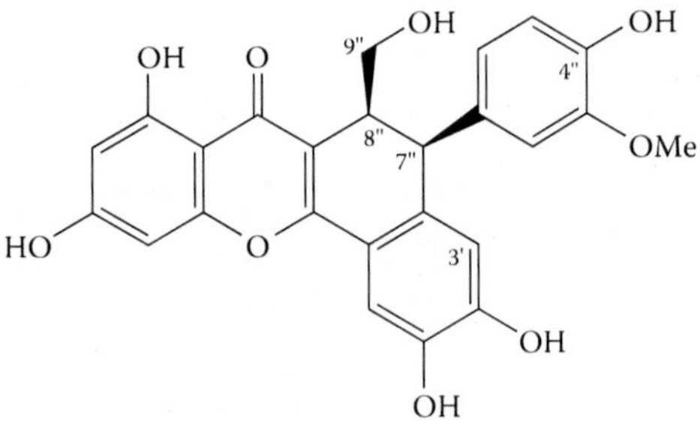

e.g. Neohydnocarpin
Flavonolignans (102)
VO1600

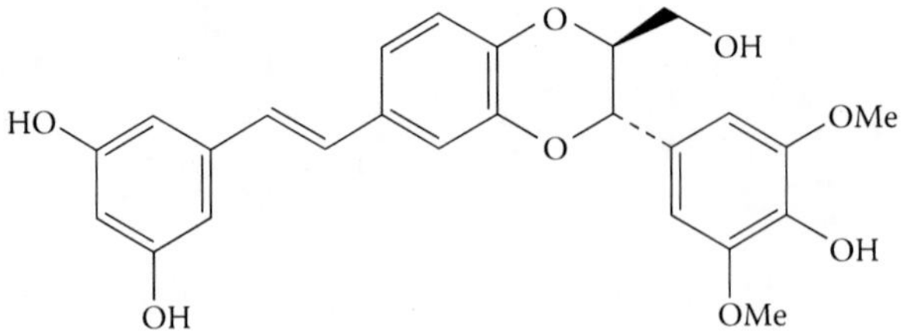

e.g. Aiphanol
Stilbenolignans
VO1650

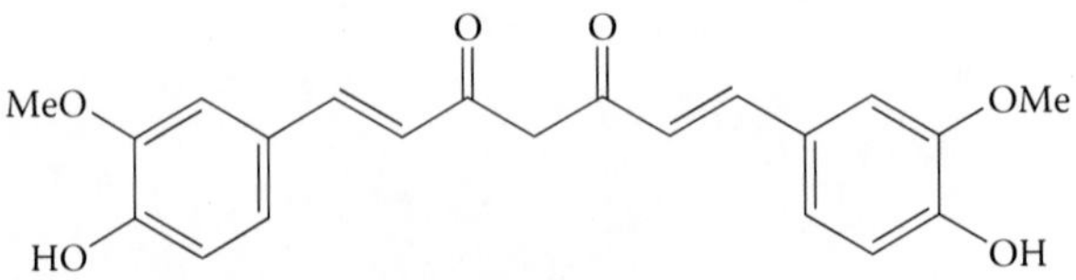

e.g. Curcumin
Diarylheptanoids
VO8000-VO8050

(Continued)

TABLE 6.1 (*Continued*)
Natural Product Parent Skeletons

Polycyclic aromatic natural products (VQ)

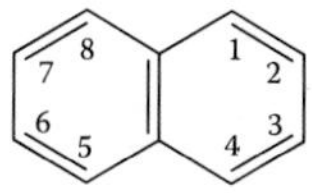

Naphthalenes (522)
VQ2000

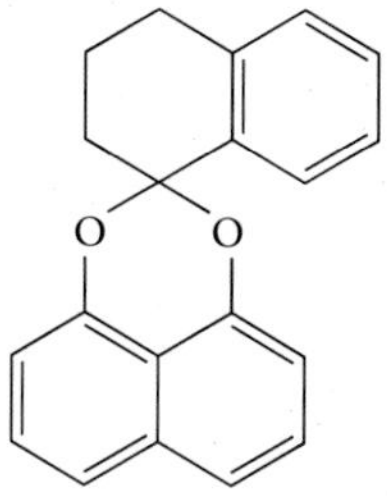

Spirobisnaphthalenes (104)
VQ2050

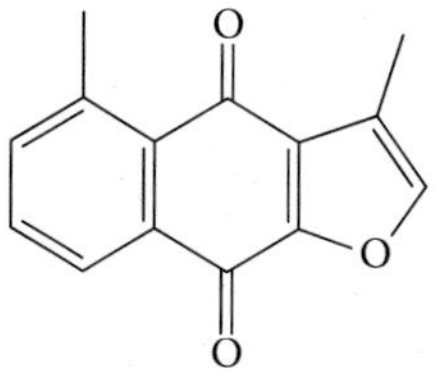

e.g. Maturinone
Furonaphthalenes (203)
VQ2100

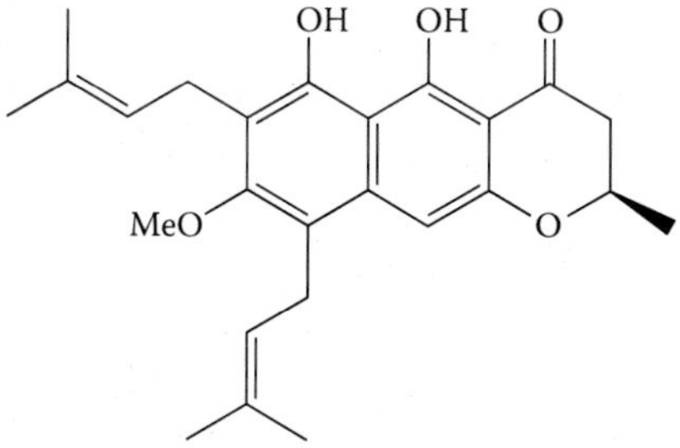

e.g. Adenaflorin A
Pyranonaphthalenes (397)
VQ2200

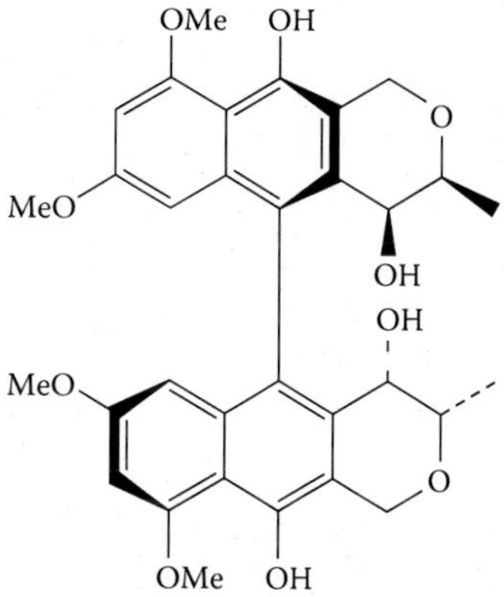

e.g. Bioxanthracene
Binaphthyls (219)
VQ2500

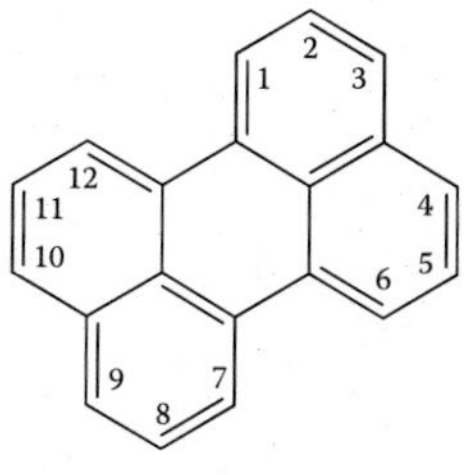

Perylenes (80)
VQ2600

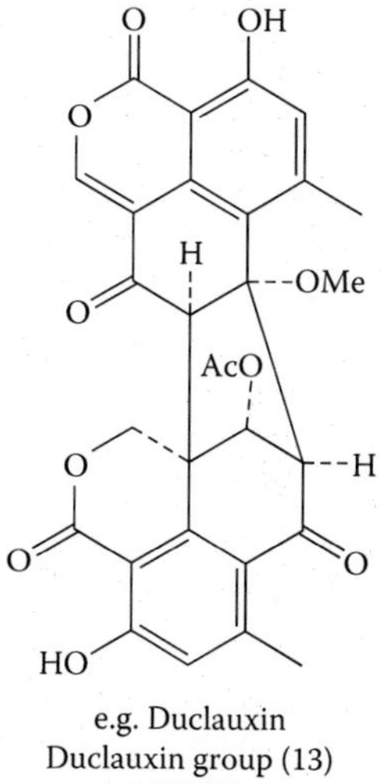

e.g. Duclauxin
Duclauxin group (13)
VQ2700

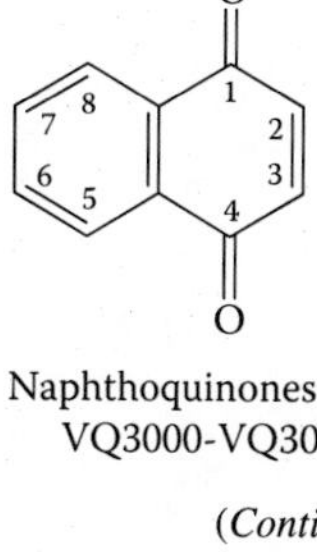

Naphthoquinones (771)
VQ3000-VQ3060

(*Continued*)

TABLE 6.1 (*Continued*)
Natural Product Parent Skeletons

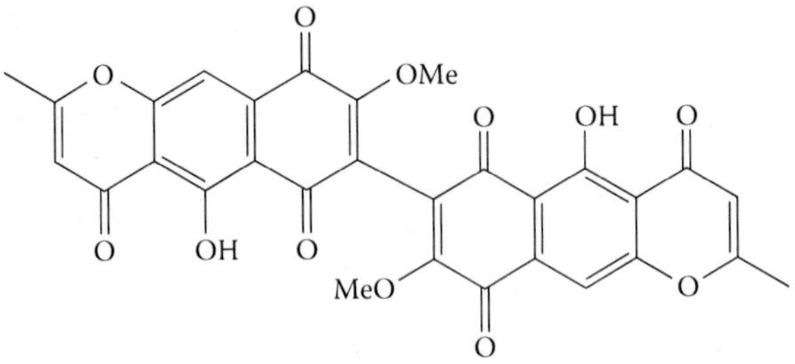

e.g. Aurofusarin
Benzochromanquinones (106)
VQ3090

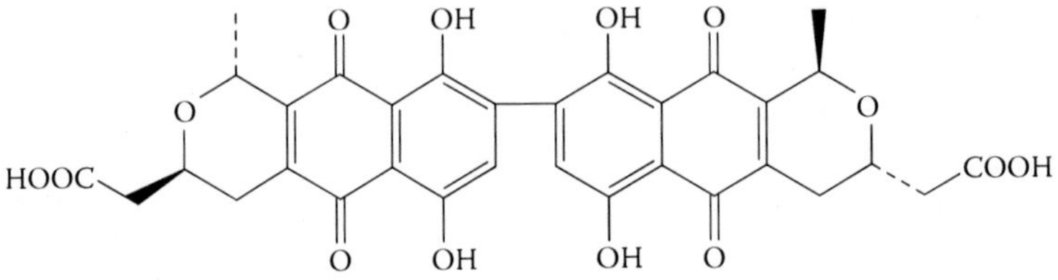

e.g. Actinorhodin
Benzoisochromanquinones (171)
VQ3100

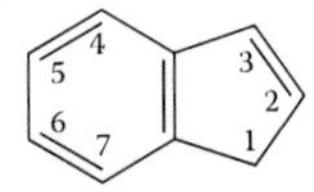

Indenes/indanes (72)
VQ3300

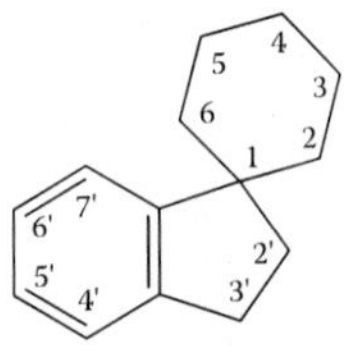

Indan-1-spiro-
cyclohexanes (18)
VQ3400

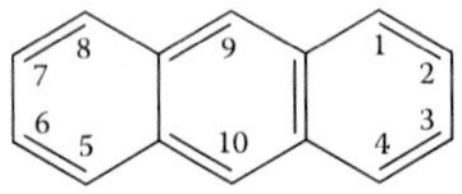

Anthracenes (564)
VQ3450

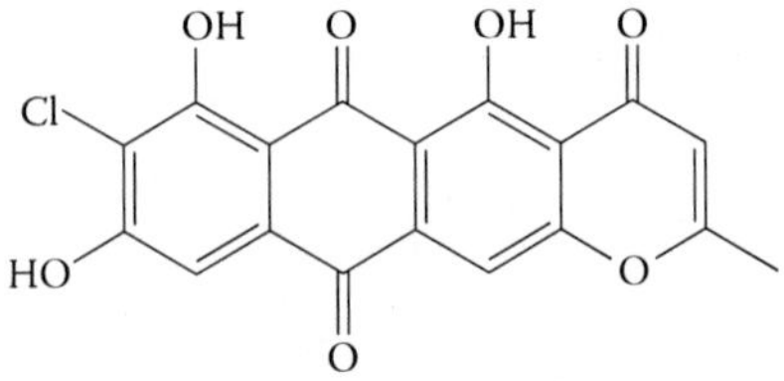

e.g. Topopyrone B
Pyranoanthracenes (9)
VQ3700

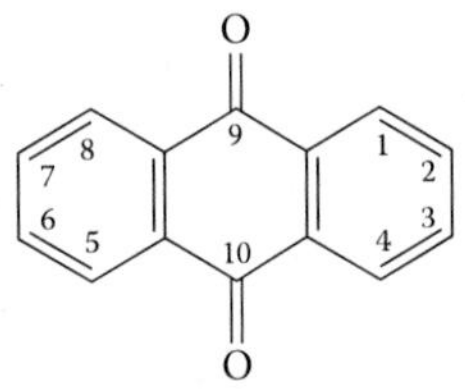

9,10-Anthraquinones (1177)
VQ4000-VQ4080
1,2- and 1,4-Anthraquinones (53)
VQ4100

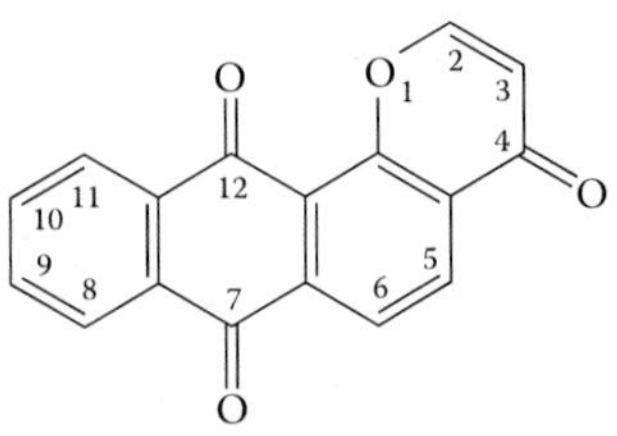

Pluramycins (82)
VQ4200

(*Continued*)

TABLE 6.1 (*Continued*)
Natural Product Parent Skeletons

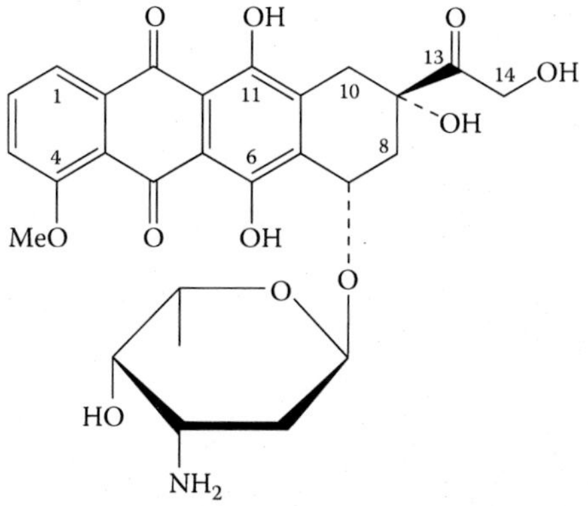

e.g. Adriamycin
Anthracyclinones (549)
VQ4300

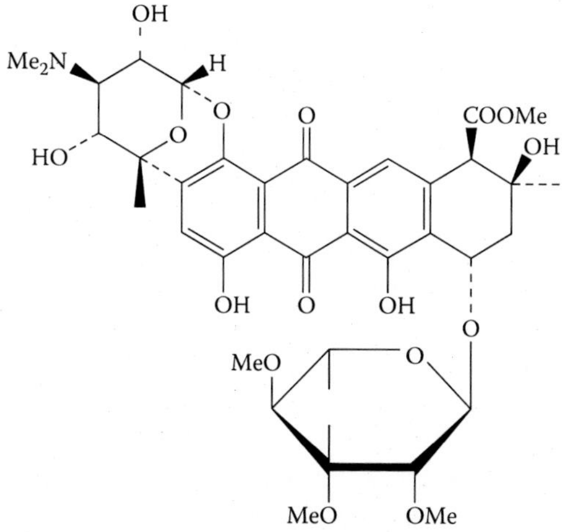

e.g. Nogalamycin
Nogarols (19)
VQ4320

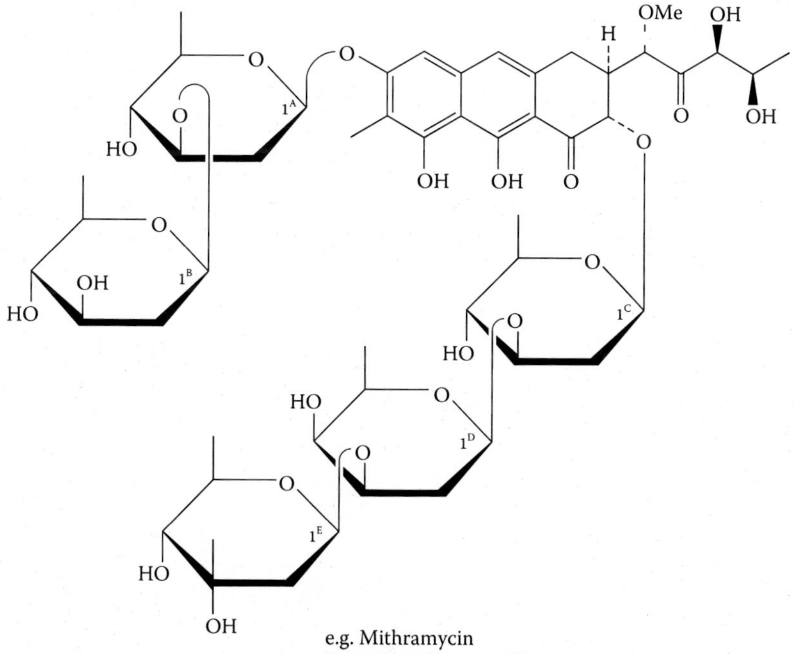

e.g. Mithramycin
Aureolic acid antibiotics (71)
VQ4330

(*Continued*)

TABLE 6.1 (*Continued*)
Natural Product Parent Skeletons

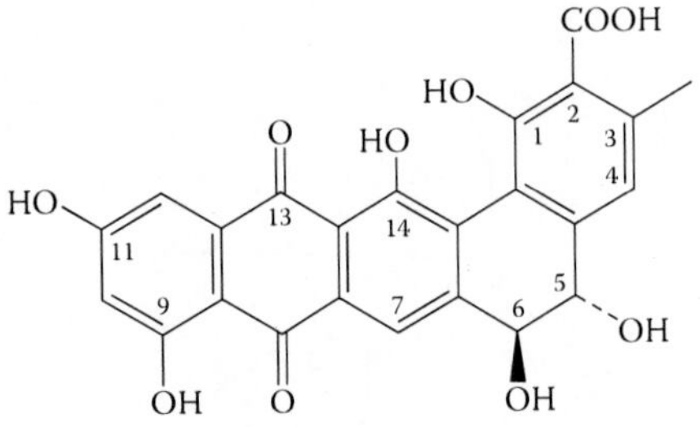

e.g. Pradimicin M
Benzoanthracyclinones (84)
VQ4350

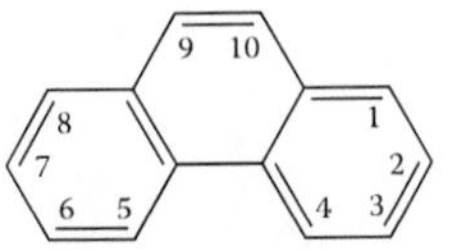

Phenanthrenes (432)
VQ4800

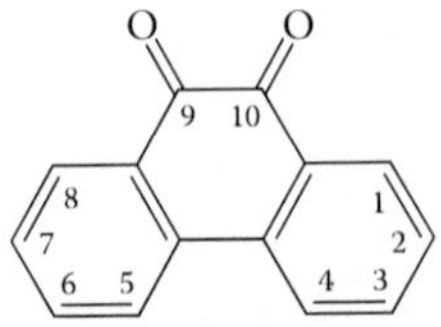

9,10-Phenanthraquinones (10)
VQ5000
1,4-Phenanthraquinones (34)
VQ5100

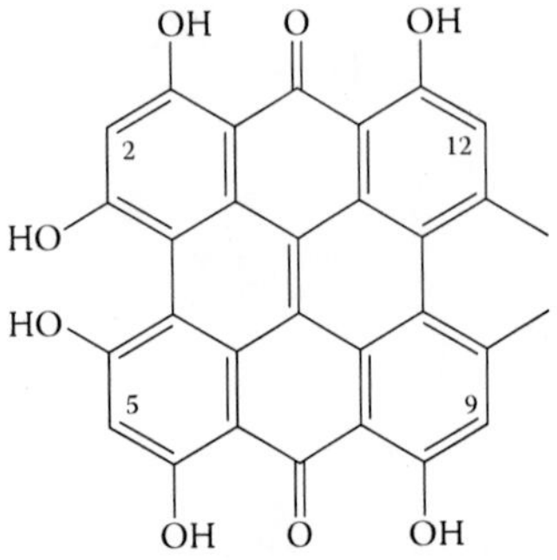

e.g. Hypericin
Extended quinones (53)
VQ6000

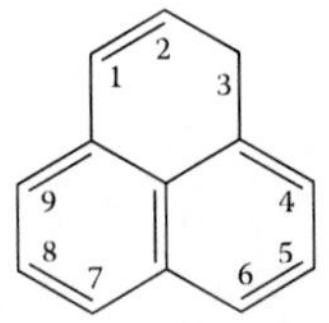

Phenalenes (122)
VQ7500

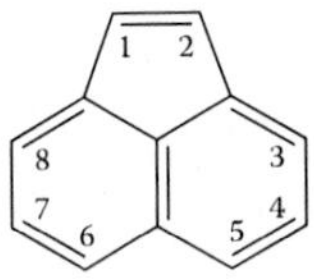

Acenaphthylenes (25)
VQ7600

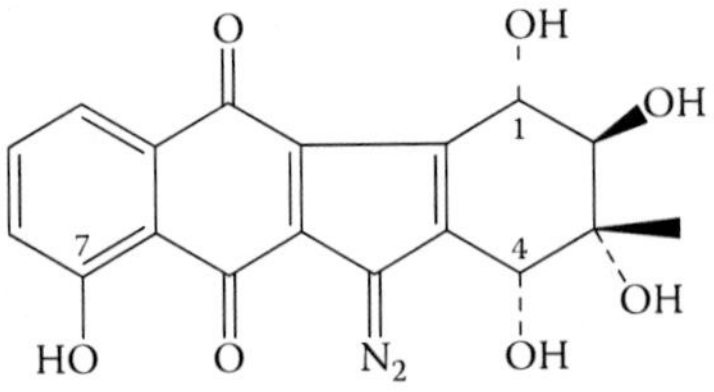

Fluorenes (43)
VQ7700

e.g. Kinamycin F
Kinamycins (20)
VQ7800

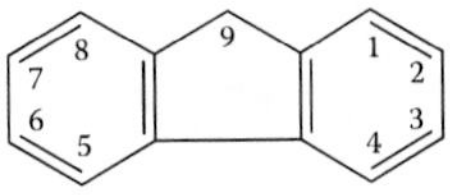

(*Continued*)

TABLE 6.1 (*Continued*)
Natural Product Parent Skeletons

Terpenoids (VS)

2,6-Dimethyloctane, CAS
Regular acyclic
monoterpenoids (772)
VS0100

Irregular acyclic monoterpenoids (102)
VS0150

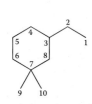

3-Ethyl-1,1-dimethyl-
cyclohexane, CAS
Ochtodanes (46)
VS0220

4,7-Dimethylcyclo-
penta[*c*]pyran, CAS
Iridoids (1706)
VS0400

Secoiridoids (573)
VS0440

Other cyclopentane monoterpenoids (33)
VS0450

1-Methyl-4-(1-methylethyl)-
cyclohexane, CAS
p-Menthanes (802)
VS0500

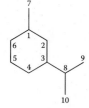

m-Menthanes (16)
VS0520

o-Menthanes (15)
VS0540

Other cyclohexane monoterpenoids (97)
VS0600

(Continued)

TABLE 6.1 (*Continued*)
Natural Product Parent Skeletons

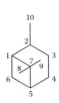

Cycloheptane monoterpenoids (30)
VS0700

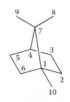

1,7,7-Trimethylbicyclo-
[2.2.1]heptane, CAS
Bornanes (72)
VS0800

1,3,3-Trimethylbicyclo-
[2.2.1]heptane, CAS
Fenchanes (17)
VS0850

2,6,6-Trimethylbicyclo-
[3.1.1]heptane, CAS
Pinanes (174)
VS0900

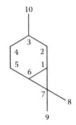

3,7,7-Trimethylbicyclo-
[4.1.0]heptane, CAS
Caranes (18)
VS0950

4-Methyl-1-(1-methylethyl)-
bicyclo[3.1.0]hexane, CAS
Thujanes (35)
VS1000

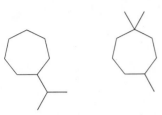

Miscellaneous bicyclic monoterpenoids (29)
VS1050

Tricyclene
1,7,7-Trimethyltricyclo-
[2.2.1.0^{2,6}]heptane, CAS
Tricyclic monoterpenoids (4)
VS1200

2,6,10-Trimethyldodecane, CAS
Farnesanes (412)
VS1300

e.g. Ngaione
Furanoid farnesane
sesquiterpenoids (139)
VS1320

(*Continued*)

TABLE 6.1 (*Continued*)
Natural Product Parent Skeletons

Irregular acyclic sesquiterpenoids (48)
VS1400

Cyclobutane and cyclopentane sesquiterpenoids (86)
VS1420, VS1430

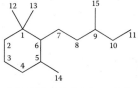

1,1,3-Trimethyl-2-(3-methyl-
pentyl)cyclohexane
Cyclofarnesanes (185)
VS1450

1,2,4-Trimethyl-3-(3-methyl-
pentyl)cyclohexane
Rearranged cyclofarnesanes (33)
VS1460

4-Methyl-2-(2-methylbutyl)-
1-(1-methylethyl)cyclohexane
Herbertianes (20)
VS1470

1-(1,5-Dimethylhexyl)-
4-methylcyclohexane, CAS
Bisabolanes (738)
VS1500

3,7-Dimethyl-7-(4-methyl-
pentyl)bicyclo-
[4.1.0]heptane, CAS
Sesquicarane

1-(1,5-Dimethylhexyl)-
4-methylbicyclo[3.1.0]-
hexane, CAS
Sesquisabinane

Cyclobisabolanes (17)
VS1550

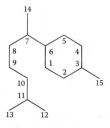

1-Ethyl-1-methyl-2,4-bis-
(1-methylethyl)cyclohexane, CAS
Elemanes (242)
VS1600

(*Continued*)

TABLE 6.1 (*Continued*)
Natural Product Parent Skeletons

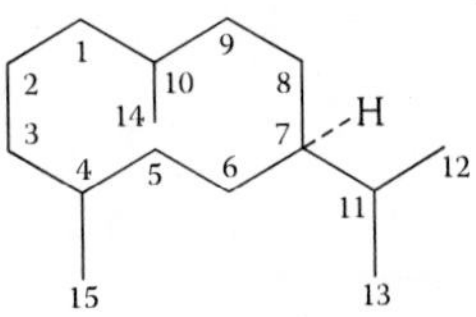

1,7-Dimethyl-4-(1-methyl-
ethyl)cyclodecane, CAS
Germacranes (327)
VS1650

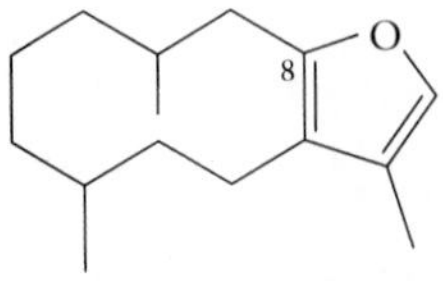

4,5,6,7,8,9,10,11-Octahydro-
3,6,10-trimethylcyclodeca[*b*]-
furan, CAS
Furanogermacranes (380)
VS1670

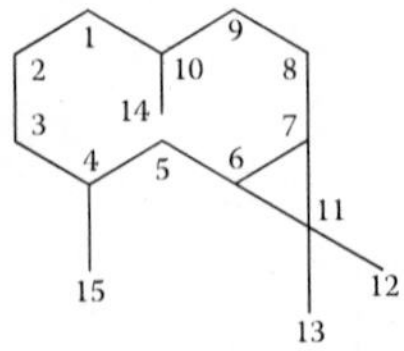

3,7,11,11-Tetramethylbicyclo-
[8.1.0]undecane, CAS
Bicyclogermacranes (59)
VS1710

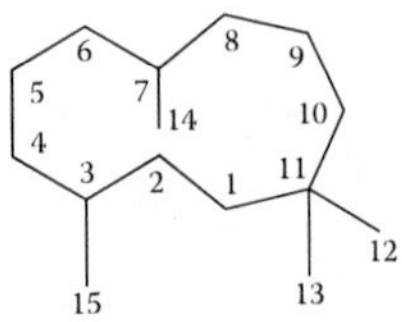

1,1,4,8-Tetramethyl-
cycloundecane, CAS
Humulanes (142)
VS1720

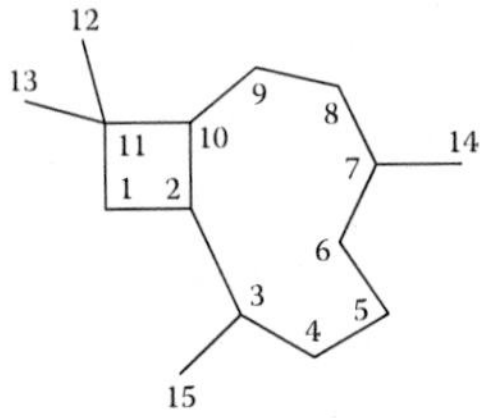

2,6,10,10-Tetramethylbicyclo-
[7.2.0]undecane, CAS
Caryophyllanes (165)
VS1730

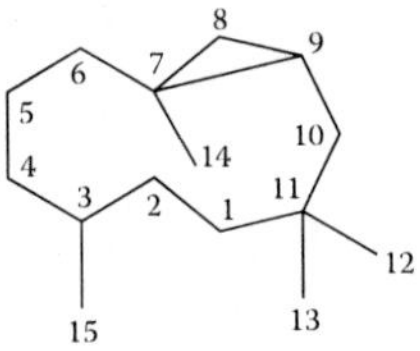

1,5,8,8-Tetramethylbicyclo-
[8.1.0]undecane, CAS
Bicyclohumulanes (4)
VS1740

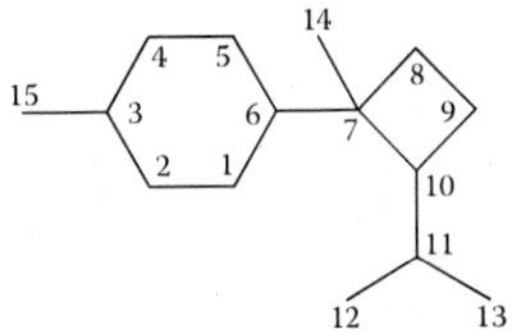

Dunnianes (2)
VS1745

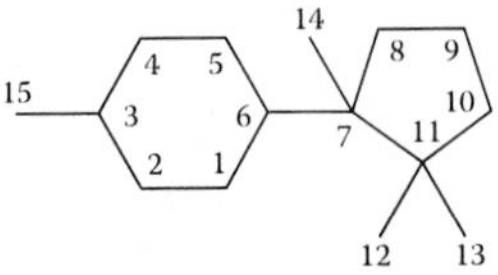

Cuparanes (96)
VS1750

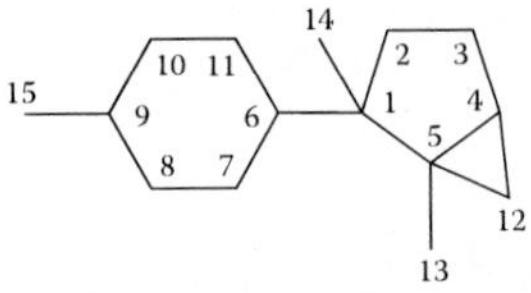

1,2-Dimethyl-2-(4-methylcyclohexyl)-
bicyclo[3.1.0]hexane, CAS
Cyclolauranes (14)
VS1760

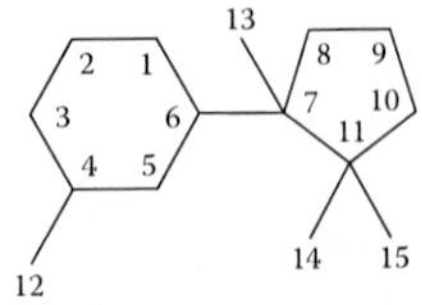

Herbertanes (25)
VS1800

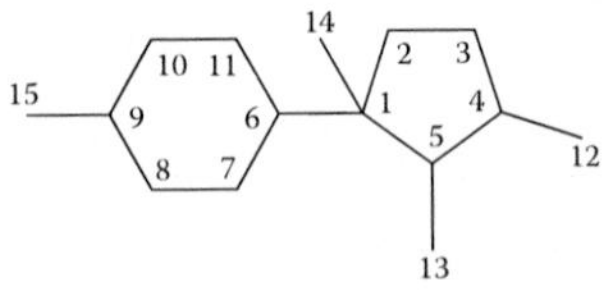

Lauranes (70)
VS1850

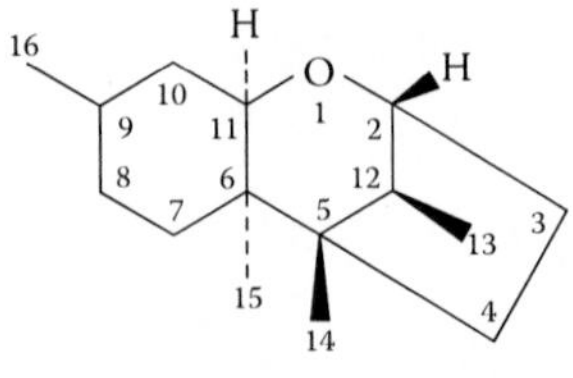

Trichothecanes (239)
VS1900

(*Continued*)

TABLE 6.1 (*Continued*)
Natural Product Parent Skeletons

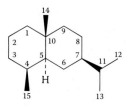

Decahydro-1,4*a*-dimethyl-
7-(1-methylethyl)-
naphthalene, CAS
Eudesmanes (1200)
VS1950, VS1990, VS2000

Dihydro-β-agarofuran
2,2,5*a*,9-Tetramethyl-
2*H*-3,9*a*-methano-
1-benzoxepin, CAS
Agarofurans (713)
VS1980

Emmotins (19)
VS2010

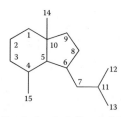

Octahydro-3*a*,7-dimethyl-1-
(2-methylpropyl)-1*H*-indene, CAS
Oppositanes (34)
VS2020

Farfugins (2)
VS2040

Cycloeudesmanes (163)
VS2050

Decahydro-1,4*a*-dimethyl-8-
(1-methylethyl)naphthalene, CAS
Gorgonanes (11)
VS2060

Decahydro-1,8*a*-dimethyl-
7-(1-methylethyl)-
naphthalene, CAS
Eremophilanes (371)
VS2100

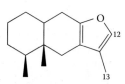

4,4*a*,5,6,7,8,8*a*,9-Octahydro-
3,4*a*,5-trimethylnaphtho-
[2,3-*b*]furan, CAS
Furanoeremophilanes (859)
VS2110

(*Continued*)

TABLE 6.1 (*Continued*)
Natural Product Parent Skeletons

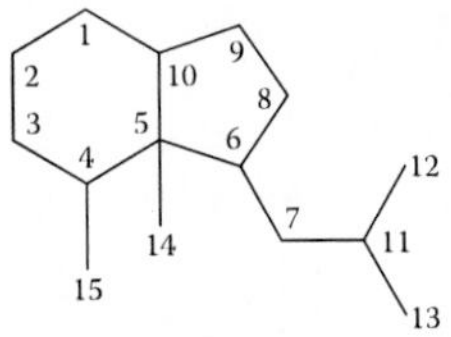

Octahydro-7,7*a*-dimethyl-
1-(2-methylpropyl)-1*H*-indene, CAS
Chiloscyphanes (14)
VS2140

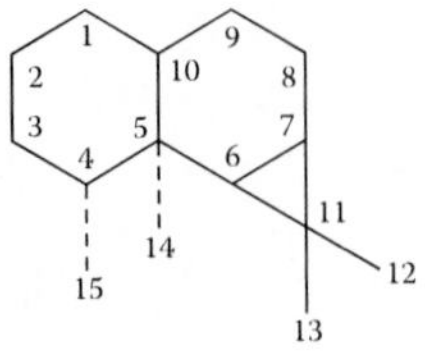

Decahydro-1,1,7,7*a*-tetramethyl-
1*H*-cyclopropa[*a*]naphthalene, CAS
Aristolanes (70)
VS2150

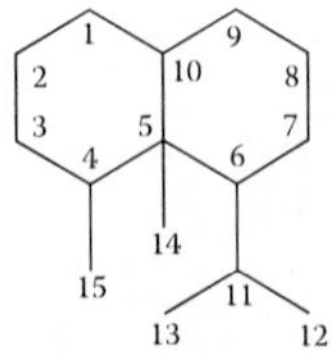

Decahydro-1,8*a*-dimethyl-
8-(1-methylethyl)naphthalene
Nardosinanes (97)
VS2160

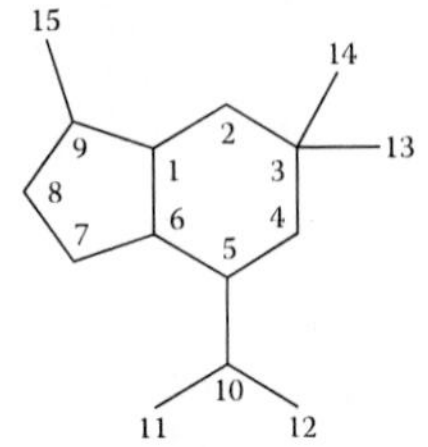

Octahydro-1,6,6-trimethyl-
4-(1-methylethyl)-1*H*-indene, CAS
Brasilanes (21)
VS2170

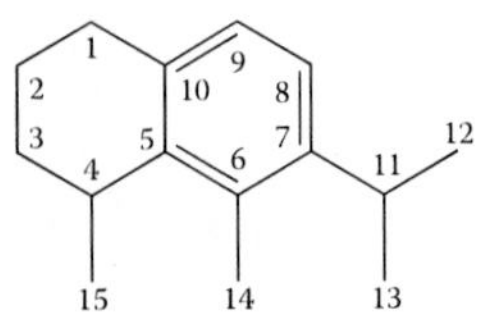

Cacalols (136)
VS2180

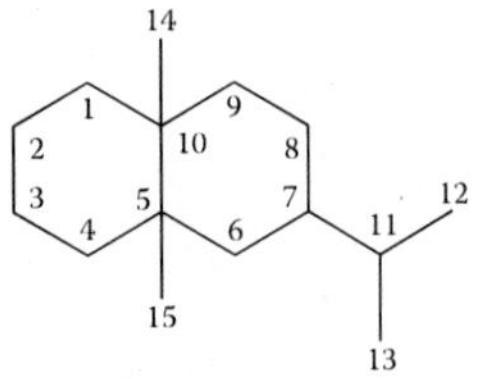

Decahydro-4*a*,8*a*-dimethyl-
2-(1-methylethyl)naphthalene, CAS
Valeranes (7)
VS2200

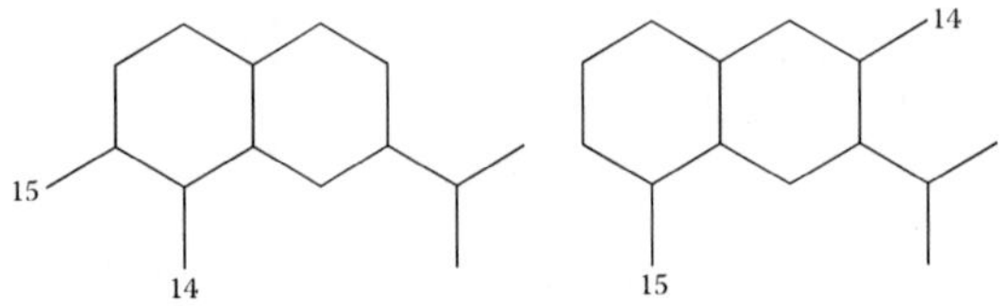

Miscellaneous rearranged eudesmanes (39)
VS2220

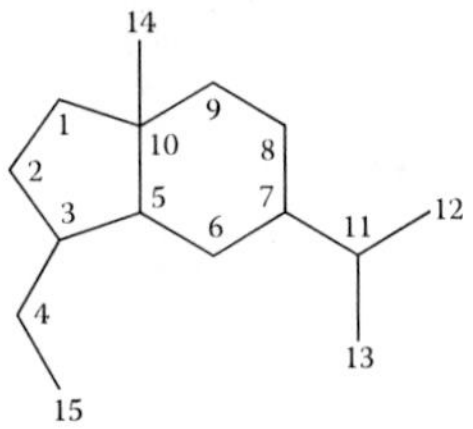

Iphionanes (6)
VS2225

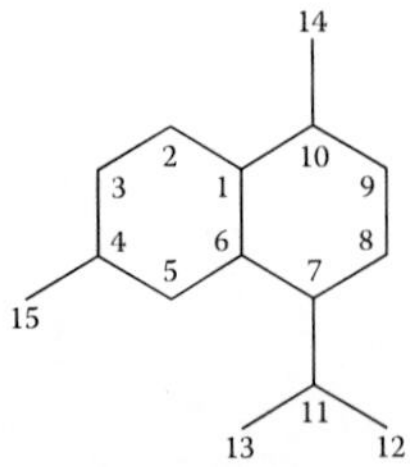

Decahydro-1,6-dimethyl-
4-(1-methylethyl)naphthalene, CAS
Cadinanes (764)
Stereoisomers called muurolanes,
bulgaranes and amorphanes.
Dehydro derivatives called
calamenenes and cadalenes
VS2250

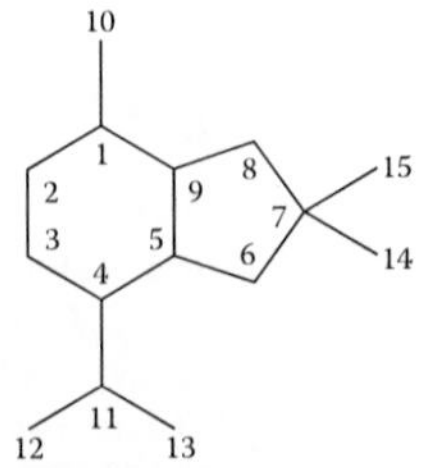

Octahydro-2,2,4-trimethyl-
7-(1-methylethyl)-
1*H*-indene, CAS
Alliacanes (13)
VS2270

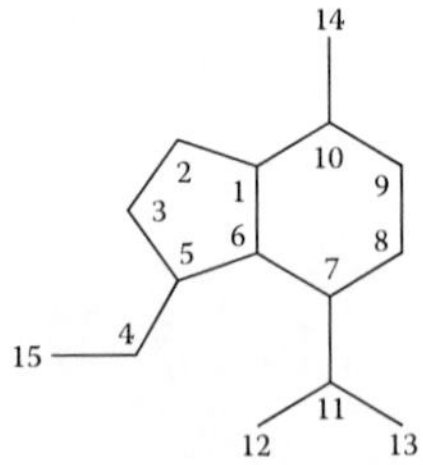

1-Ethyloctahydro-4-methyl-
7-(1-methylethyl)-1*H*-indene
Oplopanes (85)
VS2280

(Continued)

TABLE 6.1 (*Continued*)
Natural Product Parent Skeletons

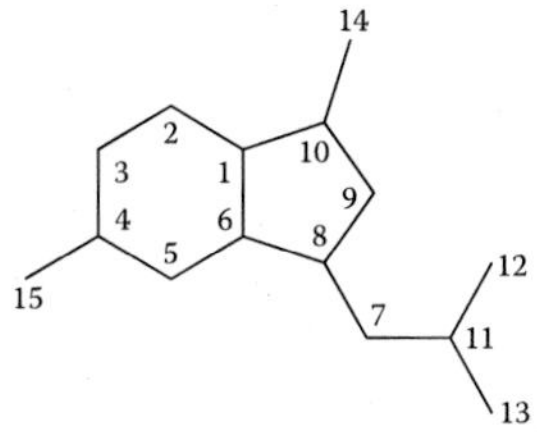

Octahydro-1,5-dimethyl-
3-(2-methylpropyl)-1*H*-indene
Mutisianthols (2)
VS2290

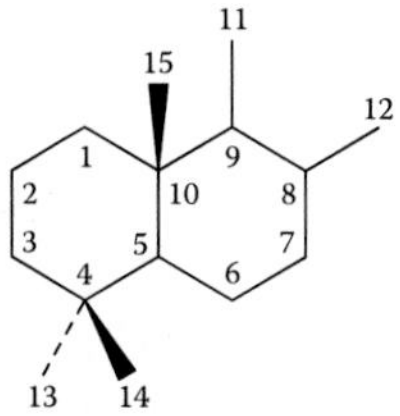

Decahydro-1,1,4*a*,5,6-
pentamethylnaphthalene
Drimanes (433)
VS2300, VS2320

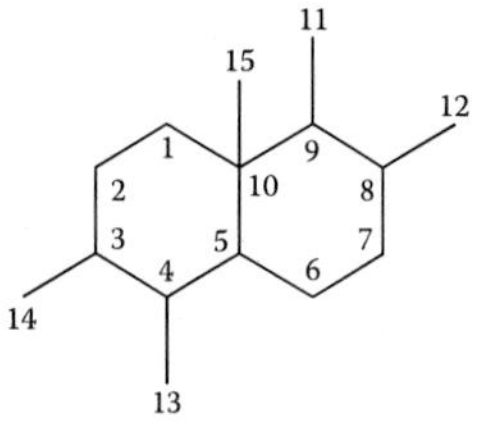

Decahydro-1,2,4*a*,5,6-
pentamethylnaphthalene
Coloratanes (23)
VS2310

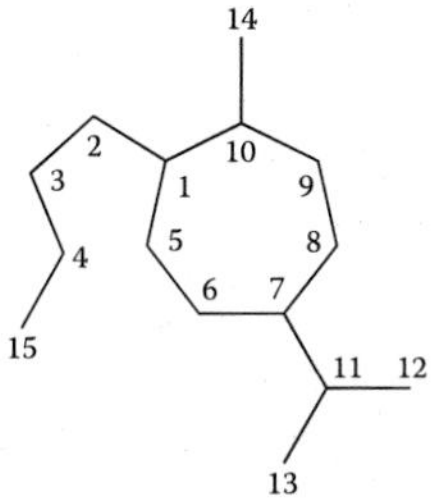

1-Butyl-2-methyl-5-
(1-methylethyl)cycloheptane
Xanthanes (91)
VS2380

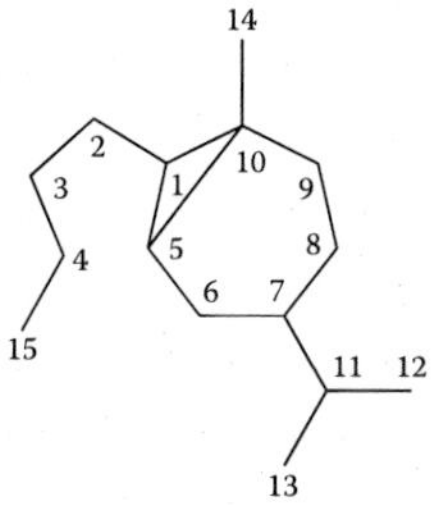

7-Butyl-1-methyl-4-(1-methyl-
ethyl)bicyclo[4.1.0]heptane
Carabranes (16)
VS2390

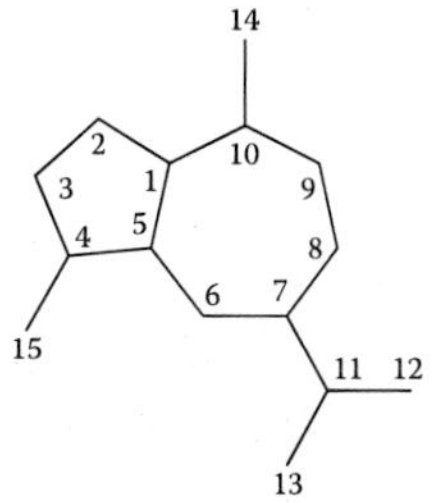

Decahydro-1,4-dimethyl-
7-(1-methylethyl)azulene, CAS
Guaianes (2330)
VS2400, VS2410, VS2420, VS2430, VS2440

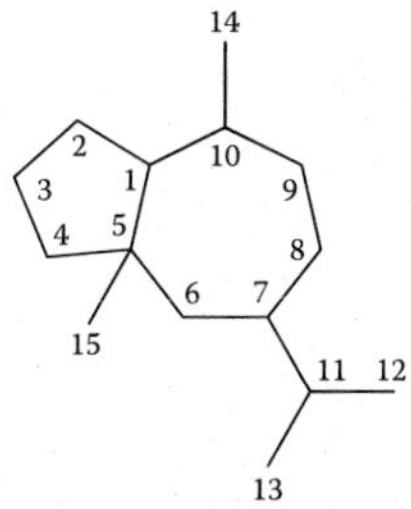

Decahydro-4,8*a*-dimethyl-
7-(1-methylethyl)azulene, CAS
Pseudoguaianes (353)
VS2450

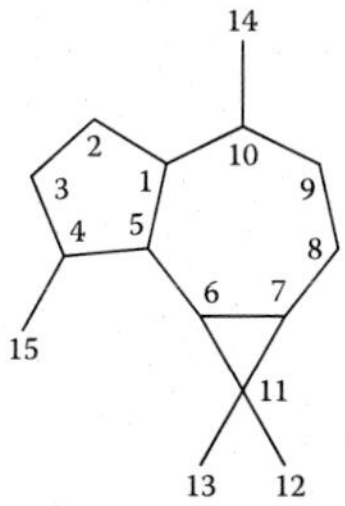

Decahydro-1,1,4,7-tetramethyl-
1*H*-cycloprop[*e*]azulene, CAS
Aromadendranes (174)
VS2500

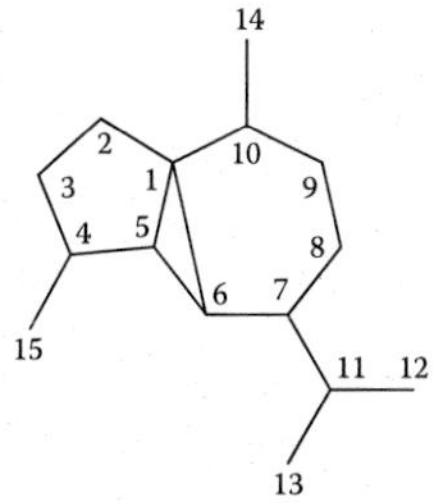

Octahydro-3,7-dimethyl-4-
(1-methylethyl)-1*H*-cyclopenta-
[1,3]cyclopropa[1,2]benzene, CAS
Cubebanes (22)
VS2600

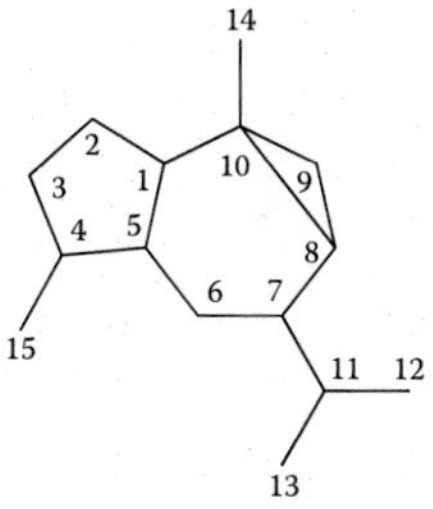

Ivaxillaranes (8)
VS2620

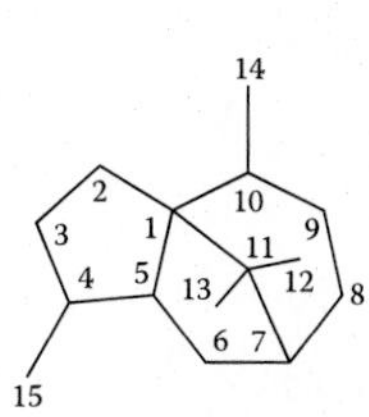

Patchoulanes (27)
VS2650

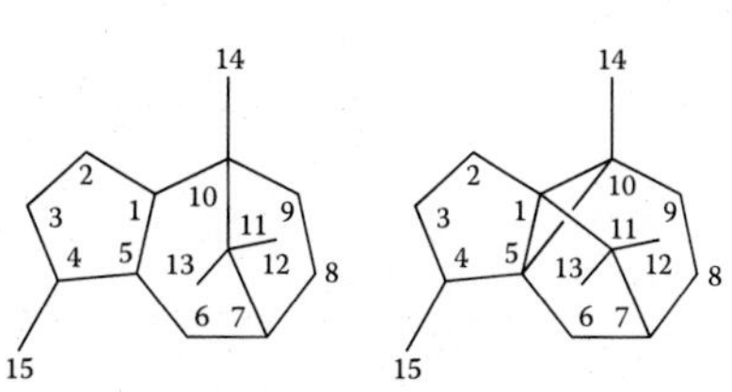

Rearranged patchoulanes (21)
VS2660

(*Continued*)

TABLE 6.1 (*Continued*)
Natural Product Parent Skeletons

Octahydro-1,4-dimethyl-
7-(2-methylpropyl)-1*H*-indene
Valerenanes (24)
VS2710

Decahydro-3,3,5,7*b*-tetramethyl-
1*H*-cycloprop[*e*]azulene
Africananes (50)
VS2750

Lippifolianes (5)
VS2760

Decahydro-2,5,9,9-tetramethyl-
1*H*-benzocycloheptene, CAS
Himachalanes (47)
VS2780

2,6,6,9-Tetramethyltricyclo-
[5.4.0.0^{2,8}]undecane, CAS
Longipinanes (96)
VS2800

Decahydro-4,8,8,9-tetramethyl-
1,4-methanoazulene, CAS
Longifolanes (5)
VS2850

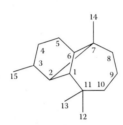

Decahydro-1,5,5,8*a*-tetramethyl-1,4-methanoazulene, CAS
Longibornanes (4)
VS2900

5-Ethyloctahydro-1,3*a*,4,7*a*-
tetramethyl-1*H*-indene
Pinguisanes (57)
VS3000

Octahydro-1,2,3*a*,7,7,7*a*-
hexamethyl-1*H*-indene
Thapsanes (26)
VS3050

Octahydro-2,3*a*,4-trimethyl-
2-(1-methylethyl)-1*H*-indene
Fukinanes (65)
VS3080

Octahydro-1,4,7*a*-trimethyl-
5-(1-methylethyl)-1*H*-indene
Picrotoxanes (109)
VS3100

(*Continued*)

TABLE 6.1 (*Continued*)
Natural Product Parent Skeletons

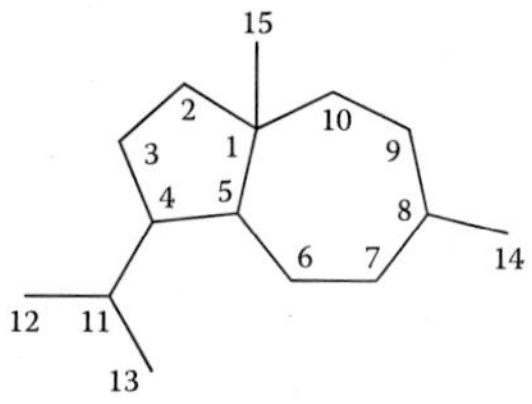

Decahydro-3*a*,6-dimethyl-
1-(1-methylethyl)azulene
Daucanes (235)
VS3180

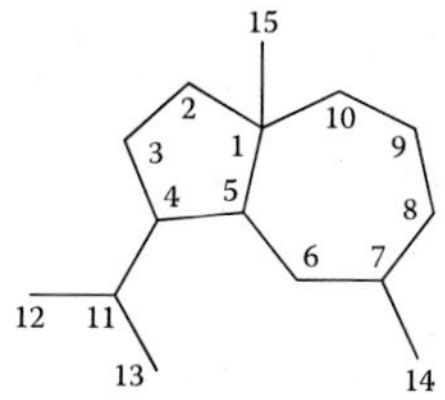

Decahydro-3*a*,7-dimethyl-
1-(1-methylethyl)azulene
Isodaucanes (41)
VS3190

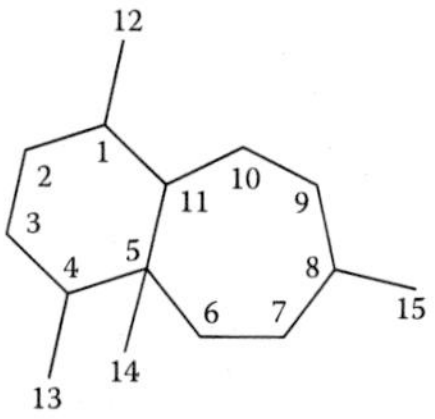

Decahydro-1,4,7,9*a*-tetra-
methyl-1*H*-benzocycloheptane
Perforanes (11)
VS3200

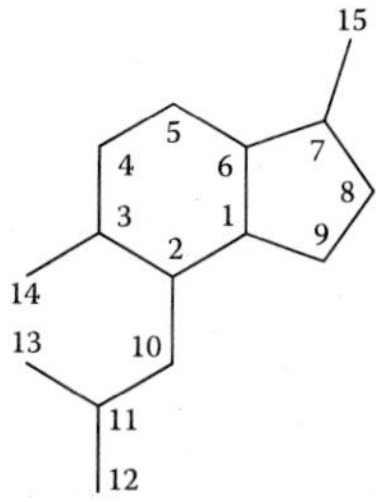

Octahydro-1,5-dimethyl-
4-(2-methylpropyl)-1*H*-indene
Pacifigorgianes (11)
VS3350

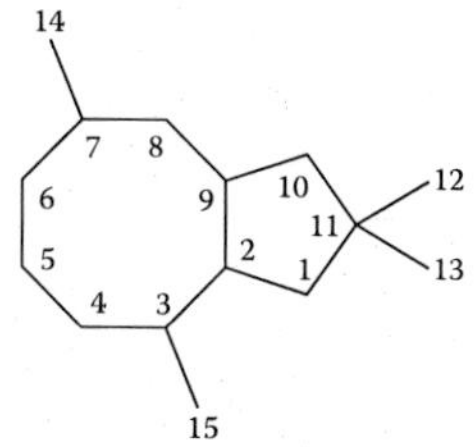

Decahydro-2,2,4,8-tetramethyl-
1*H*-cyclopentacyclooctene, CAS
Asteriscanes (29)
VS3380

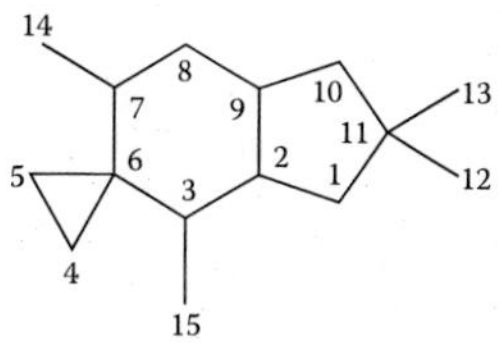

Decahydro-2',2',4',6'-tetra-
methylspiro[cyclopropane-
1,5'-[5*H*]indene], CAS
Illudanes (53)
VS3400

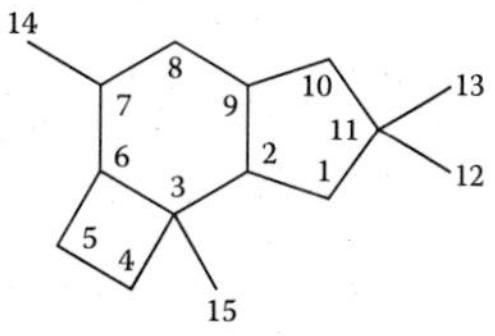

Decahydro-3,6,6,7*b*-tetra-
methyl-1*H*-cyclobut[*e*]-
indene, CAS
Protoilludanes (131)
VS3420

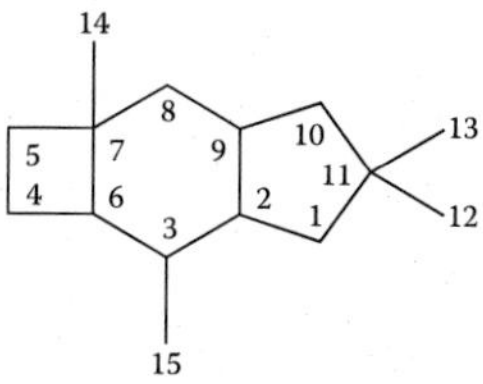

Decahydro-2*a*,5,5,7-tetra-
methyl-1*H*-cyclobut[*f*]-
indene, CAS
Sterpuranes (19)
VS3430

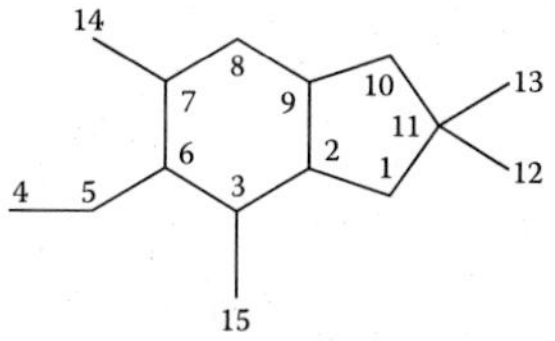

5-Ethyloctahydro-2,2,4,6-tetra-
methyl-1*H*-indene, CAS
Illudalanes (209)
VS3440

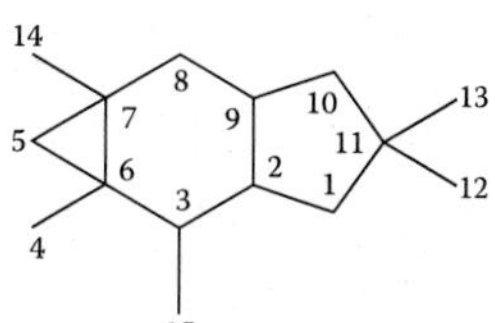

Isolactaranes (35)
VS3470

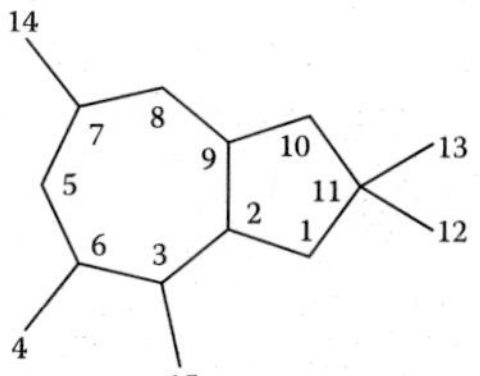

Merulanes (1)
VS3475

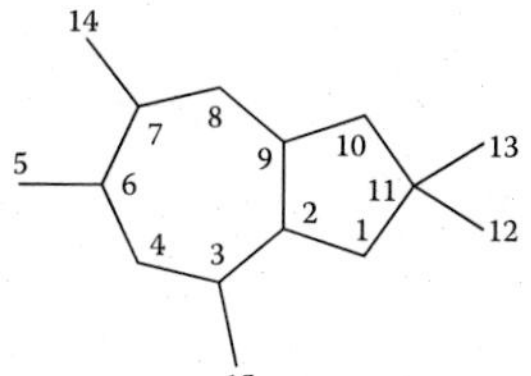

Lactaranes (73)
VS3480

(*Continued*)

TABLE 6.1 (*Continued*)
Natural Product Parent Skeletons

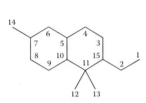

Tremulanes (19)
VS3490

Marasmanes (34)
VS3500

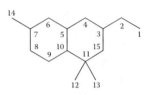

3-Ethyldecahydro-
1,1,6-trimethylnaphthalene
Furodysins (7)
VS3550

2-Ethyldecahydro-
1,1,6-trimethylnaphthalene
Furodysinins (31)
VS3560

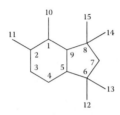

Octahydro-1,1,3,3,4,5-
hexamethyl-1*H*-indene
Botrydials (66)
VS3600

6,10-Dimethyl-2-(1-methyl-
ethyl)spiro[4.5]decane, CAS
Spirovetivanes (66)
VS3700

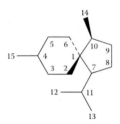

1,8-Dimethyl-4-(1-methyl-
ethyl)spiro[4.5]decane, CAS
Acoranes (59)
VS3750

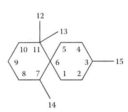

1,1,5,9-Tetramethyl-
spiro[5.5]undecane, CAS
Chamigranes (150)
VS3800

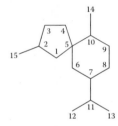

2,6-Dimethyl-9-(1-methyl-
ethyl)spiro[4.5]decane
Spiroaxanes (21)
VS3820

1,2,5,9-Tetramethylspiro-
[5.5]undecane

1,6-Dimethyl-8-(1-methyl-
ethyl)spiro[4.5]decane

Miscellaneous spirosesquiterpenoids (33)
VS3850

Octahydro-3,6,8,8-tetra-
methyl-1*H*-3*a*,7-methano-
azulene
Cedranes (63)
VS3900

(*Continued*)

TABLE 6.1 (*Continued*)
Natural Product Parent Skeletons

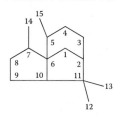

Octahydro-3,4,8,8-tetra-
methyl-1*H*-3*a*,7-methano-
azulene
Isocedranes (77)
VS3920

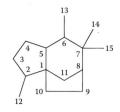

Octahydro-3,7,7,8-tetra-
methyl-1*H*-3*a*,6-methano-
azulene, CAS
Zizaanes (13)
VS3950

Octahydro-3,7,8,8-tetra-
methyl-1*H*-3*a*,6-methano-
azulene
Prezizaanes (129)
VS3960

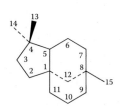

Decahydro-1,1,7-trimethyl-
3*a*,7-methano-3*aH*-cyclo-
pentacyclooctene, CAS
Clovanes (12)
VS4000

Decahydro-1,5,8,8-tetra-
methyl-1*H*-cyclopenta-
cyclooctene
Precapnellanes (6)
VS4200

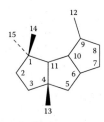

Decahydro-3,3,4,7*a*-tetra-
methyl-1*H*-cyclopenta[*a*]-
pentalene, CAS
Capnellanes (27)
VS4250

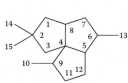

Decahydro-2,2,3*b*,4-tetra-
methyl-1*H*-cyclopenta[*a*]-
pentalene
Hirsutanes (53)
VS4300

Decahydro-2,3,3*b*,4-tetra-
methyl-1*H*-cyclopenta[*a*]-
pentalene
Rearranged hirsutanes (13)
VS4310

Decahydro-1,4,7,7-tetra-
methylcyclopenta[*c*]-
pentalene
Pentalenanes

Decahydro-1,4,6,7-tetra-
methylcyclopenta[*c*]-
pentalene
Rearranged pentalenanes

Pentalenanes and abeopentalenanes (20)
VS4400

Decahydro-1,4,4,5*a*-tetra-
methylcyclopenta[*c*]-
pentalene
Silphinanes (30)
VS4450

(*Continued*)

TABLE 6.1 (*Continued*)
Natural Product Parent Skeletons

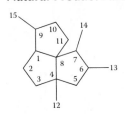

Decahydro-1,4,5,6a-tetra-
methylcyclopenta[c]-
pentalene
Silphiperfolianes (35)
VS4460

Presilphiperfolianes (19)
VS4470

Decahydro-1,3a,4,5a-tetra-
methylcyclopenta[c]pentalene
Isocomanes (16)
VS4500

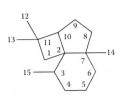

Decahydro-2,2,4a,8-tetra-
methylcyclobut[c]indene, CAS
Panasinsanes (4)
VS4630

Tetrahydro-1,1,3,4-tetramethyl-
1H,4H-3a,6a-propanopentalene
Modhephanes (10)
VS4700

Quadranes (8)
VS4750

1,7-Dimethyl-7-(4-methyl-
pentyl)bicyclo[2.2.1]-
heptane, CAS
Campherenanes (13)
VS4770

1,7-Dimethyl-7-(4-methyl-
pentyl)tricyclo[2.2.1.0^{2,6}]-
heptane
α-Santalanes (38)
VS4780

2,3-Dimethyl-2-(4-methyl-
pentyl)bicyclo[2.2.1]heptane
β-Santalanes (23)
VS4790

Octahydro-4,8-dimethyl-7-
(1-methylethyl)-1,4-methano-
1H-indene
Sativanes

1,6,7,8-Tetramethyl-4-
(1-methylethyl)bicyclo-
[3.2.1]octane
Helminthosporanes

Sativanes and helminthosporanes (22)
VS4800

(*Continued*)

TABLE 6.1 (*Continued*)
Natural Product Parent Skeletons

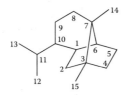
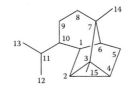

Octahydro-1,7*a*-dimethyl-5-
(1-methylethyl)-1,4-methano-
1*H*-indene, CAS
Copacamphanes

Octahydro-1,7*a*-dimethyl-5-
(1-methylethyl)-1,2,4-
metheno-1*H*-indene, CAS
Cyclocopacamphanes

Copacamphanes and cyclocopacamphanes (16)
VS4820

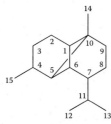

Octahydro-7*a*,8-dimethyl-5-
(1-methylethyl)-1,4-methano-
1*H*-indene, CAS
Sinularanes

Octahydro-1,4-dimethyl-7-
(1-methylethyl)-1,2,4-
metheno-1*H*-indene, CAS
Cyclosinularanes

Sinularanes and cyclosinularanes (4)
VS4850

1,3-Dimethyl-8-(1-methylethyl)-
tricyclo[4.4.0.0²·⁷]decane, CAS
Copaanes (28)
VS4960

Decahydro-1,6,6*a*-trimethyl-
1,2*a*-methano-2*aH*-cyclo-
propa[*b*]naphthalene, CAS
Ishwaranes (11)
VS5000

Decahydro-1,4,6-trimethyl-
4,7-ethanoazulene, CAS
Rotundanes (11)
VS5020

Decahydro-2,4*a*,8,8-tetramethyl-
cyclopropa[*d*]naphthalene, CAS
Thujopsanes (16)
VS5040

Decahydro-3*a*,6-dimethyl-1-
(1-methylethyl)cyclobuta[1,2:3,4]-
dicyclopentene, CAS
Bourbonanes (15)
VS5050

Decahydro-3*a*,4,7,8*a*-tetra-
methyl-4,8-methanoazulene, CAS
Gymnomitranes (34)
VS5070

(Continued)

TABLE 6.1 (*Continued*)
Natural Product Parent Skeletons

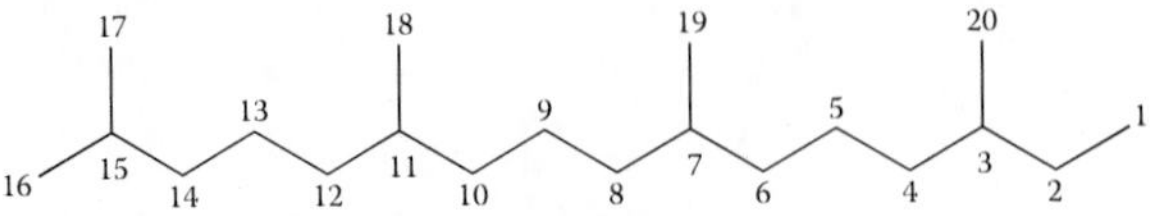

2,6,10,14-Tetramethylhexadecane
Phytanes (416)
VS5350

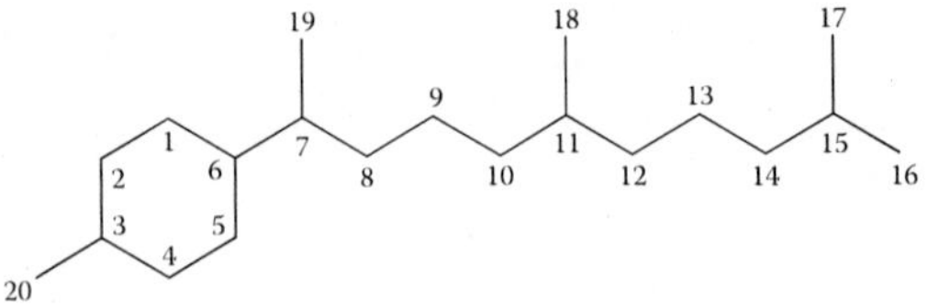

1-Methyl-4-(1,5,9-trimethyldecyl)cyclohexane
Prenylbisabolanes (19)
VS5380

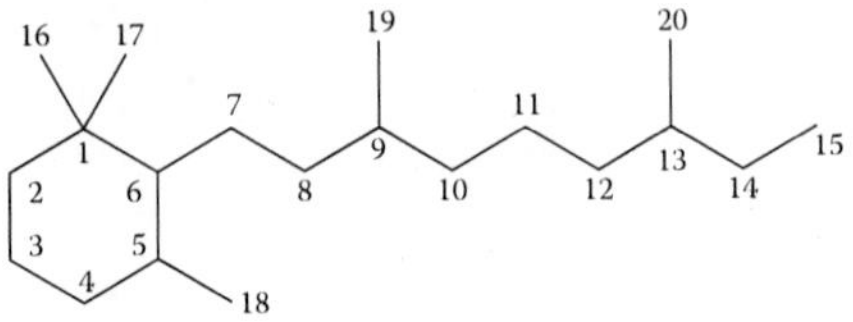

1,1,3-Trimethyl-2-(3,7-dimethylnonyl)cyclohexane
10,15-Cyclophytanes (118)
VS5390

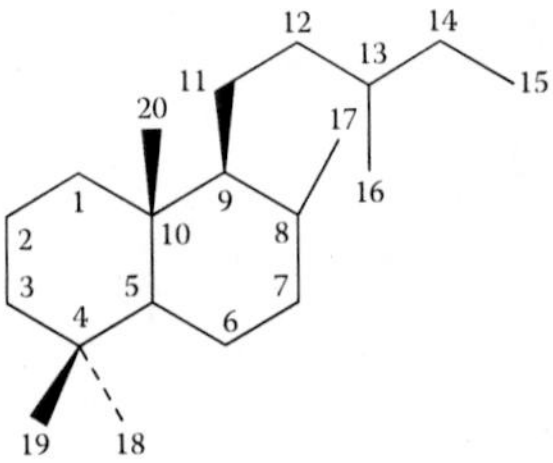

Decahydro-1,1,4a,6-tetra-
methyl-5-(3-methyl-
pentyl)naphthalene, CAS
Labdanes (1964)
VS5400

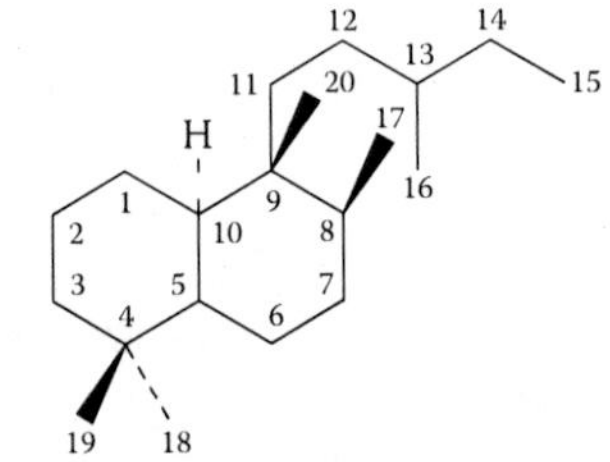

Decahydro-1,1,5,6-tetra-
methyl-5-(3-methyl-
pentyl)naphthalene
Halimanes (153)
VS5460

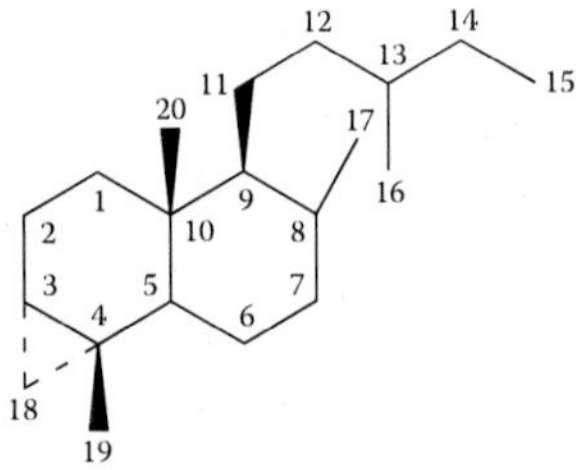

Rearranged labdane
diterpenoids (64)
VS5470

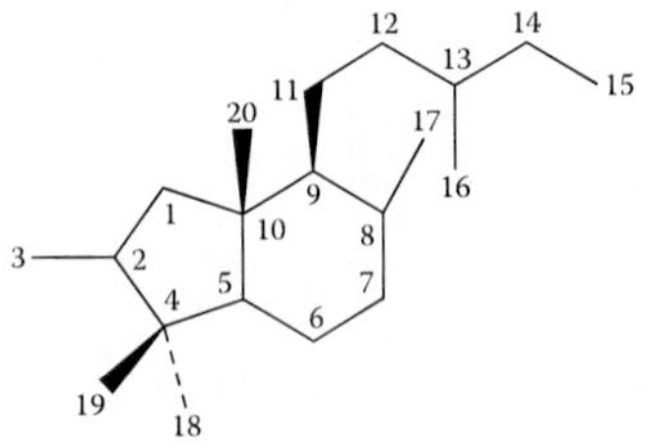

Octahydro-1,1,2,3a,5-
pentamethyl-4-(3-methyl-
pentyl)-1*H*-indene
Colensanes (5)
VS5480

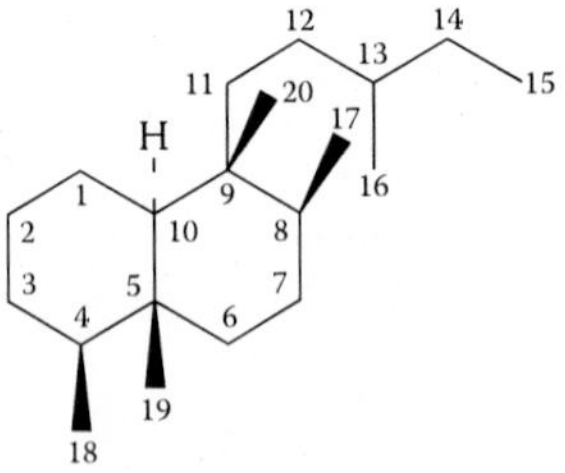

Decahydro-1,2,4a,5-
tetramethyl-1-(3-methyl-
pentyl)naphthalene, CAS
Clerodanes (1906)
VS5500

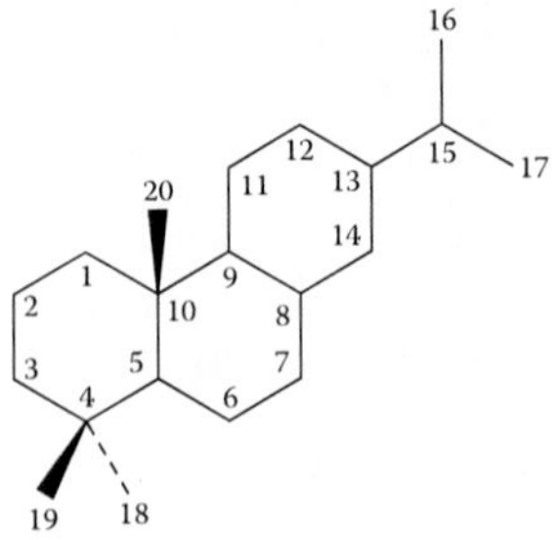

Tetradecahydro-1,1,4a-trimethyl-
7-(1-methylethyl)phenanthrene, CAS
Abietanes (1269)
VS5550, VS5560, VS5570, VS5580, VS5600

(*Continued*)

TABLE 6.1 (*Continued*)
Natural Product Parent Skeletons

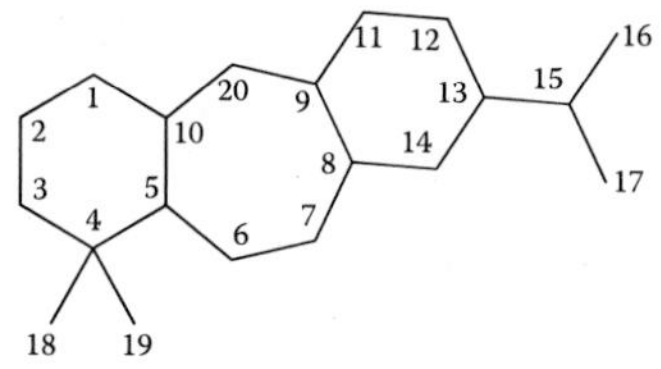

Icetexanes (60)
VS5565

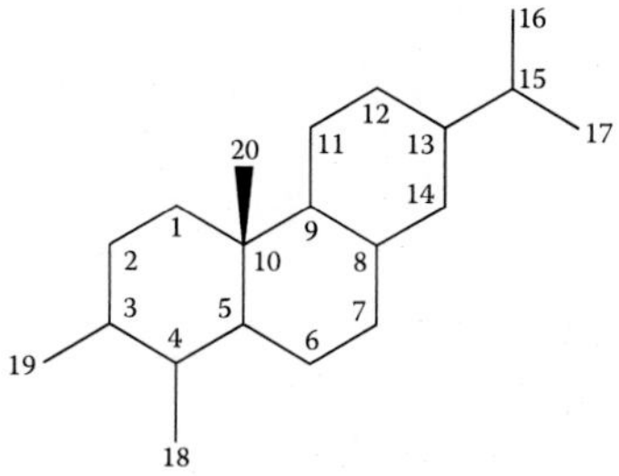

19(4⟶3)-Abeoabietanes (181)
VS5590

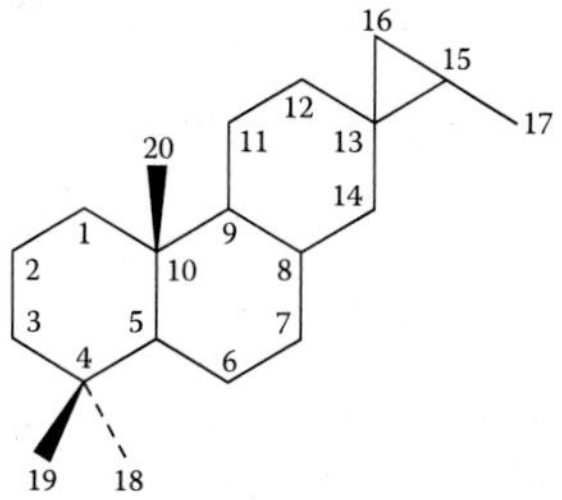

13,16-Cycloabietanes (47)
VS5620

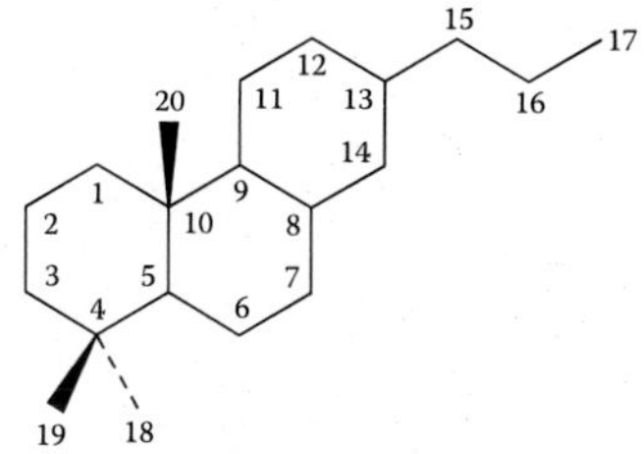

Tetradecahydro-1,1,4*a*-
trimethyl-7-propyl-
phenanthrene
17(15⟶16)-Abeoabietanes (80)
VS5630

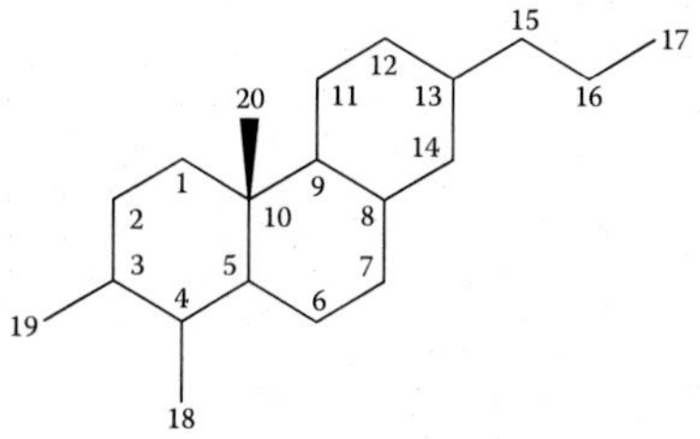

Tetradecahydro-1,2,4*a*-
trimethyl-7-propyl-
phenanthrene
17(15⟶16),19(4⟶3)-
Bisabeoabietanes (15)
VS5635

Abeoabietanes

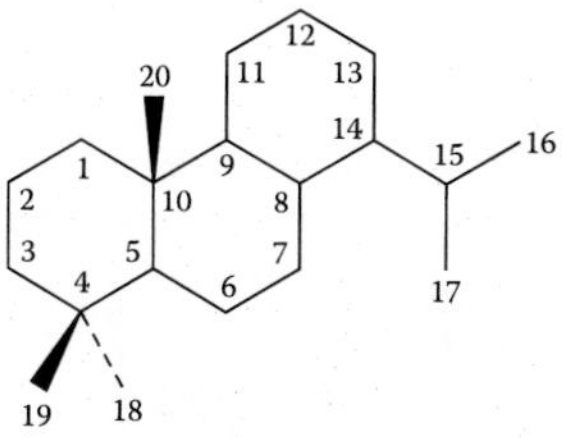

Tetradecahydro-1,1,4*a*-
trimethyl-8-(1-methylethyl)-
phenanthrene
Totaranes (45)
VS5650

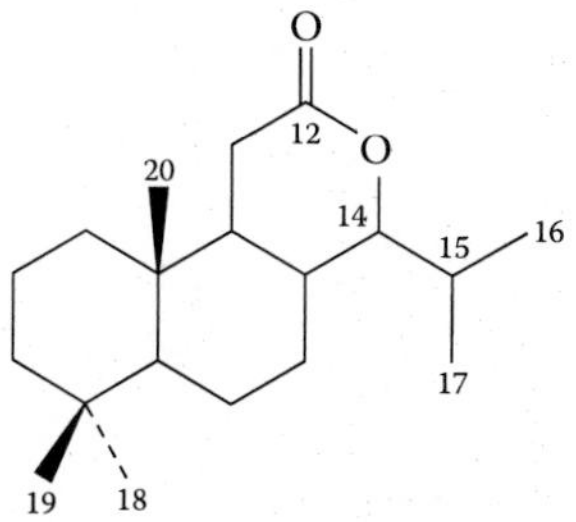

Nagilactones (103)
VS5660

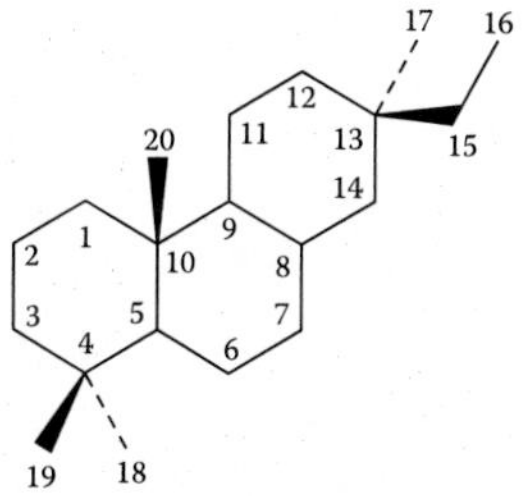

7-Ethyltetradecahydro-
1,1,4*a*,7-tetramethyl-
phenanthrene
Pimaranes (209)
VS5700

(*Continued*)

TABLE 6.1 (*Continued*)
Natural Product Parent Skeletons

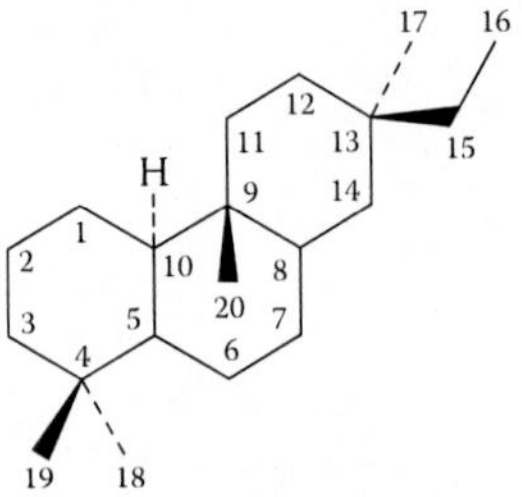

7-Ethyltetradecahydro-
1,1,4*b*,7-tetramethyl-
phenanthrene
Rosanes (97)
VS5710

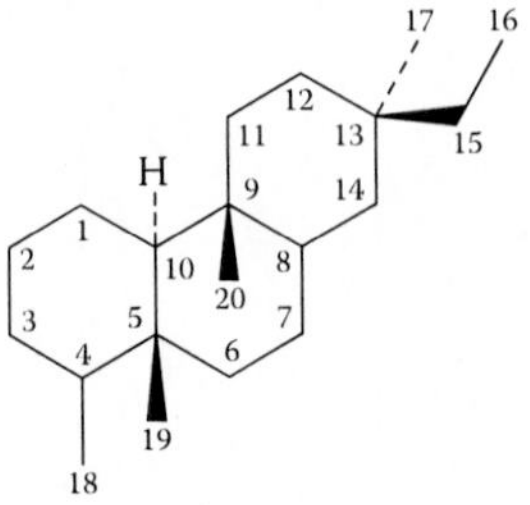

7-Ethyltetradecahydro-
1,4*b*,7,10*a*-tetramethyl-
phenanthrene
Erythroxylanes (65)
VS5720

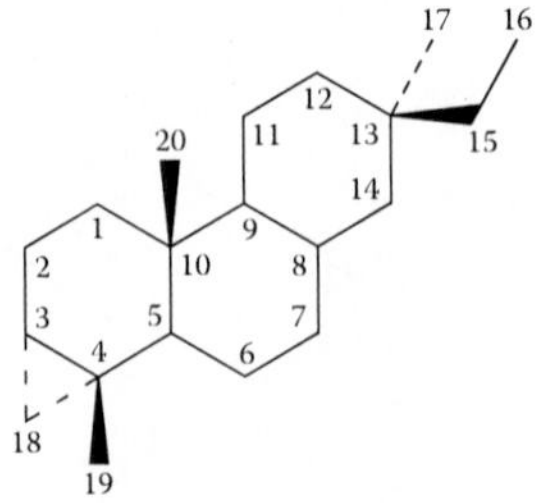

5-Ethyltetradecahydro-
1*a*,5,7*b*-trimethyl-1*H*-
cyclopropa[*a*]phenanthrene
Pargueranes (24)
VS5730

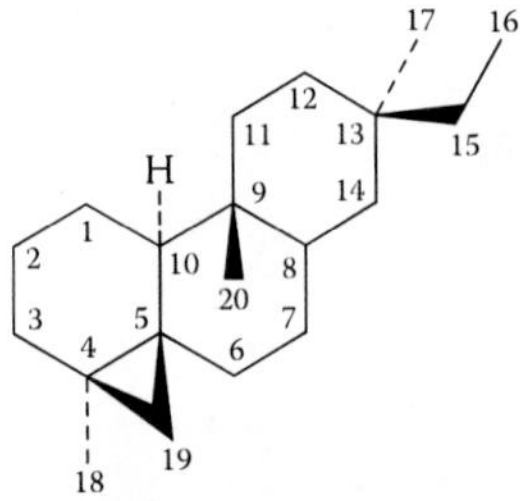

8-Ethyltetradecahydro-
3*a*,8,10*a*-trimethyl-
cyclopropa[*j*]phenanthrene
Devadaranes (5)
VS5740

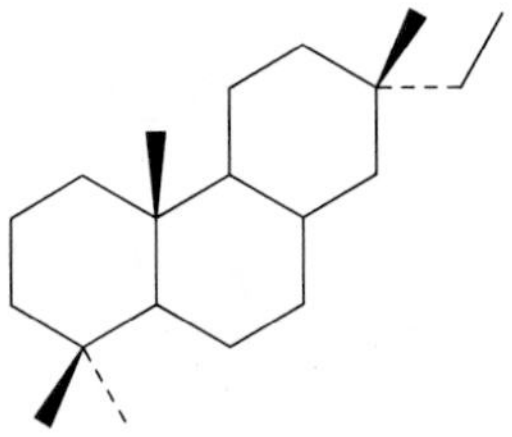

Isopimaranes (441)
VS5750

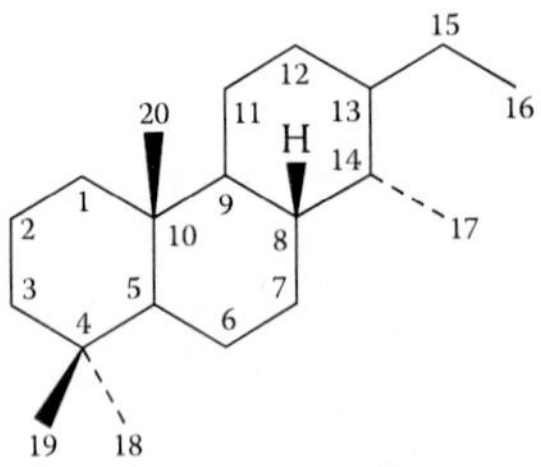

7-Ethyltetradecahydro-
1,1,4*a*,8-tetramethyl-
phenanthrene
Cassanes

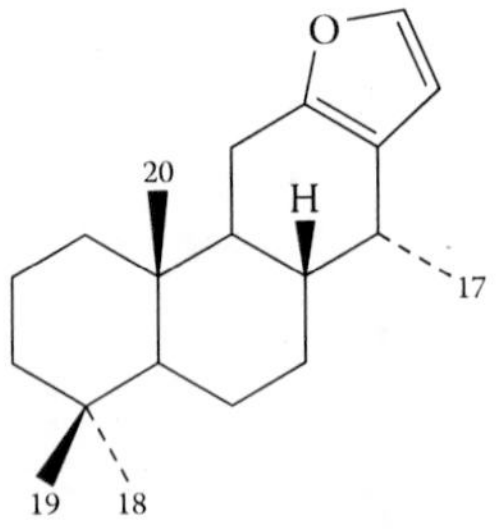

Vouacapanes

Cassanes and vouacapanes (352)
VS5800

(*Continued*)

TABLE 6.1 (*Continued*)
Natural Product Parent Skeletons

8-Ethyltetradecahydro-
1,1,4*a*,7-tetramethyl-
phenanthrene
Cleistanthanes

Isocleistanthanes

Cleistanthanes and isocleistanthanes (72)
VS5850

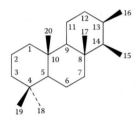

Tetradecahydro-1,1,4*a*,7,8,8*a*-
hexamethylphenanthrene
Isocopalanes

Spongianes

Isocopalanes and spongianes (155)
VS5950

ent-Kaurane
= Kaurane, CAS
Kauranes (1518)
VS6000

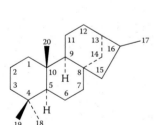

5α,9α,10β-Kaurane, CAS
Phyllocladanes (27)
VS6010

ent-Beyerane
Beyeranes (111)
VS6050

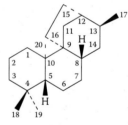

ent-Villanovane
Tetradecahydro-4,4,8,11*b*-
tetramethyl-9,11*a*-methano-
11*aH*-cyclohepta[*a*]-
naphthalene, CAS
Villanovanes (18)
VS6080

(*Continued*)

TABLE 6.1 (*Continued*)
Natural Product Parent Skeletons

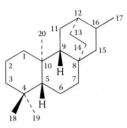

ent-Atisane
= Atisane, CAS
Atisanes (109)
VS6100

ent-Trachylobane
= Trachylobane, CAS
Trachylobanes (60)
VS6150

ent-Helvifulvane
Tetradecahydro-4,4,7*a*,9*b*-
tetramethyl-8,9*a*-methano-
9*aH*-cyclopropa[*b*]-
phenanthrene, CAS
Helvifulvanes (4)
VS6160

4,4,17-Trimethyl-9,15-
cyclo-*C*,18-dinor-
14,15-secoandrostane
Aphidicolanes (30)
VS6180

ent-Gibberellane

Gibbane, CAS

Gibberellins (162)
VS6200

Leucothols (8)
VS6220

Grayanotoxanes (113)
VS6300

1,7,11-Trimethyl-4-(1-methyl-
ethyl)cyclotetradecane, CAS
Cembranes (967)
VS6400

(*Continued*)

TABLE 6.1 (*Continued*)
Natural Product Parent Skeletons

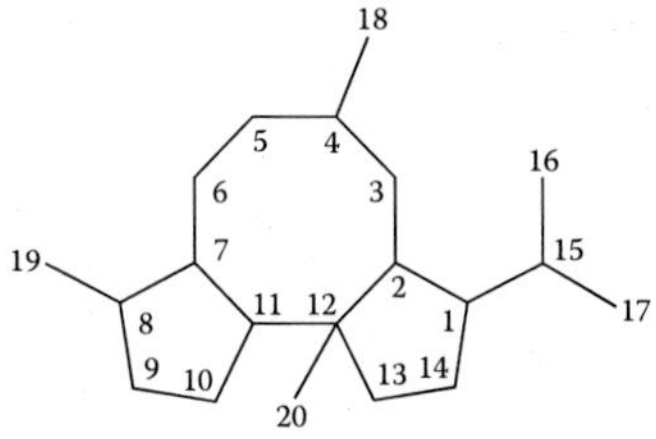

Basmanes

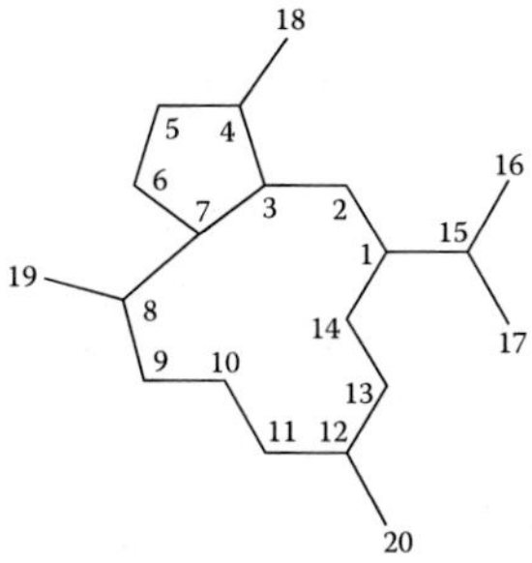

Capnosanes

Rearranged cembranes (97)
VS6420

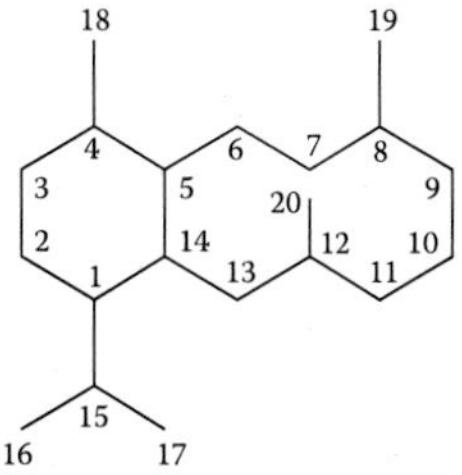

Eunicellanes (277)
VS6440

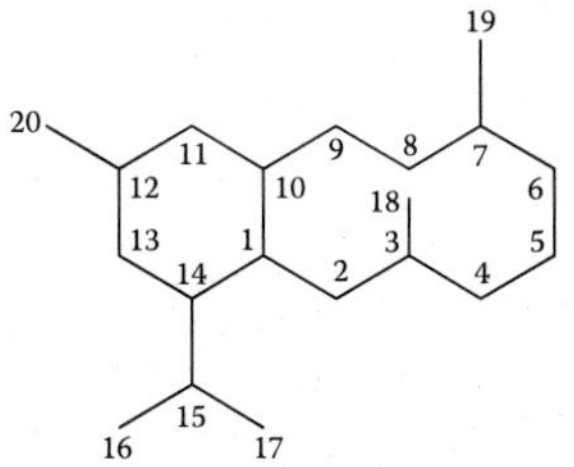

Asbestinanes (40)
VS6450

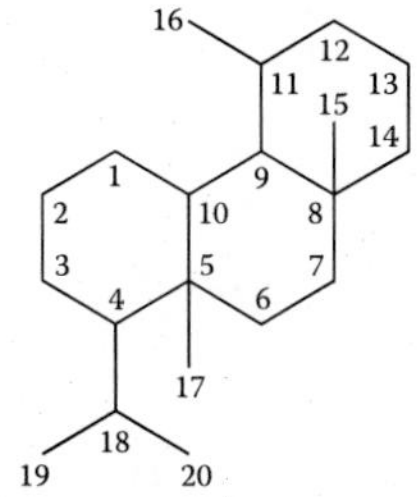

Tetradecahydro-5,8*a*,10*a*-
trimethyl-1-(1-methyl-
ethyl)phenanthrene
Sphaeranes (32)
VS6460

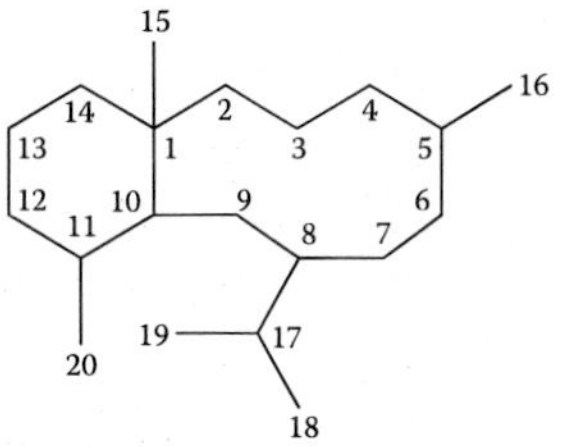

Tetradecahydro-1,4*a*,8-tri-
methyl-11-(1-methylethyl)-
benzocyclodecene
Briaranes (569)
VS6470

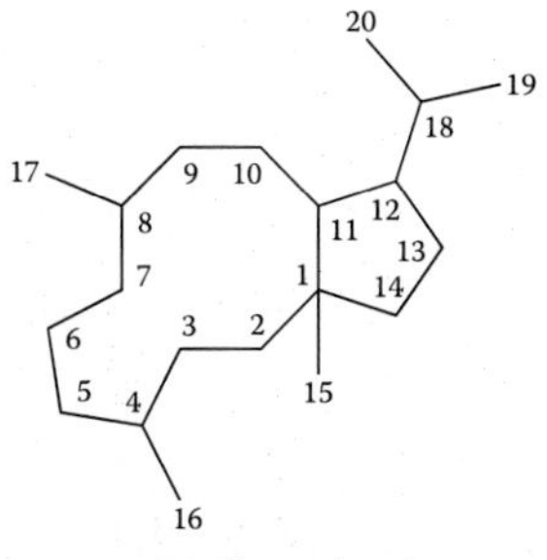

Dolabellanes (231)
VS6500

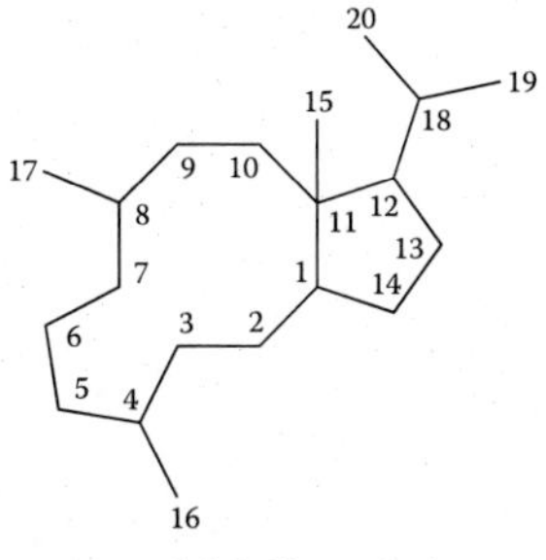

Neodolabellanes (38)
VS6510

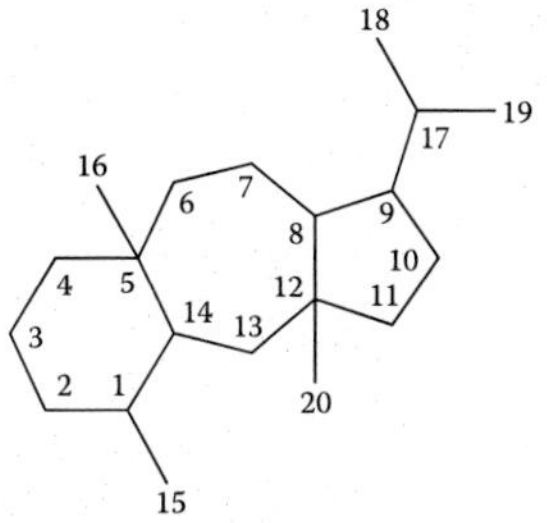

Tetradecahydro-3*a*,5,8*a*-tri-
methyl-1-(1-methylethyl)-
benz[*f*]azulene, CAS
Dolastanes (51)
VS6540

(*Continued*)

TABLE 6.1 (*Continued*)
Natural Product Parent Skeletons

Tetradecahydro-3*a*,5*a*,8-tri-
methyl-1-(1-methylethyl)-
cyclohept[*e*]indene
Cyathanes (126)
VS6560

Dodecahydro-3*a*,6,9-tri-
methyl-3-(1-methylethyl)-
benz[*e*]azulene, CAS
Sphaeroanes (7)
VS6570

Tetradecahydro-3*a*,5*a*,7*a*-tri-
methyl-1-(1-methylethyl)cyclo-
penta[*a*]cyclopropa[*g*]-
naphthalene, CAS
Verrucosanes (12)
VS6580

Neoverrucosanes

Homoverrucosanes

Homoneoverrucosanes

Neo-, homo- and homoneoverrucosanes (46)
VS6590

3,7,11,15,15-Pentamethyl-
bicyclo[12.1.0]pentadecane, CAS
Casbanes (41)
VS6600

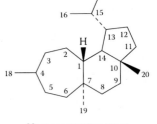

Jatrophanes (262)
VS6610

(*Continued*)

TABLE 6.1 (*Continued*)
Natural Product Parent Skeletons

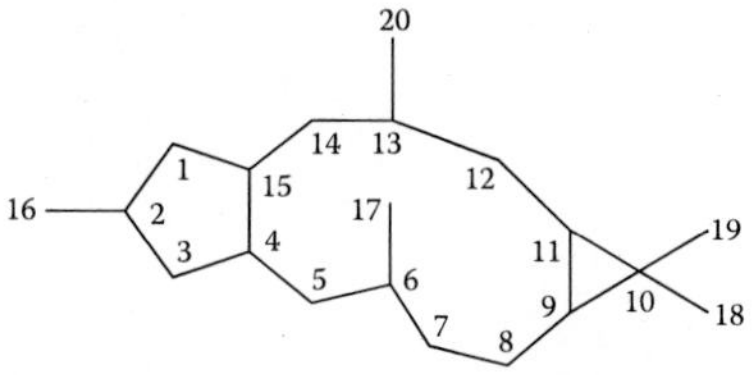

Tetradecahydro-1,1,3,6,9-penta-
methyl-1*H*-cyclopenta[*a*]-
cyclopropa[*f*]cycloundecene
Lathyranes (159)
VS6650

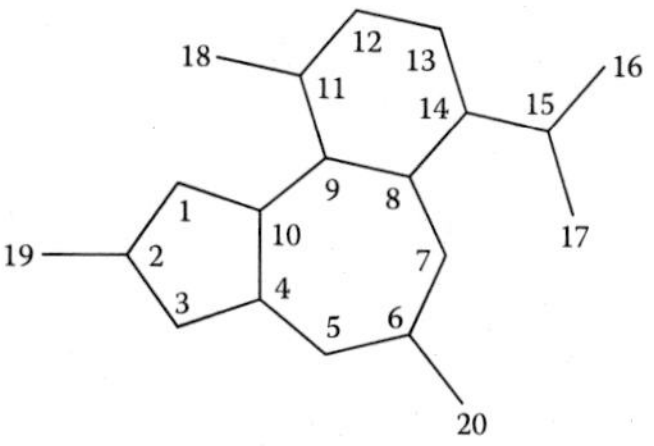

Rhamnofolanes (17)
VS6660

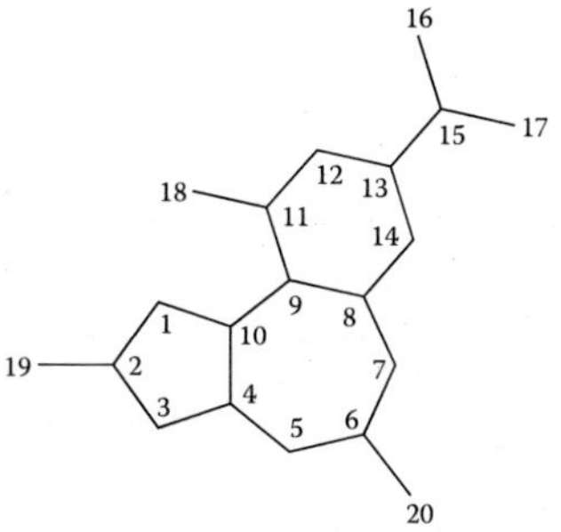

Daphnanes (216)
VS6680

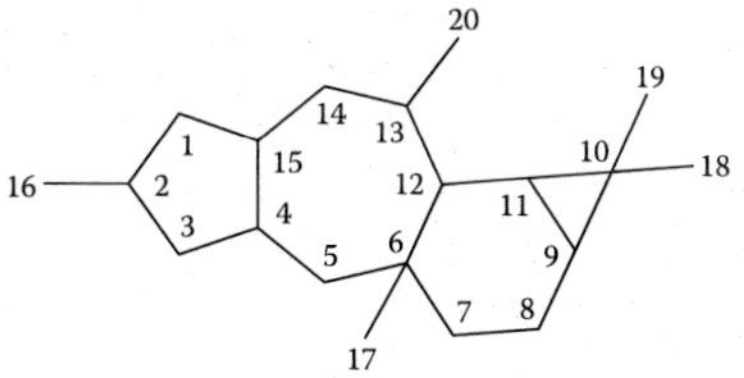

Premyrsinanes (24)
VS6685

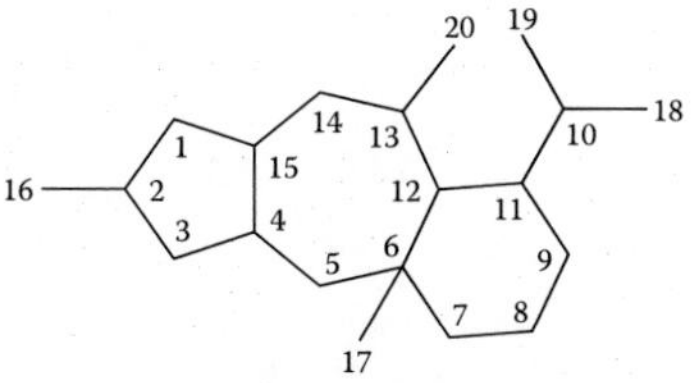

Myrsinanes (66)
VS6690

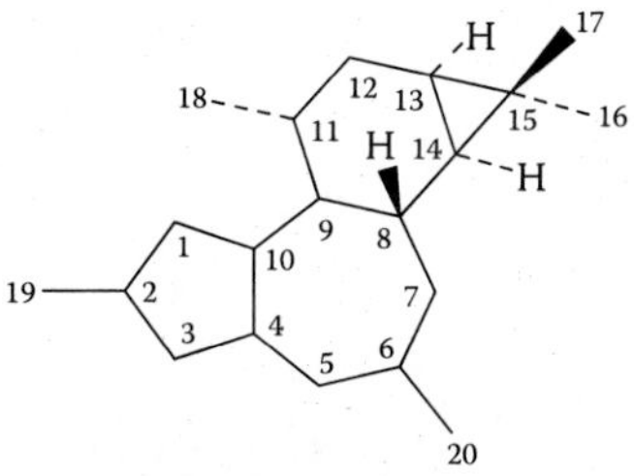

Tetradecahydro-1,1,3,6,8-
pentamethyl-1*H*-cyclo-
propa[3,4]benz[1,2-*e*]-
azulene, CAS
Tiglianes (155)
VS6700

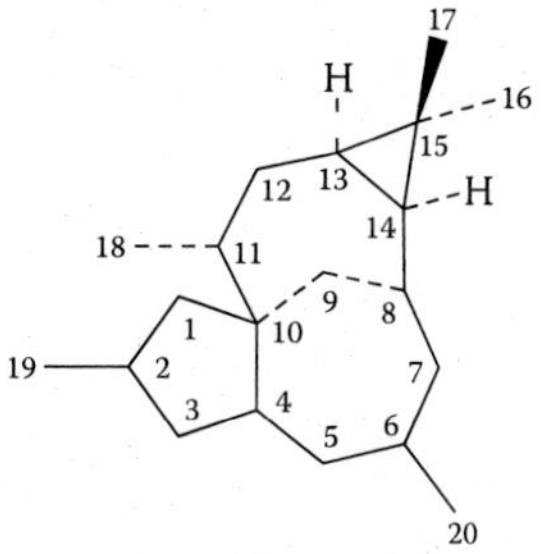

Dodecahydro-1,1,4,7,9-
pentamethyl-1*H*-2,8*a*-
methanocyclopenta[*a*]cyclo-
propa[*e*]cyclodecene, CAS
Ingenanes (106)
VS6710

(Continued)

TABLE 6.1 (*Continued*)
Natural Product Parent Skeletons

Tetradecahydro-1,1,2,5,7-
pentamethyl-1*H*-cyclopropa-
[3,4]cyclohept[1,2-*e*]indene
Jatropholanes

Tetradecahydro-2,5,10-
trimethyl-6-(1-methyl-
ethyl)cyclohept[*e*]-
indene, CAS
Crotofolanes

Jatropholanes and secojatropholanes (31)
VS6720

Tetradecahydro-1,4,9*a*-tri-
methyl-7-(1-methylethyl)dicyclo-
penta[*a,d*]cyclooctene, CAS
Fusicoccanes (84)
VS6750

Tetradecahydro-5*a*,8,10*b*-tri-
methyl-3-(1-methylethyl)-
cyclohept[*e*]indene, CAS
Valparanes (28)
VS6770

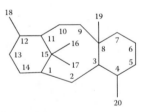

Mulinanes (27)
VS6780

Decahydro-3*a*,6-dimethyl-1-
(1,5-dimethylhexyl)cyclobuta-
[1,2:3,4]dicyclopentene, CAS
Spatanes (45)
VS6800, VS6810

4,8,12,15,15-Pentamethyl-
bicyclo[9.3.1]pentadecane, CAS
Verticillanes (120)
VS6880

Tetradecahydro-4,9,12*a*,13,13-
pentamethyl-6,10-methano-
benzocyclodecene, CAS
Taxanes (441)
VS6900

(Continued)

TABLE 6.1 (*Continued*)
Natural Product Parent Skeletons

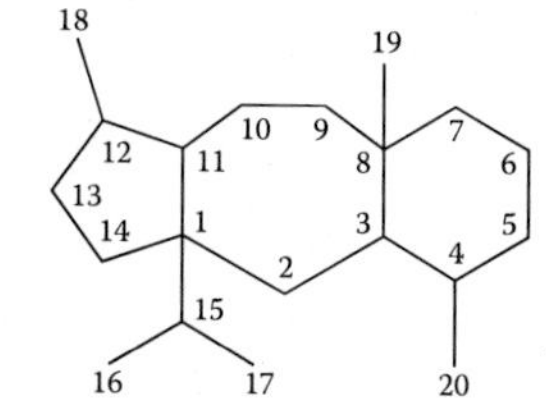

11(15→1)-Abeotaxanes (146)
VS6950

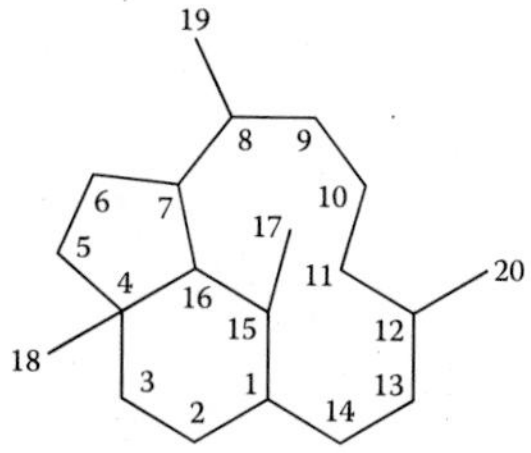

Hexadecahydro-1,4,8,12-
tetramethyl-1,11-ethanocyclo-
pentacycloundecane, CAS
Trinervitanes (37)
VS7000

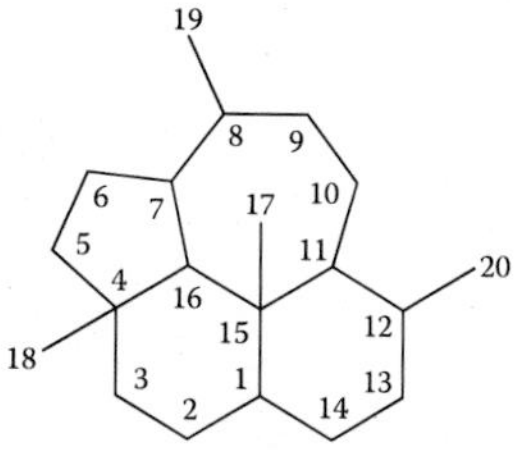

Hexadecahydro-2*a*,7,10,10*c*-
tetramethylnaphth-
[2,1,8-*cde*]azulene, CAS
Kempanes (7)
VS7010

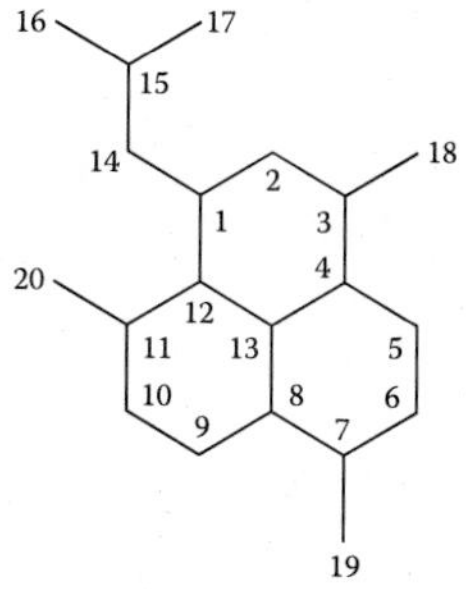

Dodecahydro-1,4,7-tri-
methyl-3-(2-methylpropyl)-
1*H*-phenalene, CAS
Amphilectanes (77)
VS7020

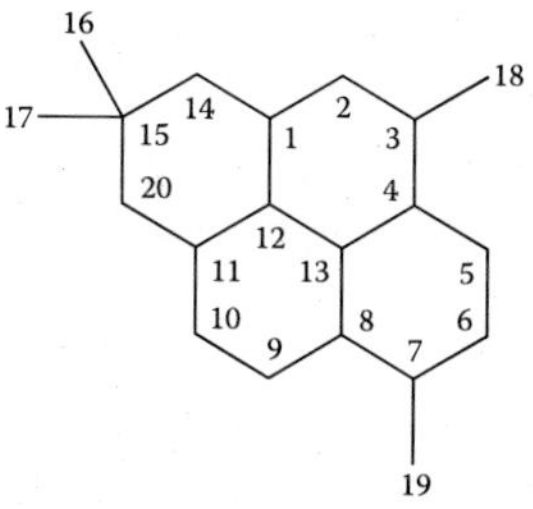

Hexadecahydro-1,4,7,7-
tetramethylpyrene
Cycloamphilectanes (10)
VS7030

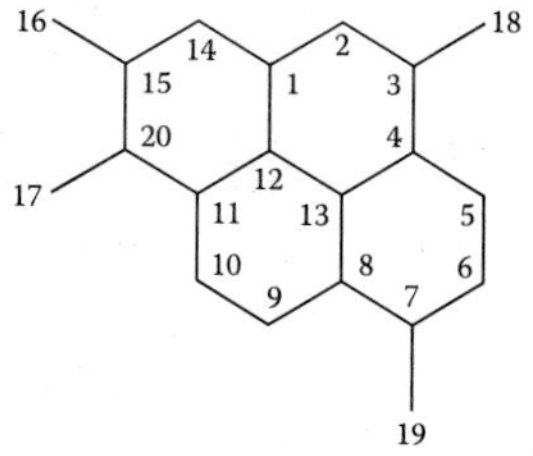

Hexadecahydro-1,2,5,8-
tetramethylpyrene, CAS
Adocianes (8)
VS7040

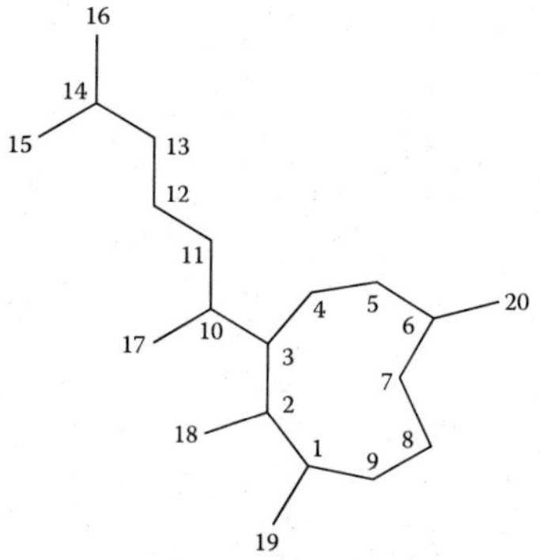

1,2,6-Trimethyl-3-
(1,5-dimethylhexyl)-
cyclononane
Xenicanes (214)
VS7100

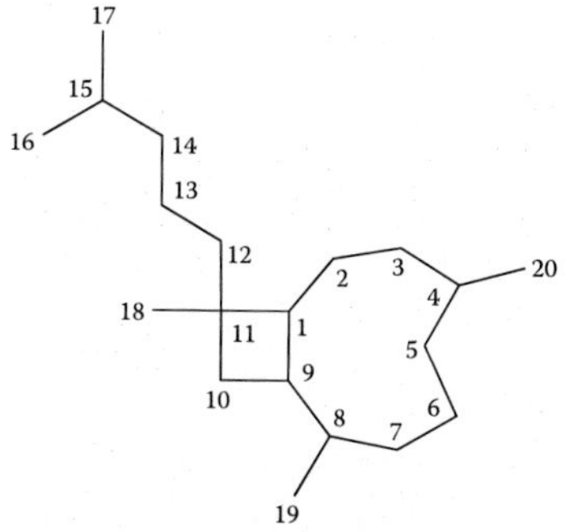

2,6,10-Trimethyl-10-
(4-methylpentyl)-
bicyclo[7.2.0]undecane
Xeniaphyllanes (37)
VS7150

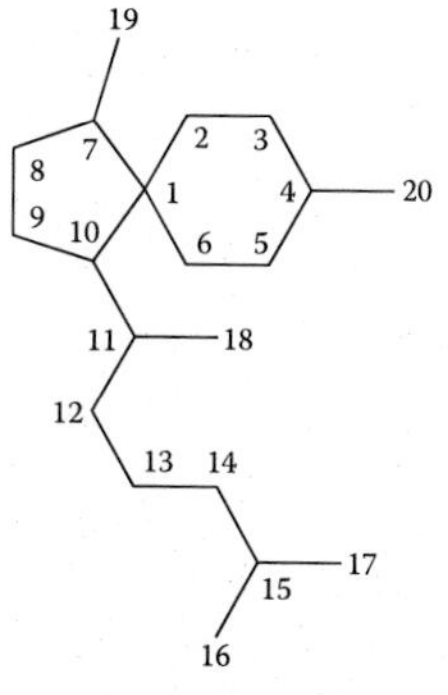

1,8-Dimethyl-4-(1,5-di-
methylhexyl)spiro[4.5]decane
Viscidanes (20)
VS7160

(*Continued*)

TABLE 6.1 (*Continued*)
Natural Product Parent Skeletons

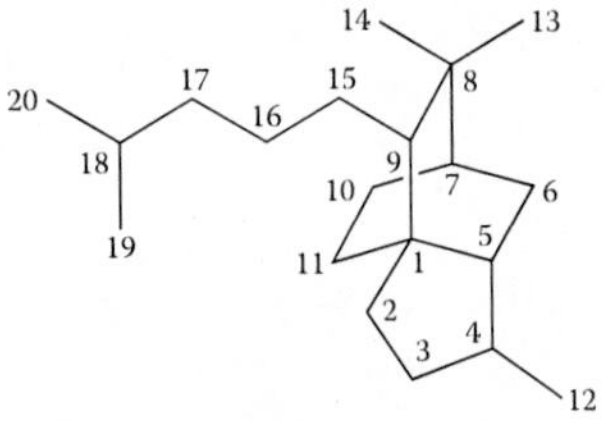

Decahydro-1,5,5-trimethyl-
4-(4-methylpentyl)-3a,6-
ethano-3aH-indene
Eremanes (3)
VS7180

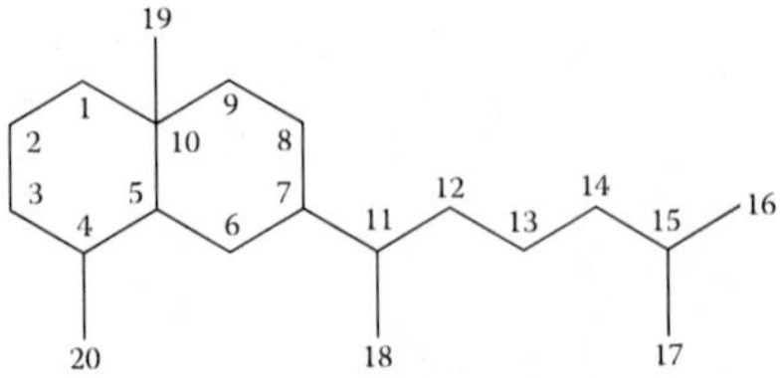

Decahydro-7-(1,5-methylhexyl)-
1,4a-dimethylnaphthalene
Prenyleudesmanes (20)
VS7190

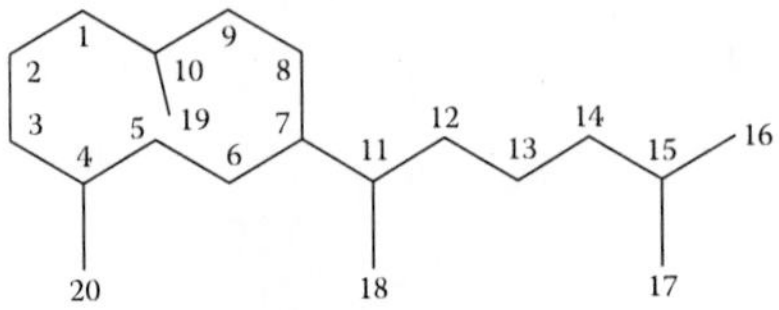

4-(1,5-Dimethylhexyl)-
1,7-dimethylcyclodecane
Prenylgermacranes (30)
VS7200

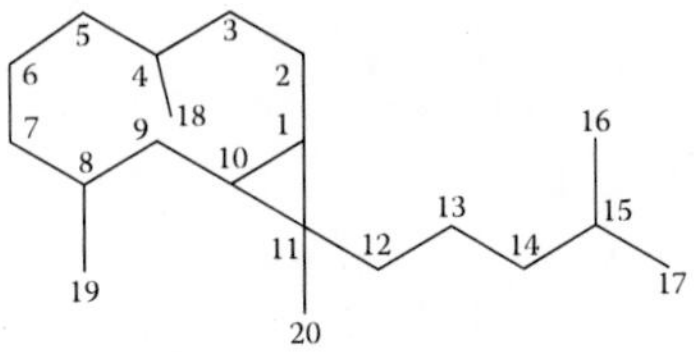

3,7,11-Trimethyl-11-(4-methyl-
pentyl)bicyclo[8.1.0]undecane
Prenylbicyclogermacranes (23)
VS7210

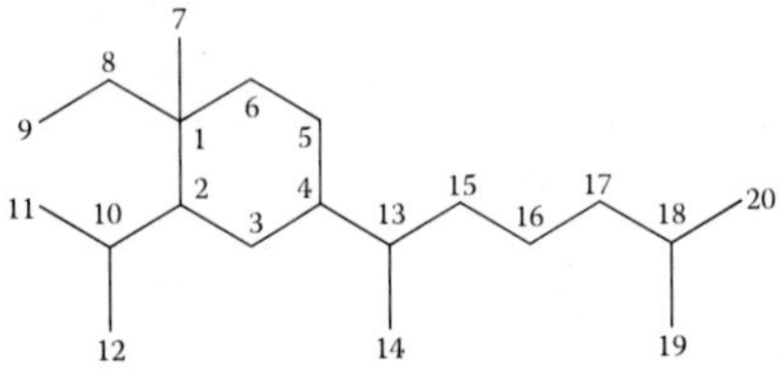

4-(1,5-Dimethylhexyl)-1-ethyl-1-
methyl-2-(1-methylethyl)cyclohexane
Lobanes (42)
VS7220

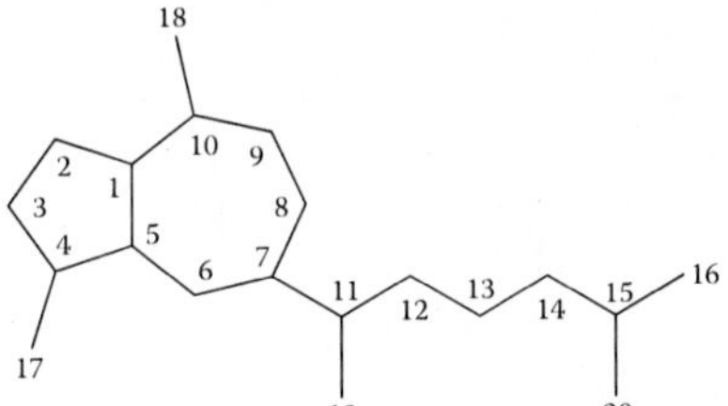

Decahydro-7-(1,5-dimethyl-
hexyl)-1,4-dimethylazulene
Pachydictyanes (72)
VS7230

(*Continued*)

TABLE 6.1 (*Continued*)
Natural Product Parent Skeletons

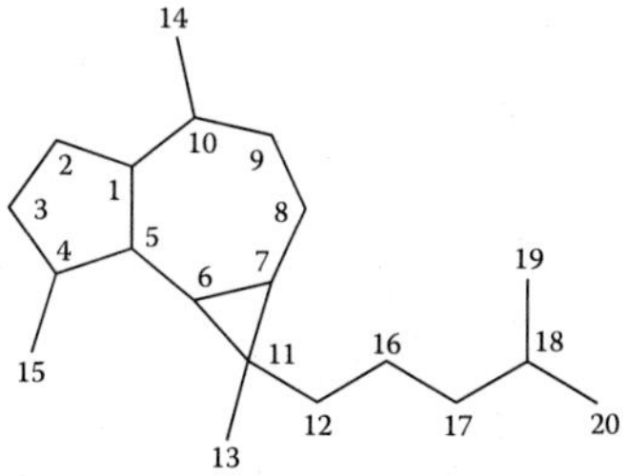

Decahydro-1,4,7-trimethyl-
1-(4-methylpentyl)-1*H*-
cycloprop[*e*]azulene
Cneorubanes (20)
VS7240

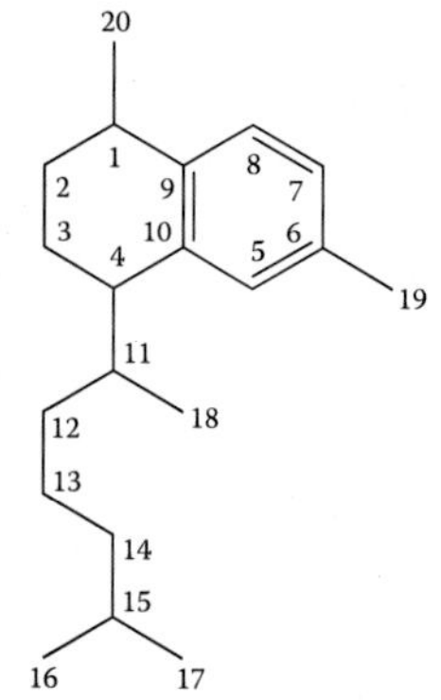

4-(1,5-Dimethylhexyl)-
1,2,3,4-tetrahydro-1,6-
dimethylnaphthalene
Serrulatanes

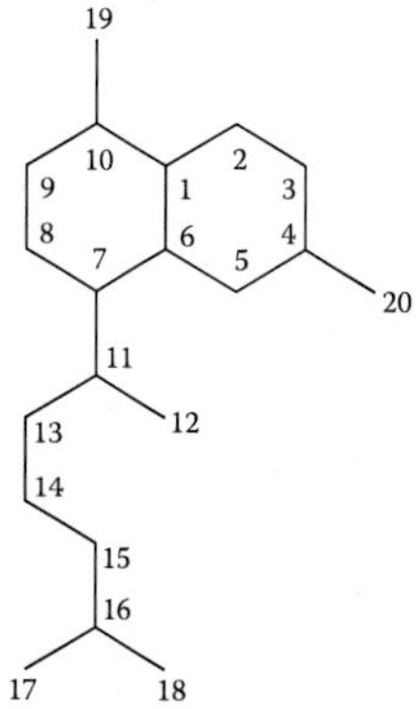

4-(1,5-Dimethylhexyl)-
decahydro-1,6-dimethyl-
naphthalene
Bifloranes

Serrulatanes and bifloranes (148)
VS7250

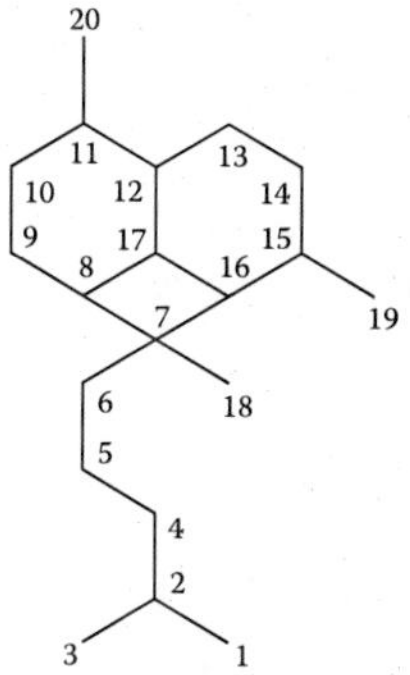

Decahydro-1,2,5-trimethyl-
1-(4-methylpentyl)-1*H*-
cyclobuta[*de*]naphthalene, CAS
Decipianes (4)
VS7260

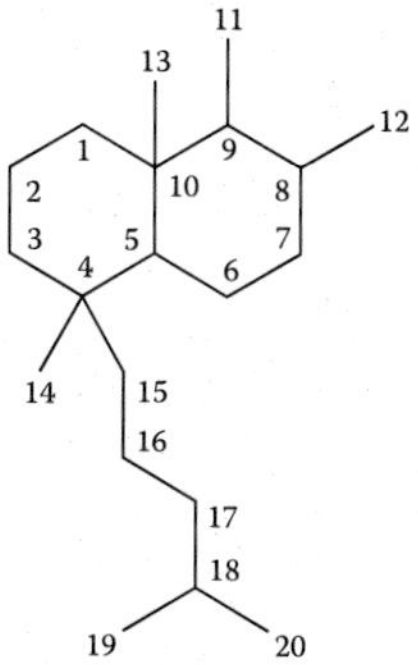

Decahydro-1,4a,5,6-tetramethyl-
1-(4-methylpentyl)naphthalene
Sacculatanes (31)
VS7270

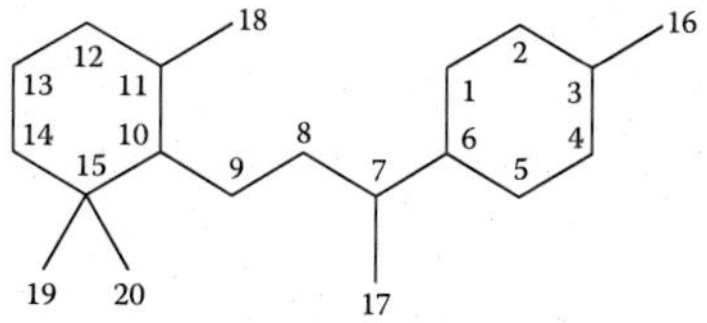

1,1,3-Trimethyl-2-[3-(4-methyl-
cyclohexyl)butyl]cyclohexane
Obtusanes (9)
VS7280

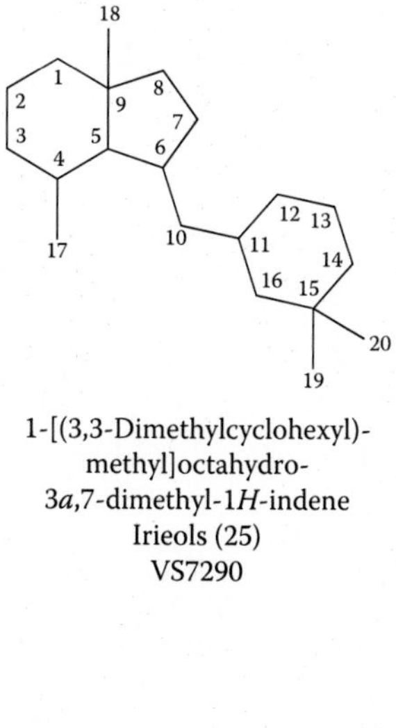

1-[(3,3-Dimethylcyclohexyl)-
methyl]octahydro-
3*a*,7-dimethyl-1*H*-indene
Irieols (25)
VS7290

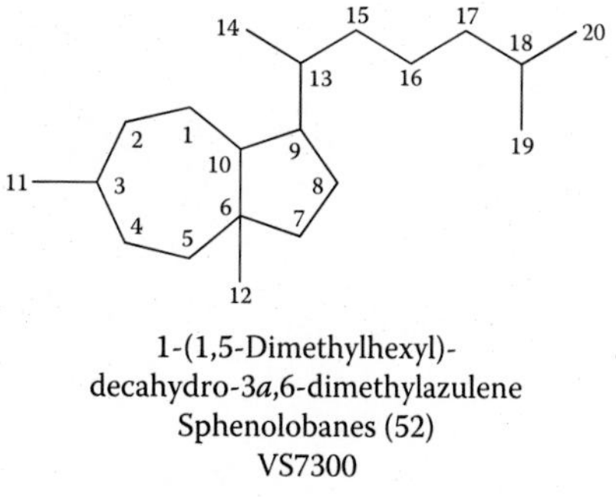

1-(1,5-Dimethylhexyl)-
decahydro-3*a*,6-dimethylazulene
Sphenolobanes (52)
VS7300

(*Continued*)

TABLE 6.1 (*Continued*)
Natural Product Parent Skeletons

2,6,10,14,18-Pentamethyleicosane
Acyclic sesterterpenoids (179)
VS7400

C_{21} Sesterterpenoid
3,7,11,15-Tetramethylheptadecane
Noracyclic sesterterpenoids (100)
VS7410

Cyclohexane sesterterpenoids (95)
VS7420

1-(1,5-Dimethylhexyl)-4,8,12-
trimethylcyclotetradecane
Cericeranes (15)
VS7440

Bicyclic sesterterpenoids (145)
VS7460

(Continued)

TABLE 6.1 (*Continued*)
Natural Product Parent Skeletons

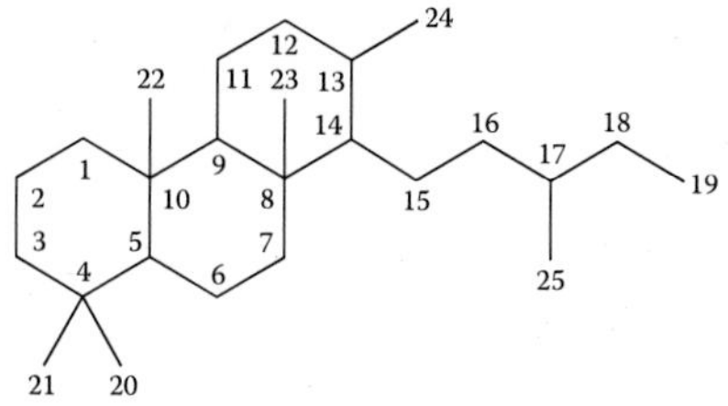

4,4,8-Trimethyl-*D*(15),24-
dinor-13,17-secocholane, CAS
Cheilanthanes (70)
VS7500

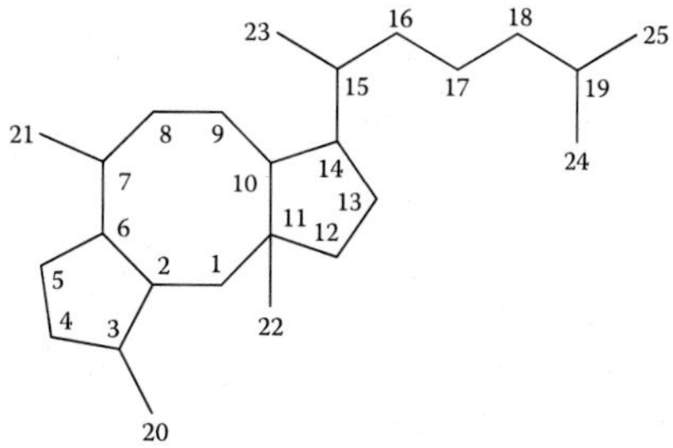

Ophiobolanes (44)
(CAS numbering differs)
VS7520

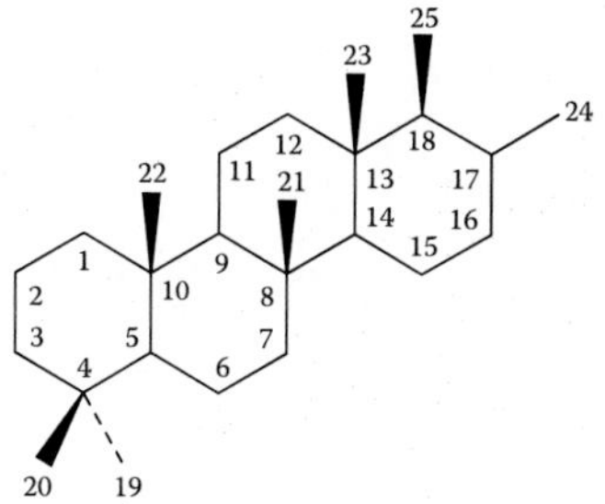

4,4,8,17,17*a*-Pentamethyl-
D-homoandrostane, CAS
Scalaranes (143)
VS7540

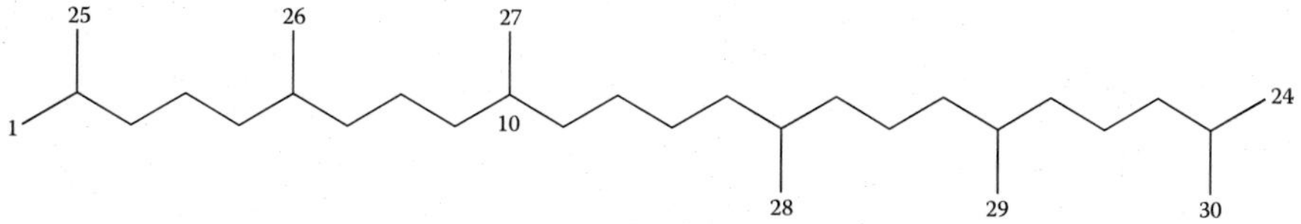

e.g. Squalane
2,6,10,15,19,23-Hexamethyltetracosane
Linear triterpenoids (166)
VS7600

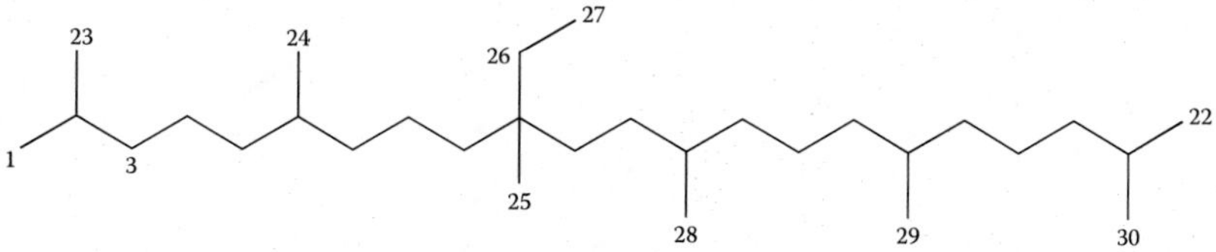

e.g. C$_{30}$-Botryococcene
10-Ethyl-2,6,10,13,17,21-hexamethyldocosane
Botryococcenes (16)
VS7620

(*Continued*)

TABLE 6.1 (*Continued*)
Natural Product Parent Skeletons

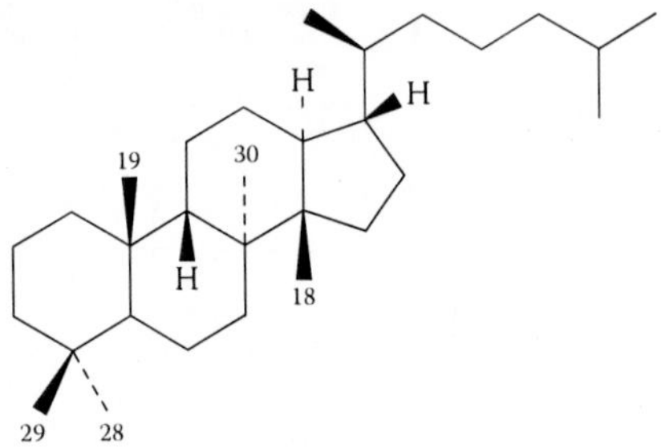

Protostanes

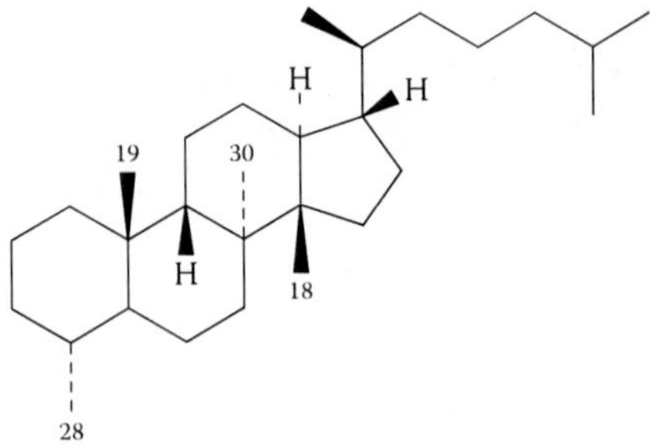

Fusidanes

Protostanes and fusidanes (91)
VS7700

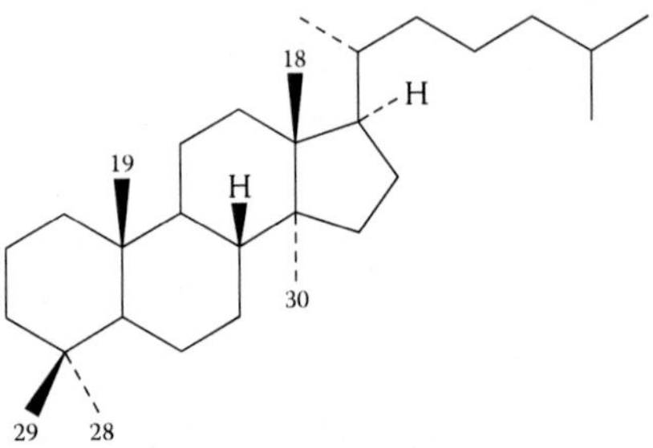

Lanostanes (1545)
VS7750

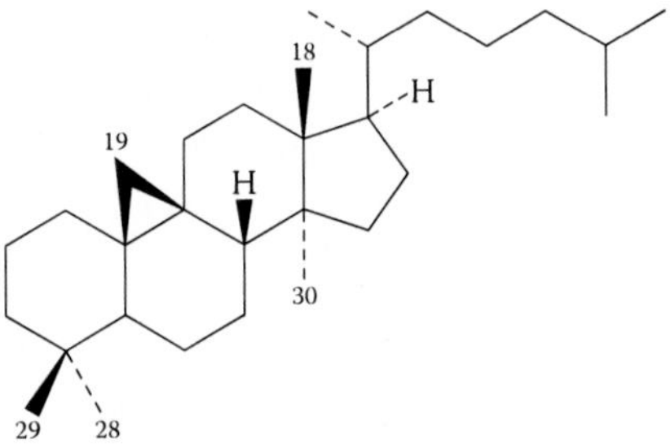

Cycloartanes (1729)
VS7800

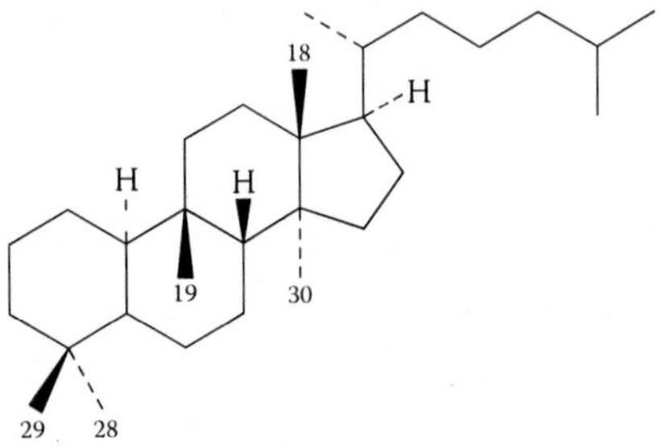

Cucurbitanes (575)
VS7900

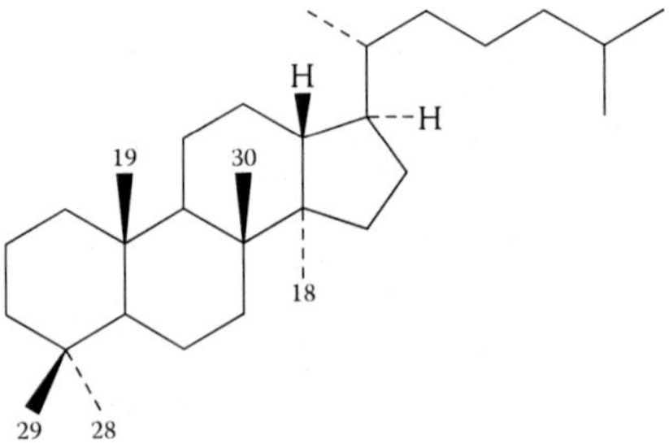

Dammaranes (1243)
VS7950

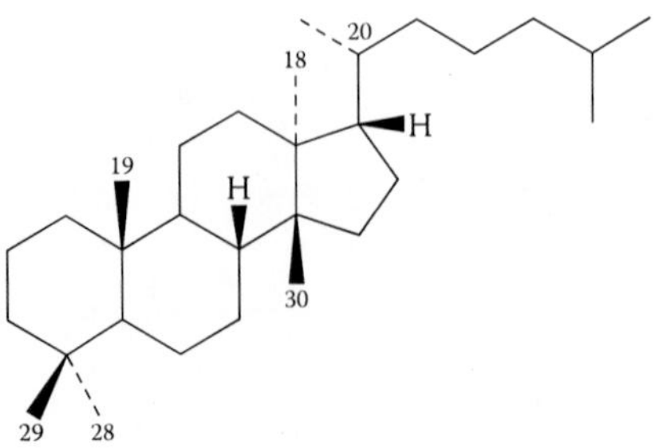

Euphanes

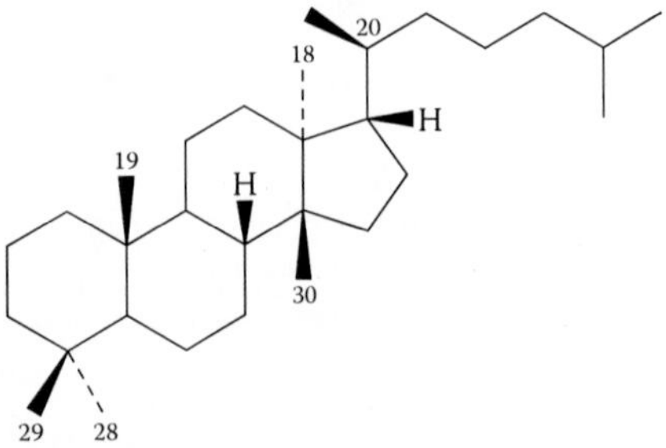

Tirucallanes

Euphanes and tirucallanes (462)
VS8000

(*Continued*)

TABLE 6.1 (*Continued*)
Natural Product Parent Skeletons

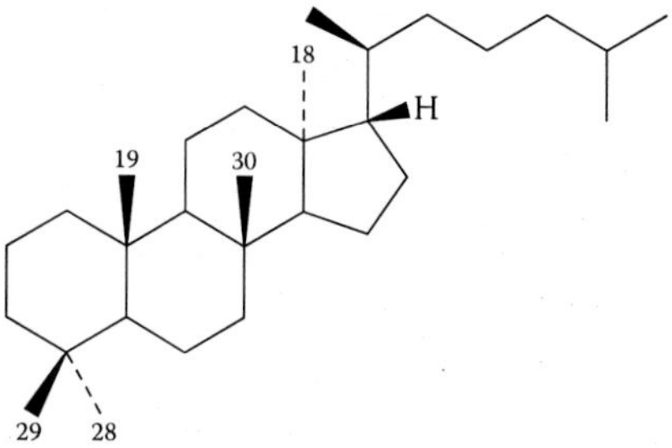

Apotirucallanes (150)
VS8050

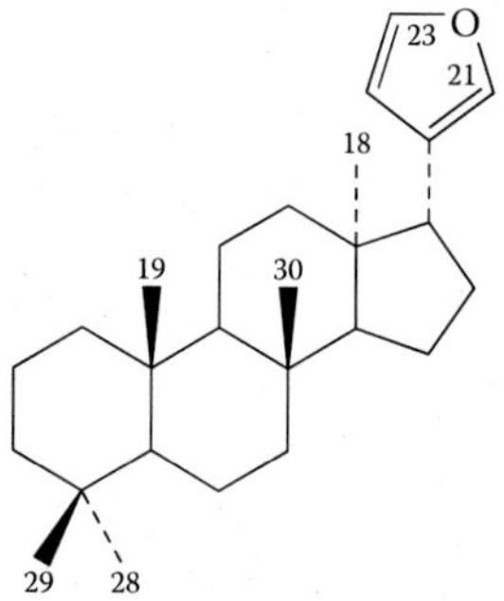

Melliacanes (297)
VS8100

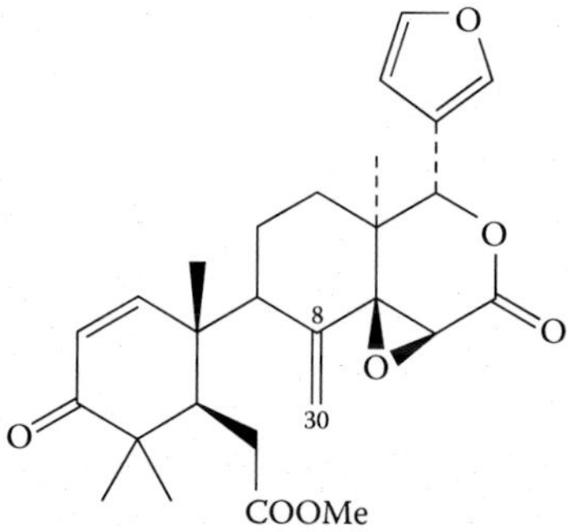

e.g. Andirobin
Ring cleaved tetranortriterpenoids (739)
VS8120

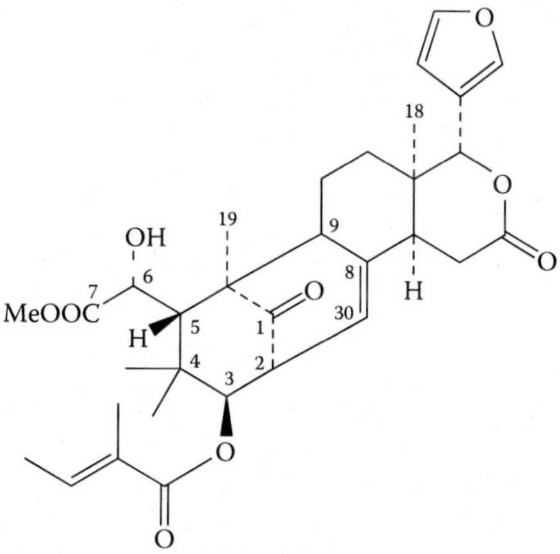

e.g. Swietenine
Rearranged tetranortriterpenoids (669)
VS8130

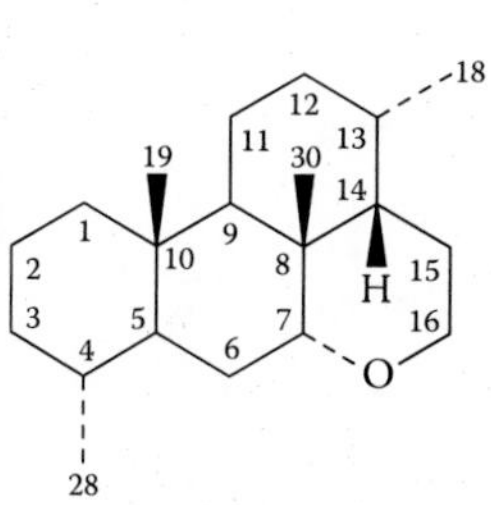

Picrasanes (C_{20})
(CAS numbering differs)

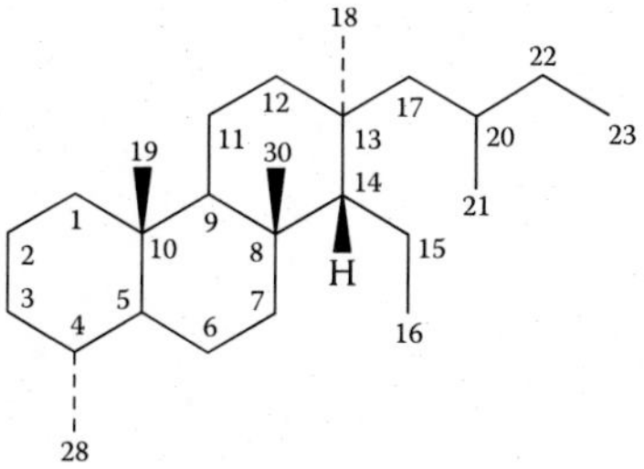

C_{25} Quassinoid
skeleton

Quassinoids (399)
VS8200, VS8205

(*Continued*)

TABLE 6.1 (*Continued*)
Natural Product Parent Skeletons

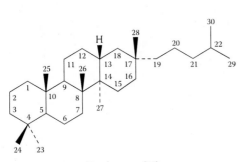

Baccharanes (36)
VS8230

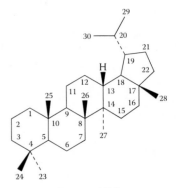

Lupanes (626)
VS8250

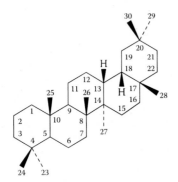

Oleananes (5770)
VS8300

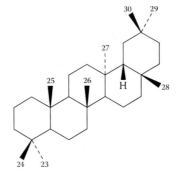

Taraxeranes (106)
VS8350, VS8360

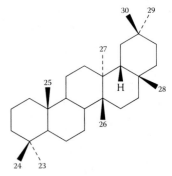

Multifloranes (62)
VS8400

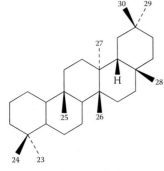

Glutinanes (28)
VS8450

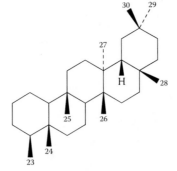

Friedelanes (459)
VS8500, VS8510

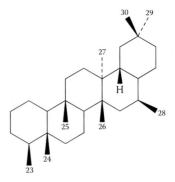

Pachysananes (4)
VS8520

(*Continued*)

TABLE 6.1 (*Continued*)
Natural Product Parent Skeletons

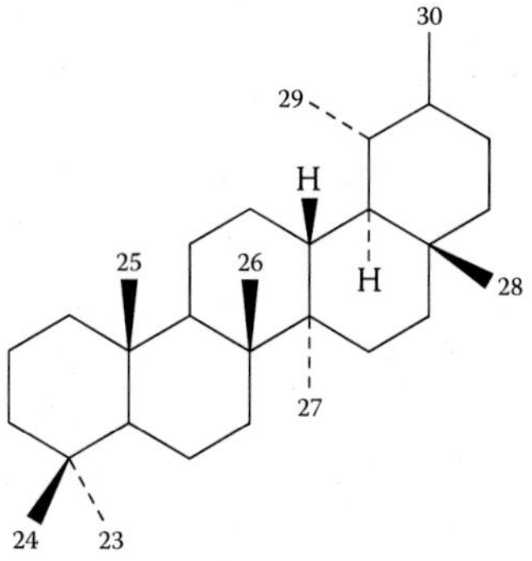

Taraxastanes (146)
VS8550

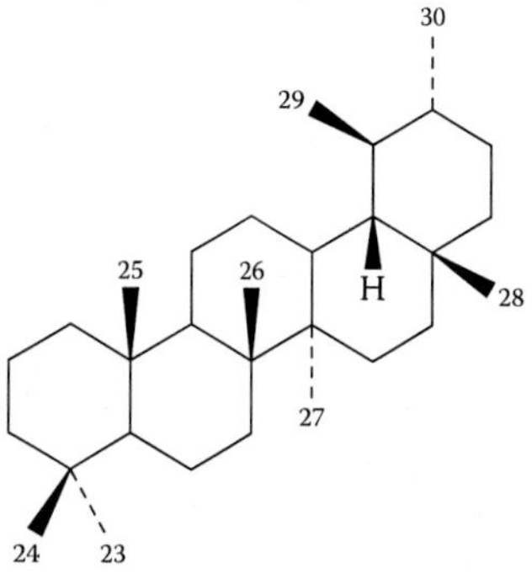

Ursanes (1354)
VS8650, VS8660

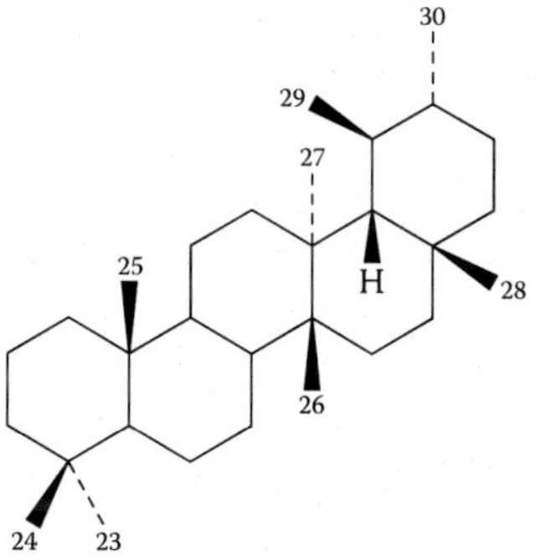

Baueranes (18)
VS8700

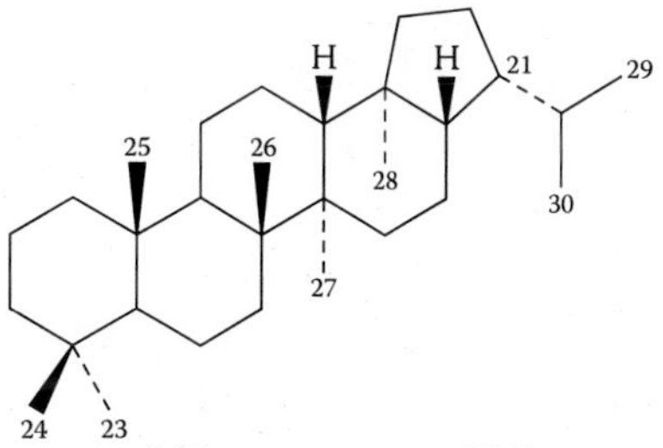

A'-Neogammacerane, CAS
Hopanes (332)
VS8720, VS8730

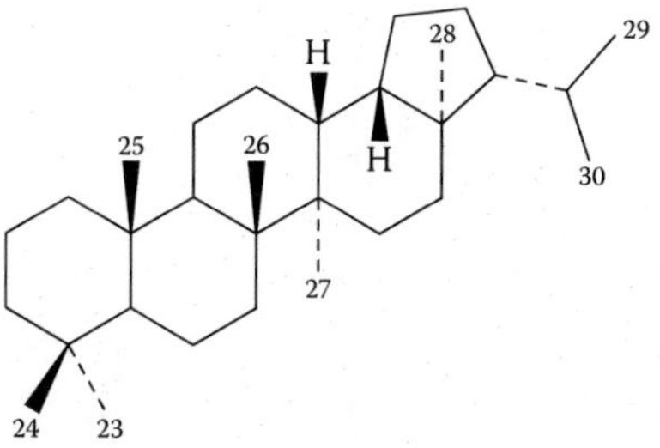

B': *A*'-Neogammacerane, CAS
Neohopanes (19)
VS8770

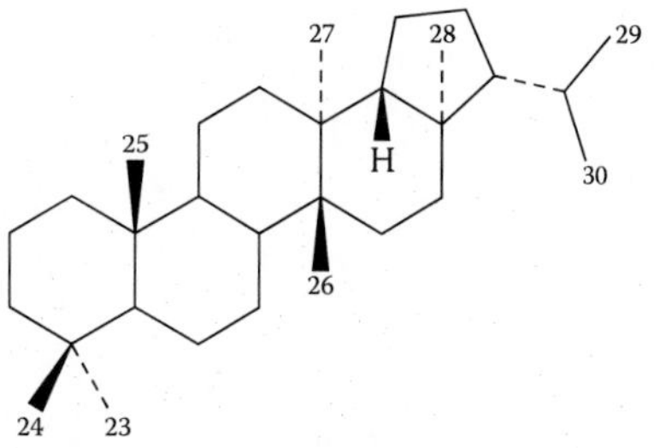

Fernanes (100)
VS8800

(*Continued*)

TABLE 6.1 (*Continued*)
Natural Product Parent Skeletons

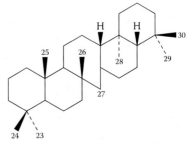

Adiananes (17)
VS8850

Filicanes (22)
VS8870

Arborinanes (52)
VS8880

Stictanes (19)
VS8900

Gammaceranes (26)
VS8950

C(14*a*)-Homo-27-
norgammacerane, CAS
Serratanes (144)
VS9000

(*Continued*)

TABLE 6.1 (*Continued*)
Natural Product Parent Skeletons

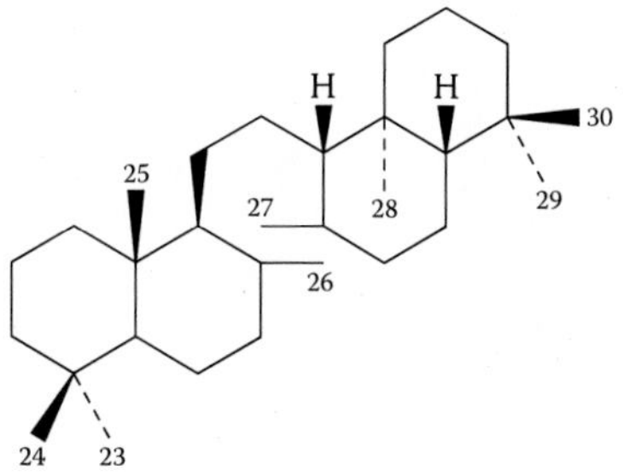

8,14-Secogammacerane, CAS
Onoceranes (39)
VS9050

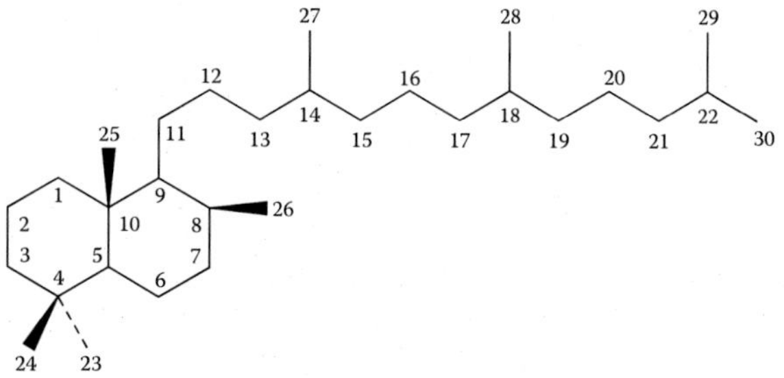

Polypodanes (14)
VS9080

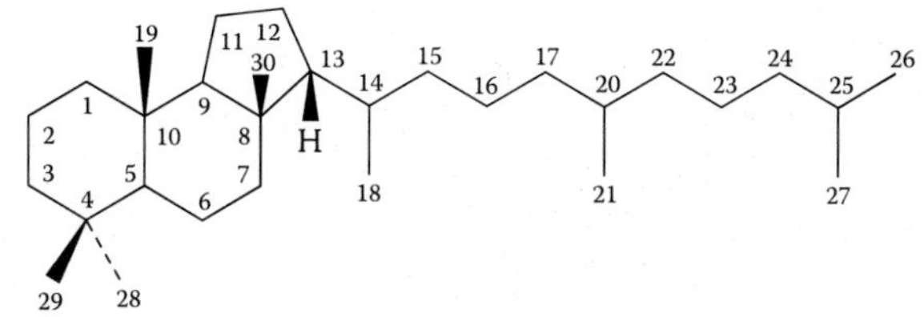

Malabaricanes (130)
VS9100

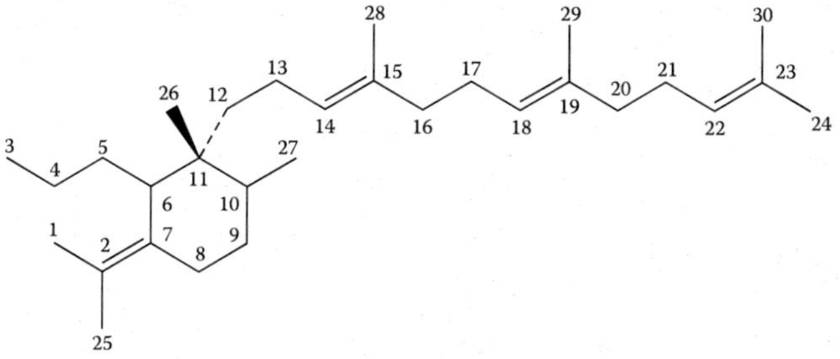

Iridal norterpenoids (92)
VS9350

(*Continued*)

TABLE 6.1 (*Continued*)
Natural Product Parent Skeletons

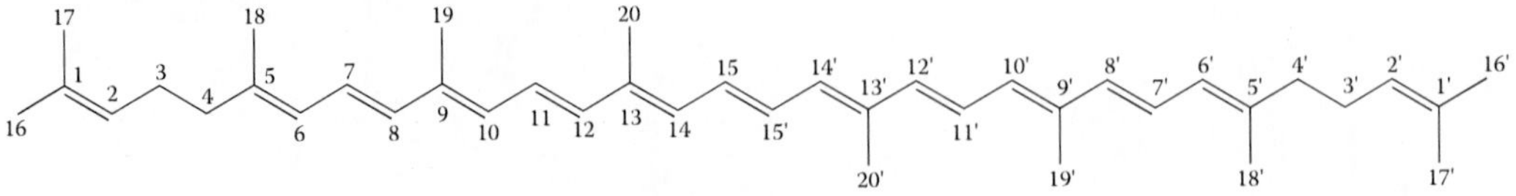

e.g. ψ,ψ-Carotene

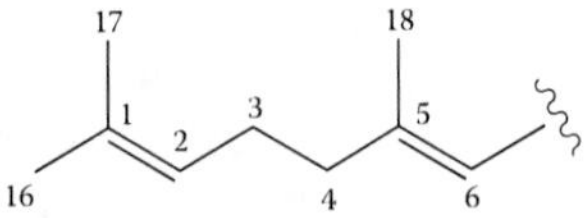

ψ

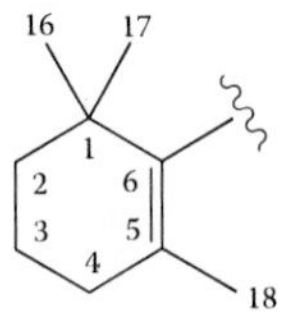

β

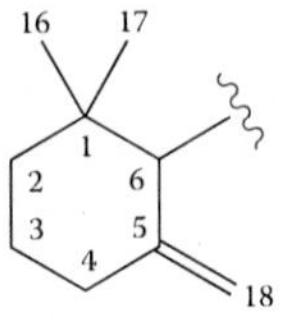

γ

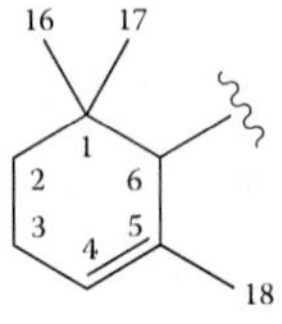

ε

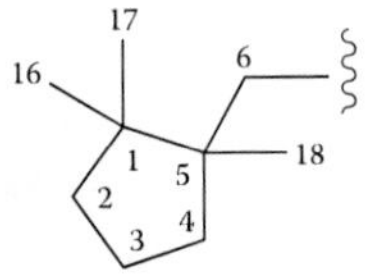

κ

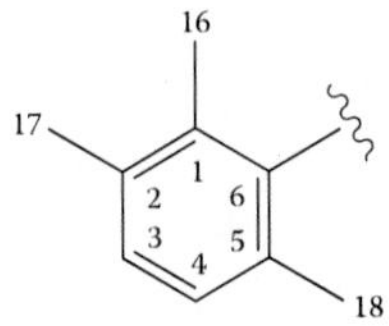

φ

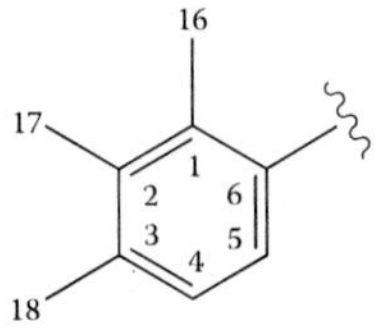

χ

Carotenoid end-groups
Tetraterpenoids (706)
VS9400

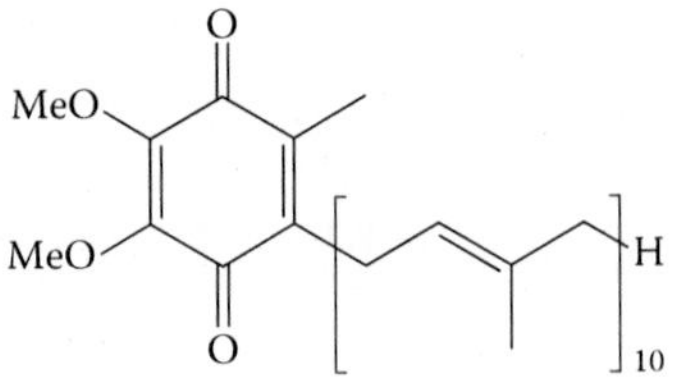

Ubiquinone 10

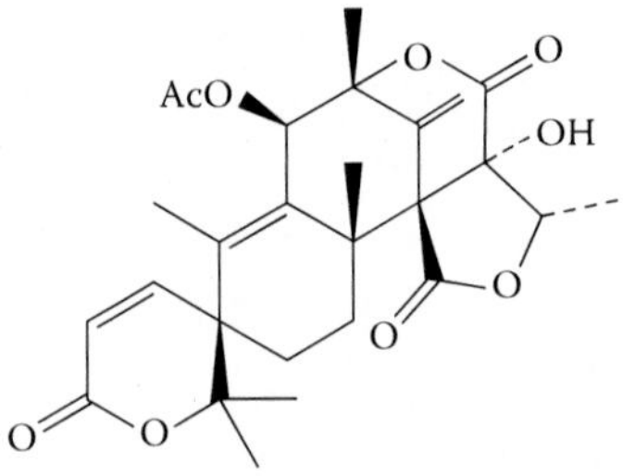

Austin

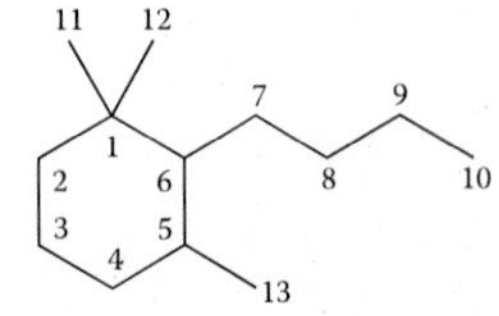

Megastigmane norterpenoids (555)
VS9450

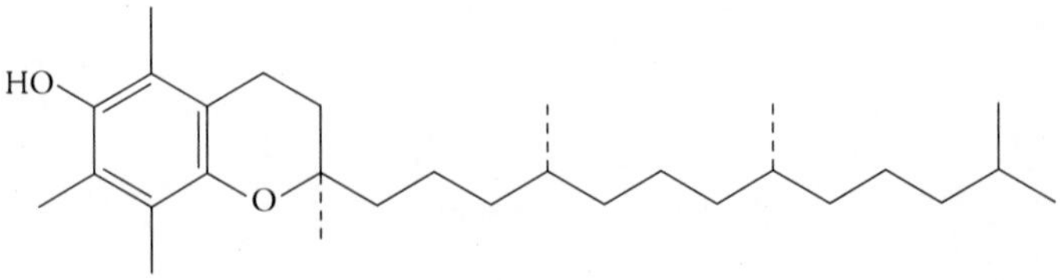

α-Tocopherol

Representative meroterpenoids (3315)
VS9900

(*Continued*)

TABLE 6.1 (*Continued*)
Natural Product Parent Skeletons

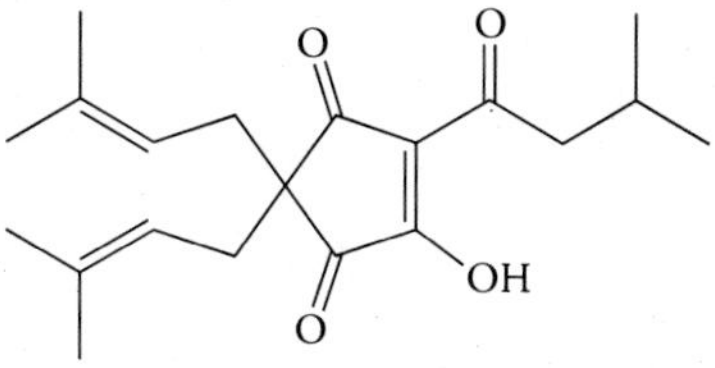

e.g. Hulupone
Hop meroterpenoids (33)
VS9910

Steroids (VT)

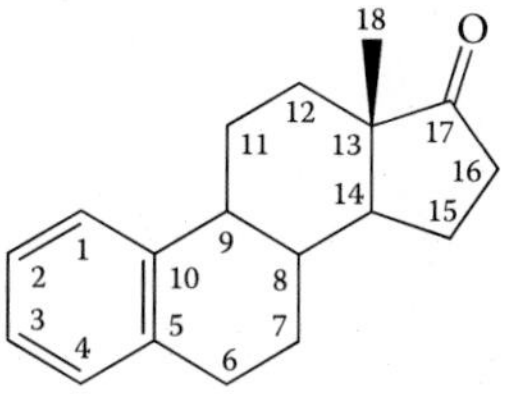

Estranes (48)
VT0100

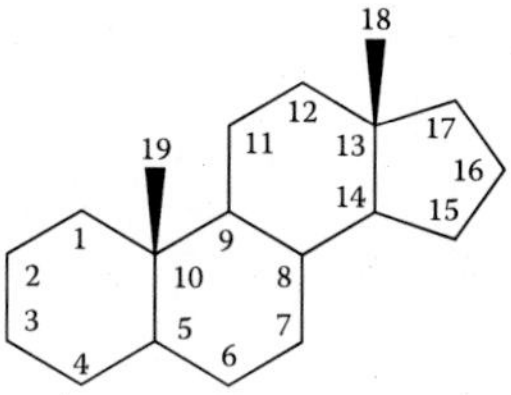

Androstanes (120)
VT0250

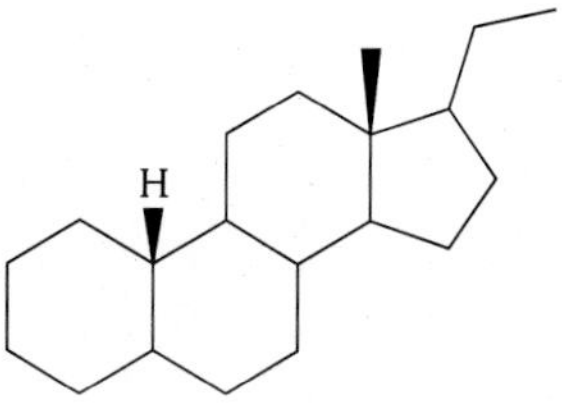

19-Norpregnanes (8)
VT0400

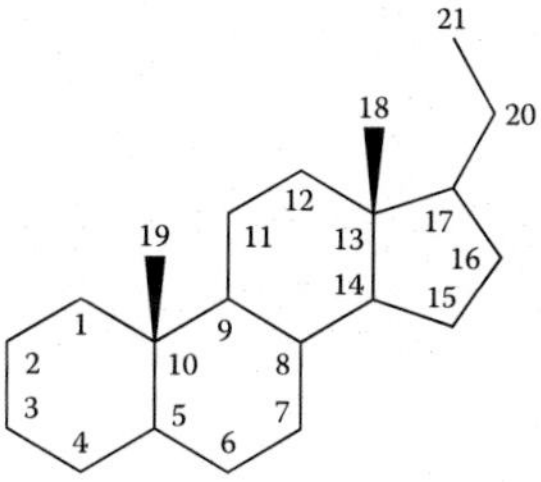

Pregnanes (2260)
VT0450

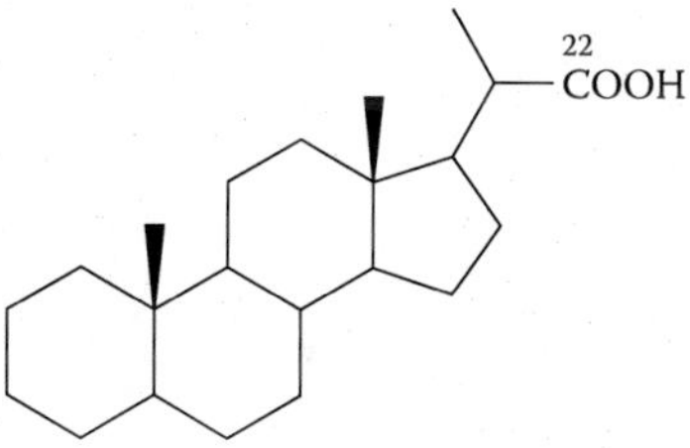

Pregnane-20-carboxylic
acids (15)
VT0550

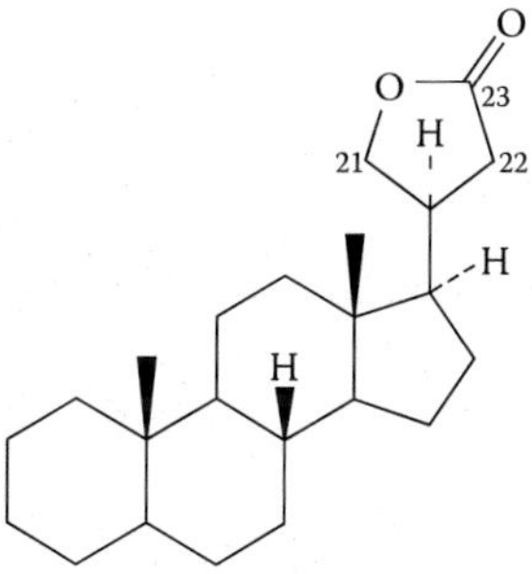

Cardanolides (953)
VT0750

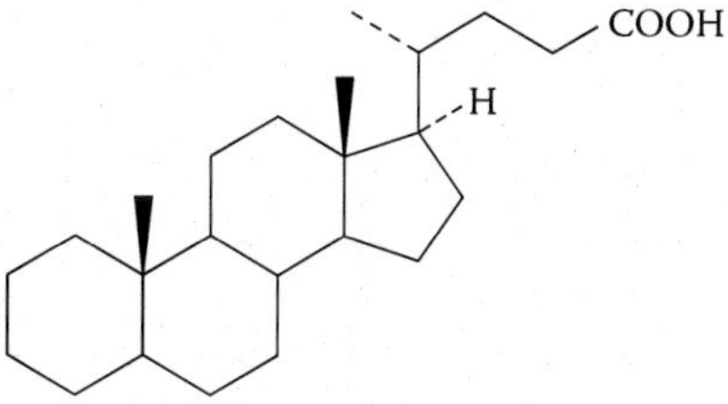

Cholan-24-oic acids (197)
VT0800

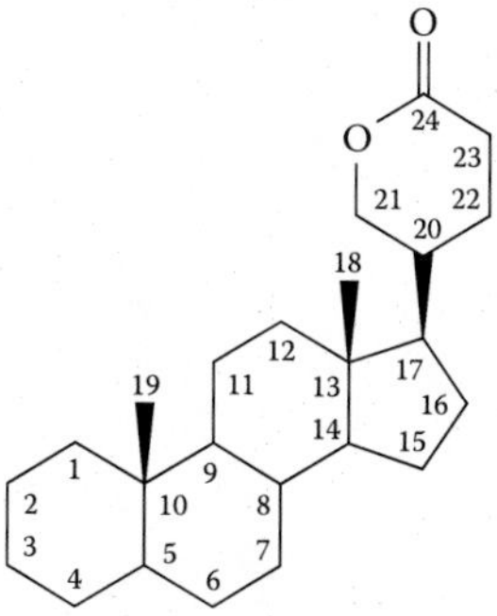

Bufanolides (295)
VT0900

(*Continued*)

TABLE 6.1 (*Continued*)
Natural Product Parent Skeletons

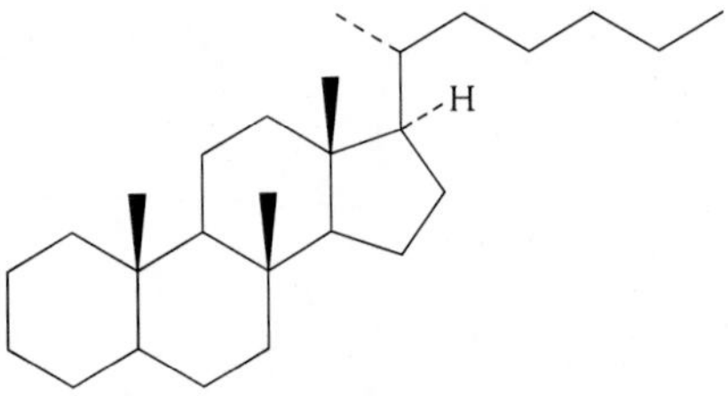

27-Norcholestanes (45)
VT1000

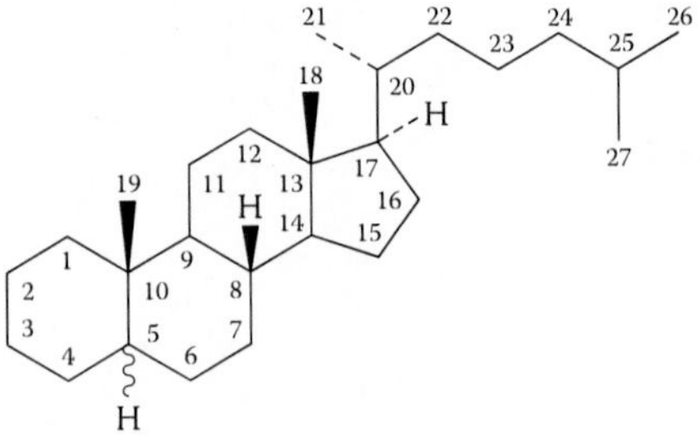

Cholestanes (1721)
VT1050, VT1100

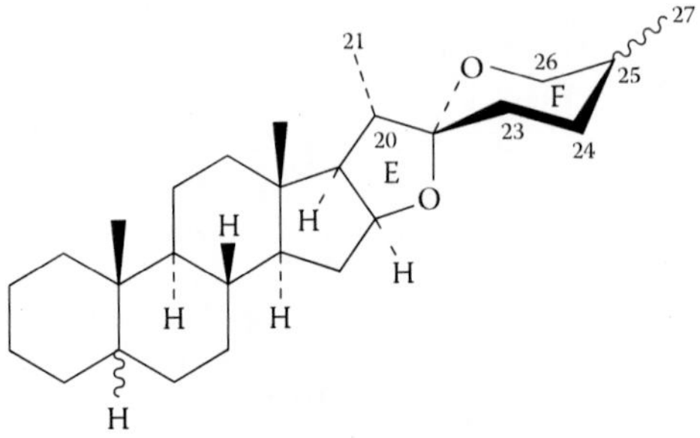

Spirostanes (1251)
VT1200

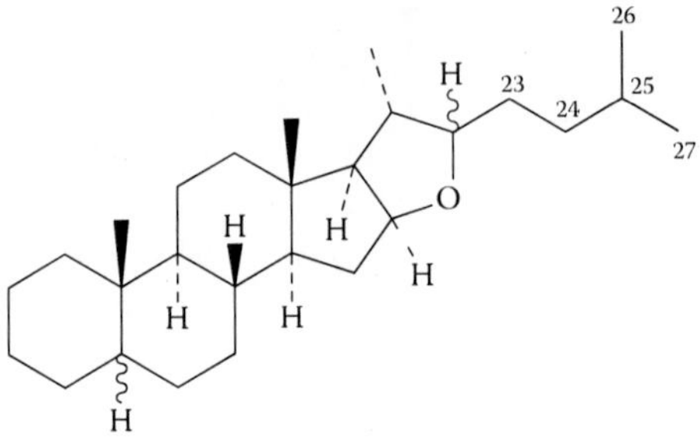

Furostanes (696)
VT1250

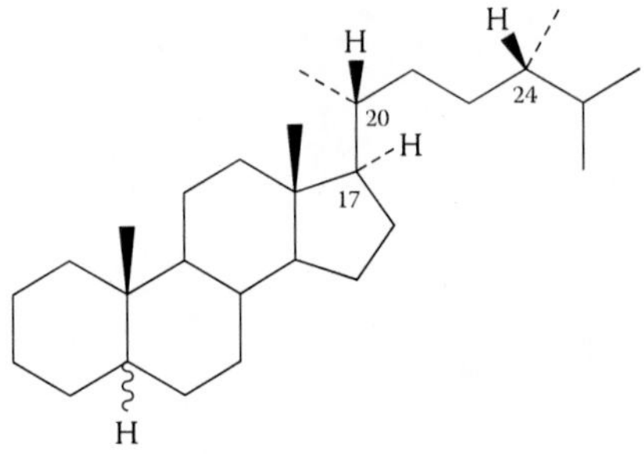

Ergostanes (1744)
24-Epimers = Campestanes
VT1300

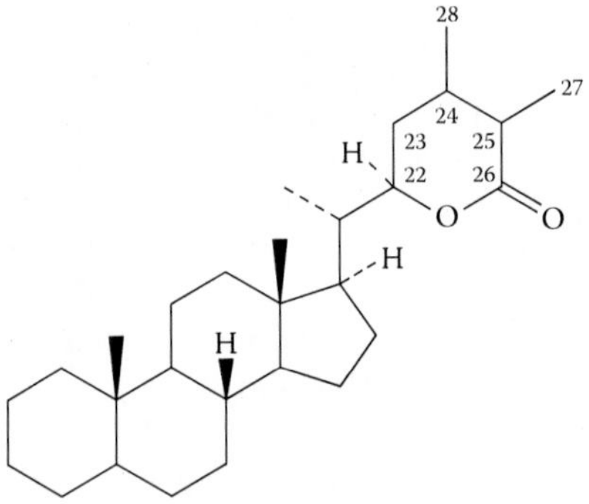

Withanolides (718)
VT1400

(*Continued*)

TABLE 6.1 (*Continued*)
Natural Product Parent Skeletons

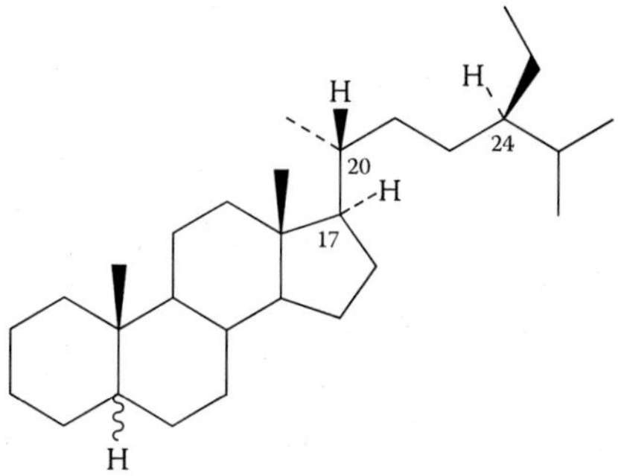

Stigmastanes (1007)
24-Epimers = Poriferastanes
VT1550

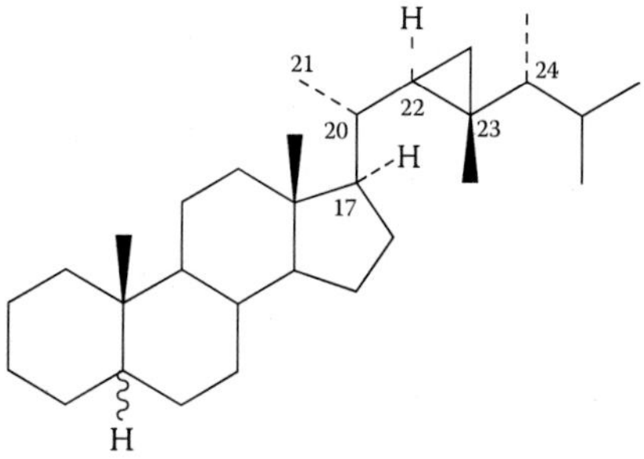

Gorgostanes (140)
VT1700

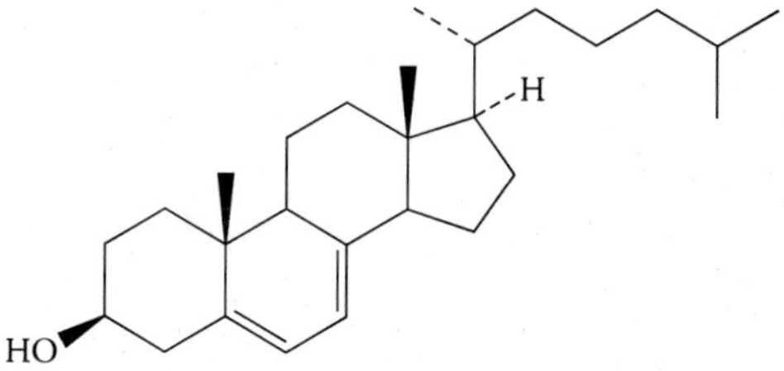

7-Dehydrocholesterol
Cholesta-5,7-dien-3β-ol

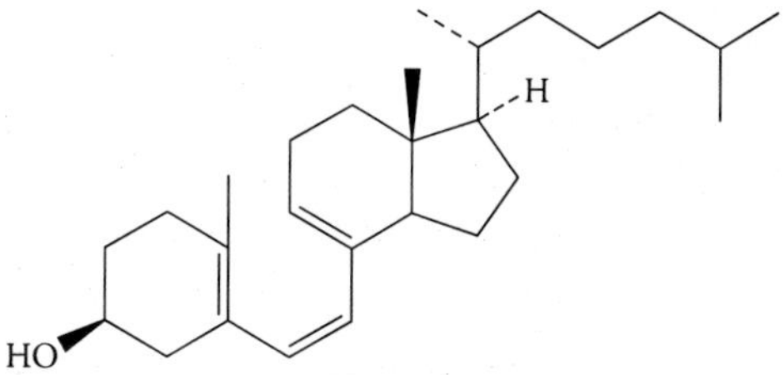

Previtamin D_3
9,10-Secocholesta-
5(10),6Z,8-trien-3S-ol

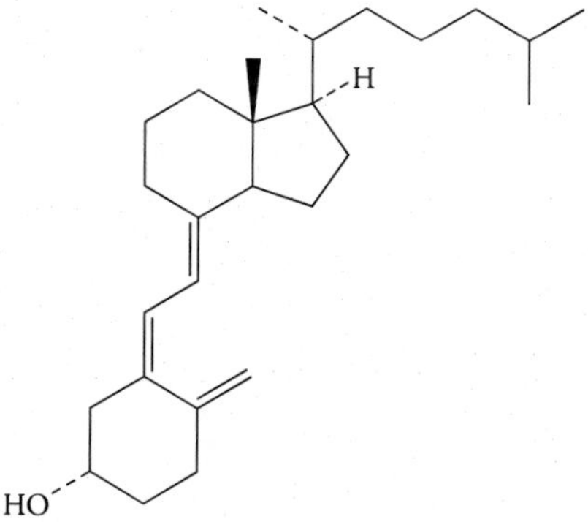

Cholecalciferol
Vitamin D_3
9,10-Secocholesta-
5Z,7E,10(19)-trien-3S-ol

Vitamin D_3 analogues (58)
VT2850

(*Continued*)

TABLE 6.1 (*Continued*)
Natural Product Parent Skeletons

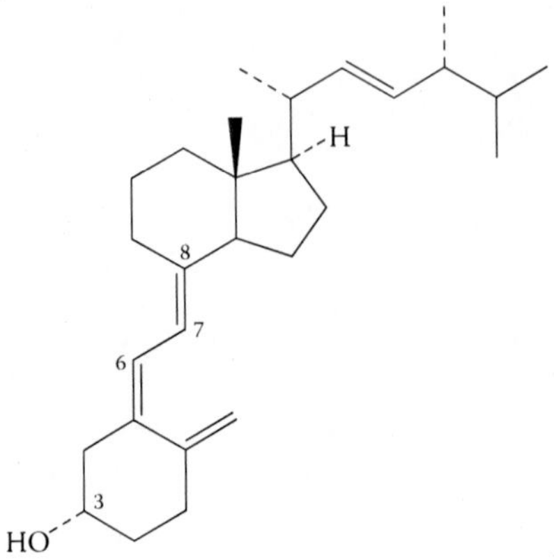

Ergocalciferol
Vitamin D$_2$
9,10-Secocholesta-
5Z,7E,10(19),22E-tetraen-3S-ol
Vitamin D$_2$ analogues (20)
VT2900

Amino acids and peptides (VV)

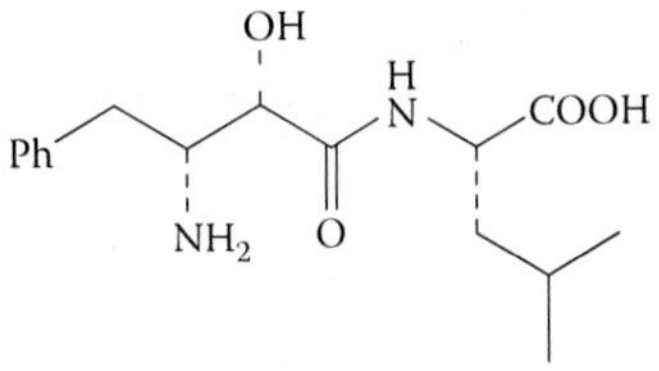

e.g. Bestatin
Bestatins (7)
VV0122

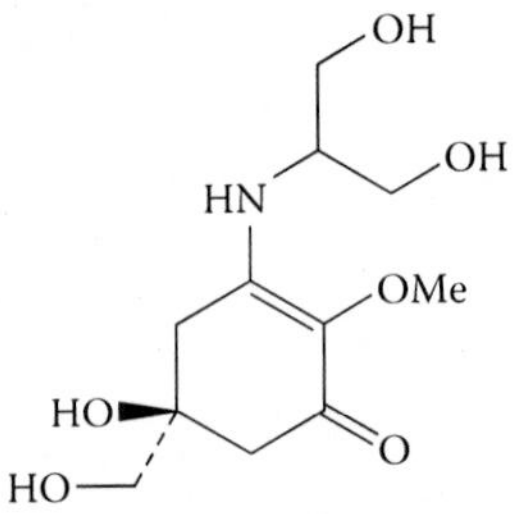

e.g. Mycosporin 1
Mycosporins (37)
VV0125

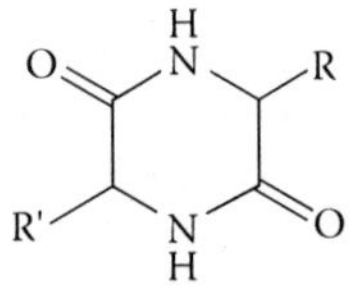

Diketopiperazines (681)
VV0150

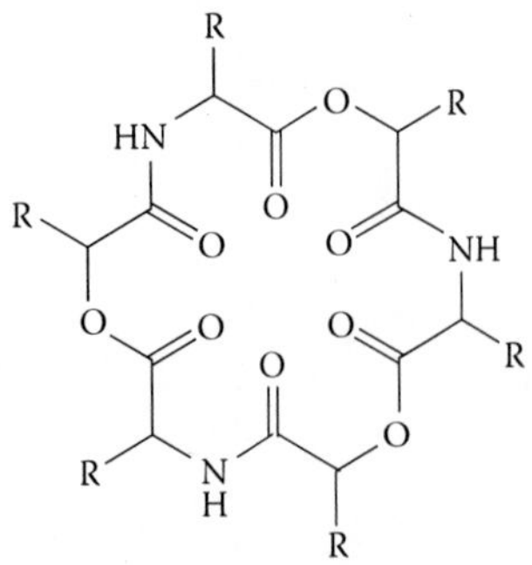

Beauvericins/Enniatins (48)
VV0505

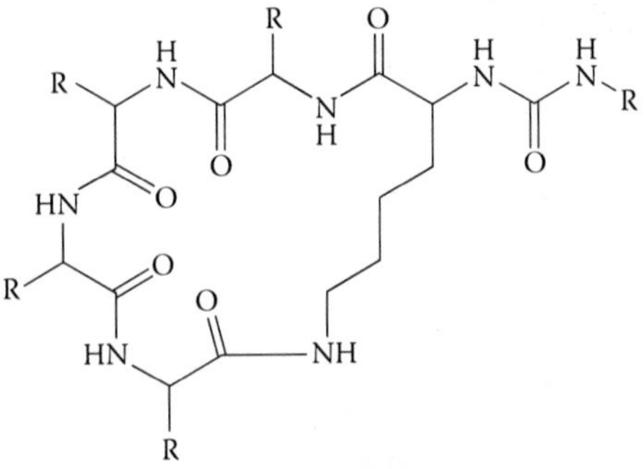

Anabaenopeptins (46)
VV0510

(*Continued*)

TABLE 6.1 (*Continued*)
Natural Product Parent Skeletons

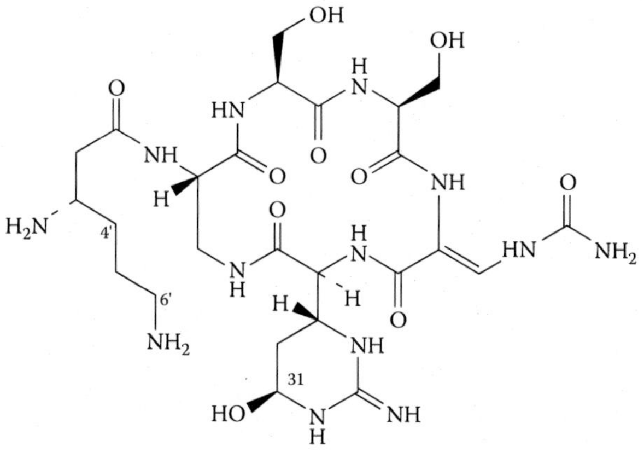

e.g. Viomycin
Tuberactinomycins (38)
VV0640

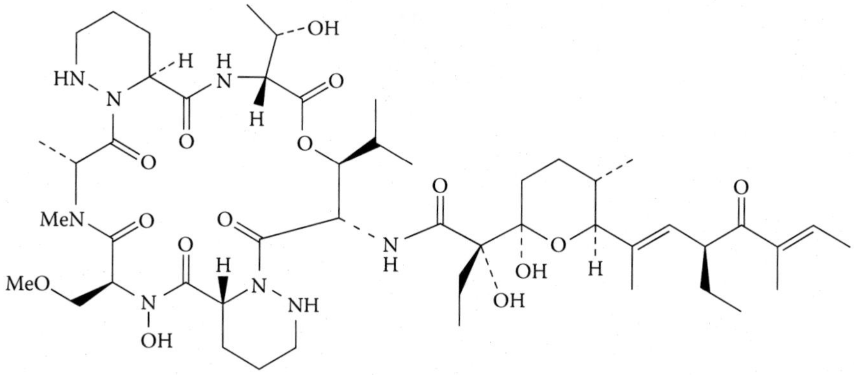

e.g. Azinothricin
Azinothricins (35)
VV0660

(*Continued*)

TABLE 6.1 (*Continued*)
Natural Product Parent Skeletons

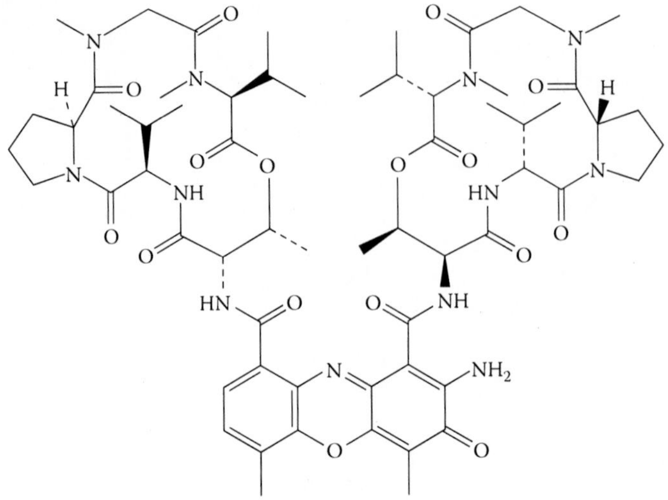

e.g. Actinomycin D
Actinomycins (54)
VV0670

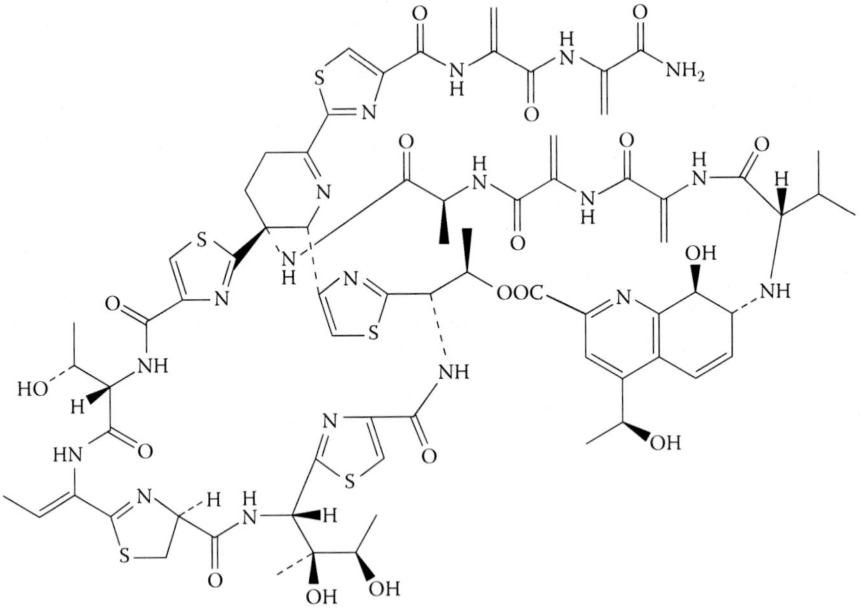

e.g. Siomycin A
Thiopeptins/Siomycins (43)
VV0680

(*Continued*)

TABLE 6.1 (*Continued*)
Natural Product Parent Skeletons

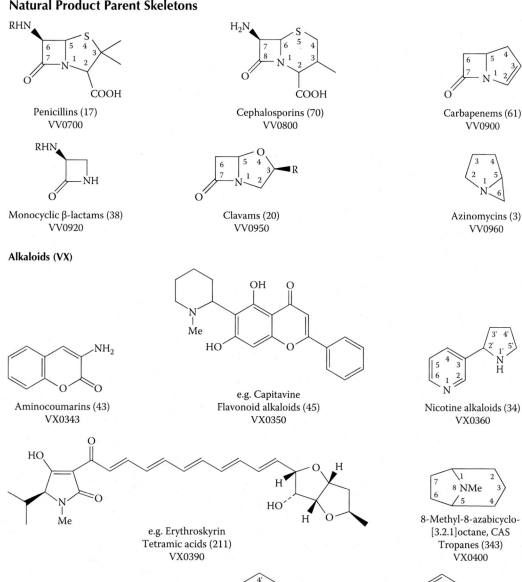

Penicillins (17)
VV0700

Cephalosporins (70)
VV0800

Carbapenems (61)
VV0900

Monocyclic β-lactams (38)
VV0920

Clavams (20)
VV0950

Azinomycins (3)
VV0960

Alkaloids (VX)

Aminocoumarins (43)
VX0343

e.g. Capitavine
Flavonoid alkaloids (45)
VX0350

Nicotine alkaloids (34)
VX0360

e.g. Erythroskyrin
Tetramic acids (211)
VX0390

8-Methyl-8-azabicyclo-
[3.2.1]octane, CAS
Tropanes (343)
VX0400

Pyrrolizidines (680)
VX0440–VX0520

Anabasine alkaloids (35)
VX0620

e.g. Sedinine
Lobelia alkaloids (60)
VX0660

(*Continued*)

TABLE 6.1 (*Continued*)
Natural Product Parent Skeletons

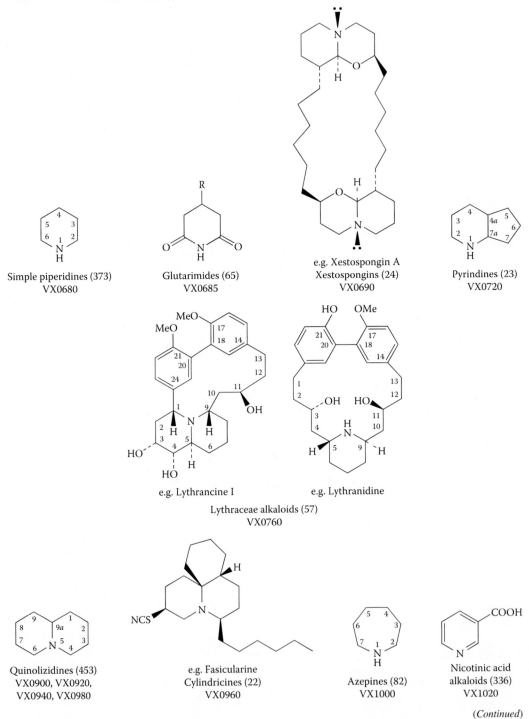

Simple piperidines (373)
VX0680

Glutarimides (65)
VX0685

e.g. Xestospongin A
Xestospongins (24)
VX0690

Pyrindines (23)
VX0720

e.g. Lythrancine I

e.g. Lythranidine

Lythraceae alkaloids (57)
VX0760

Quinolizidines (453)
VX0900, VX0920,
VX0940, VX0980

e.g. Fasicularine
Cylindricines (22)
VX0960

Azepines (82)
VX1000

Nicotinic acid
alkaloids (336)
VX1020

(*Continued*)

TABLE 6.1 (*Continued*)
Natural Product Parent Skeletons

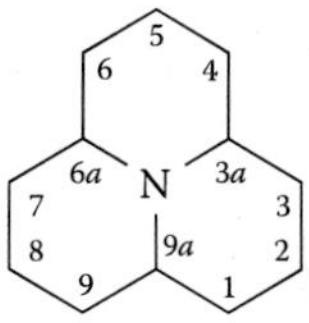

9B-Azaphenalenes (23)
VX1120

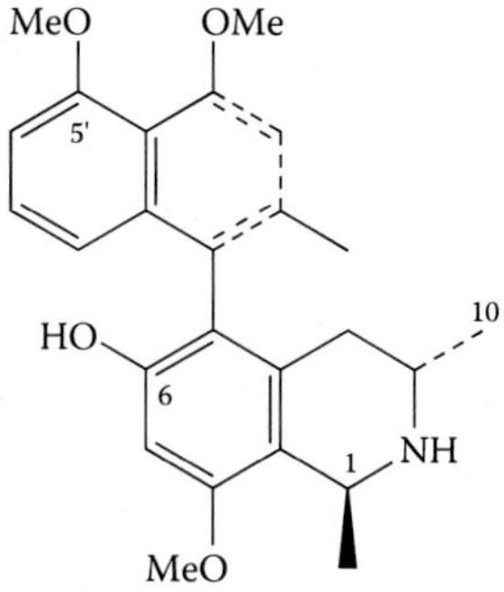

e.g. Ancistrocladine
Naphthalene-isoquinoline
alkaloids (125)
VX1140

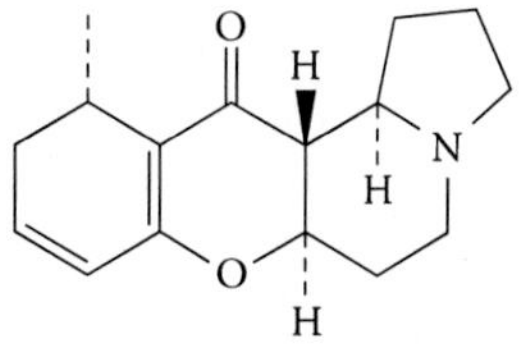

e.g. Elaeocarpiline
Elaeocarpus alkaloids (28)
VX1160

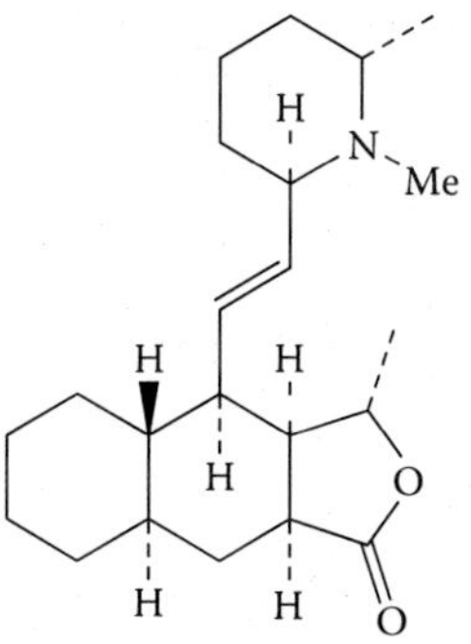

e.g. Himbacine
Galbulimima alkaloids (30)
VX1240

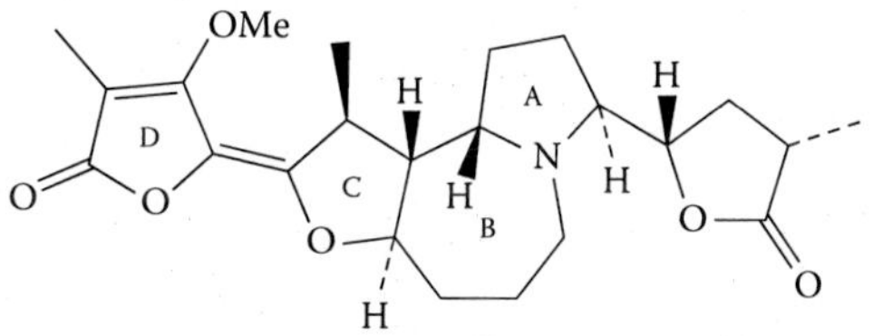

e.g. Protostemonine
Stemona alkaloids (172)
VX1260

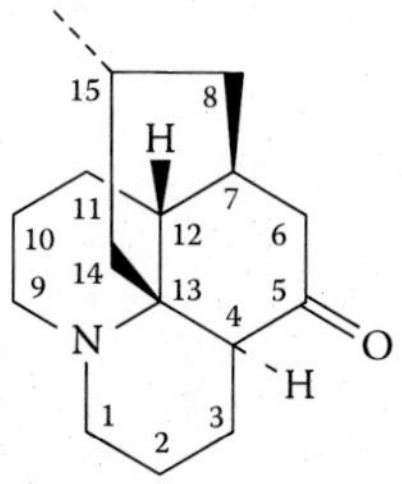

e.g. Lycopodine
Lycopodium alkaloids (338)
VX1280

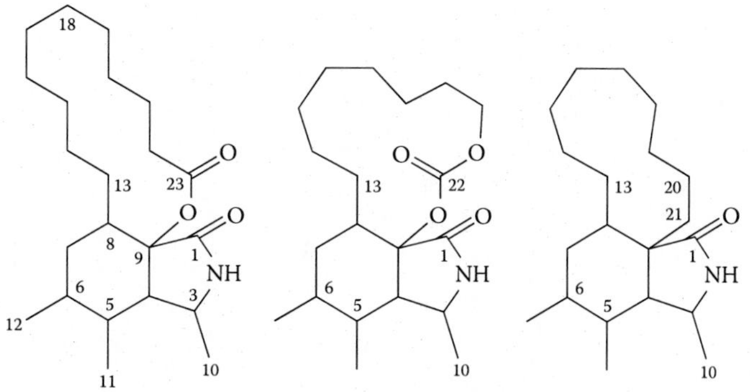

lactone

carbonate

carbocycle

Cytochalasins (233)
VX1300

(Continued)

TABLE 6.1 (*Continued*)
Natural Product Parent Skeletons

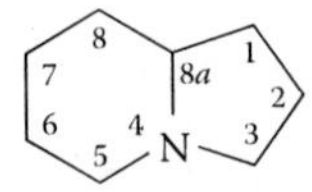

Indolizidines (161)
VX1360

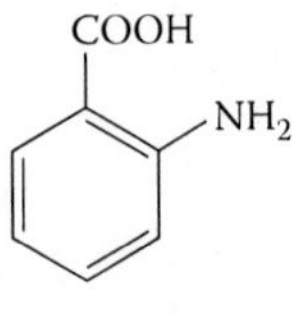

Simple anthranilic
acid alkaloids (82)
VX1460

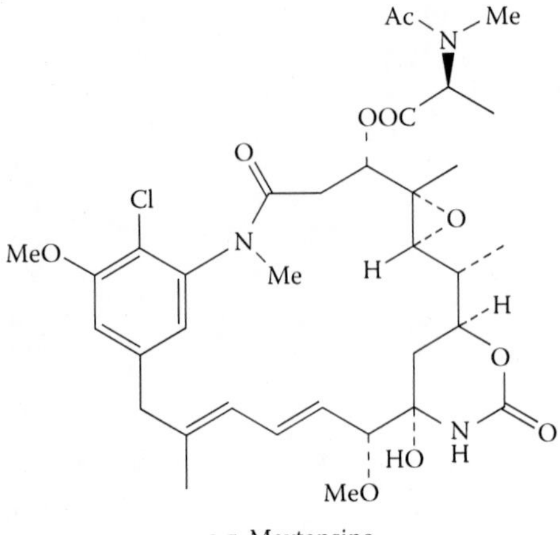

e.g. Maytansine
Maytansinoids (60)
VX1470

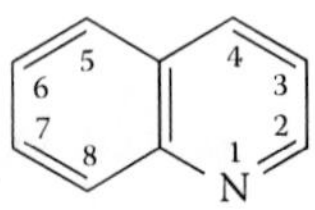

Quinolines (719)
VX1480, VX1580

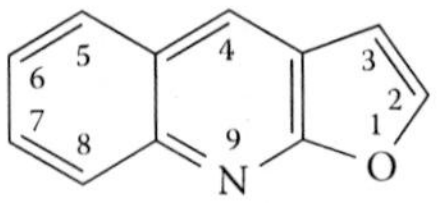

Furanoquinolines (224)
VX1520

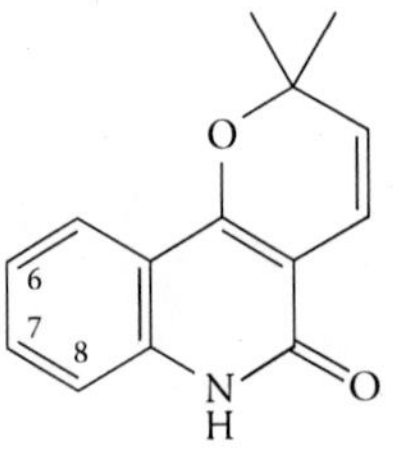

e.g. Flindersine
Pyranoquinolines (86)
VX1540

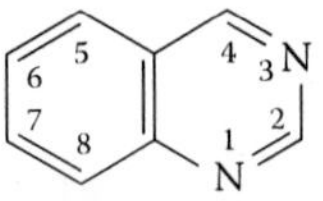

Quinazolines (144)
VX1600

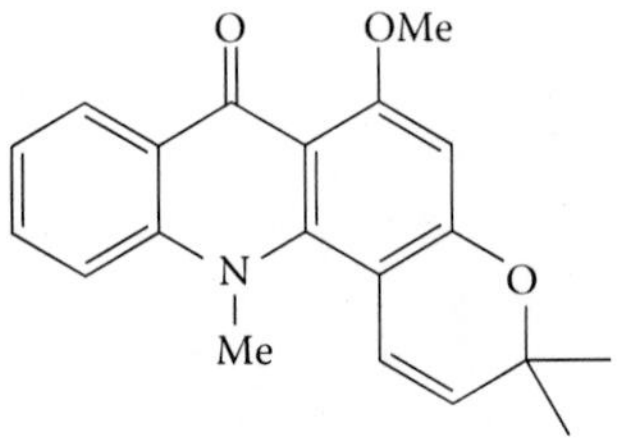

e.g. Acronycine
Acridones (237)
VX1620-VX1690

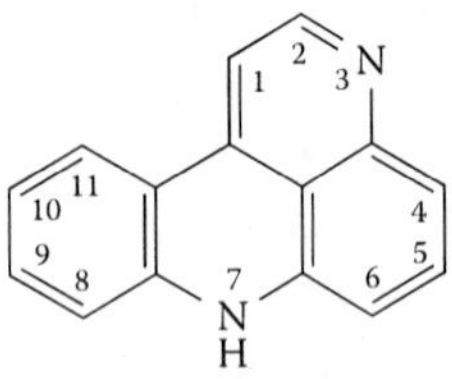

Pyrido[2,3,4-*kl*]-
acridines (98)
VX1700

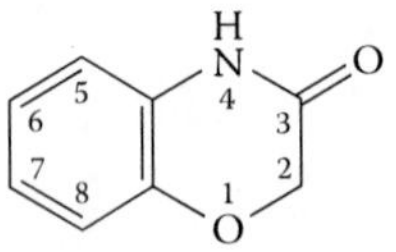

1,4-Benzoxazin-
3-ones (23)
VX1720

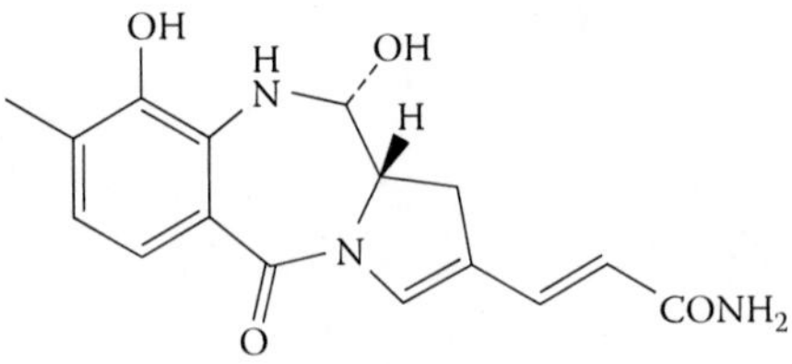

e.g. Anthramycin
Benzodiazepines (83)
VX1760

(*Continued*)

TABLE 6.1 (*Continued*)
Natural Product Parent Skeletons

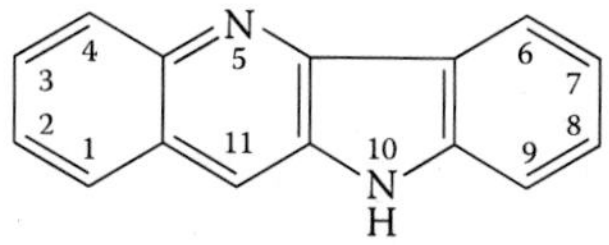

e.g. Quindoline
Cryptolepine alkaloids (19)
VX1800

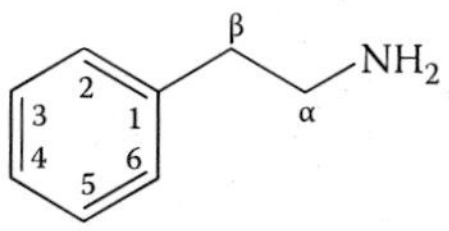

Phenethylamines (561)
VX2000, VX2005, VX2015

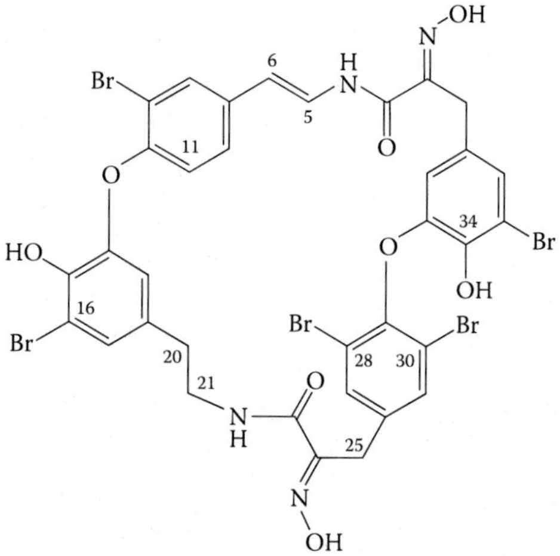

e.g. Bastadin 4
Halogenated tyrosinoids (161)
VX2008

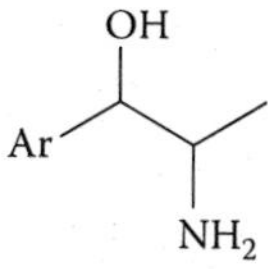

Ephedra alkaloids (15)
VX2010

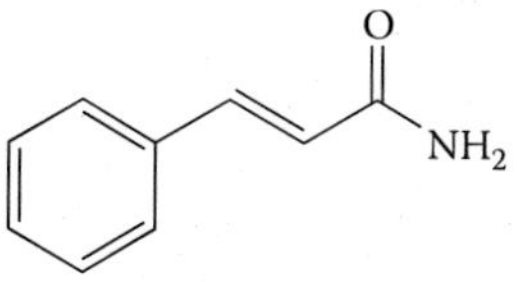

Cinnamic acid amides (381)
VX2020

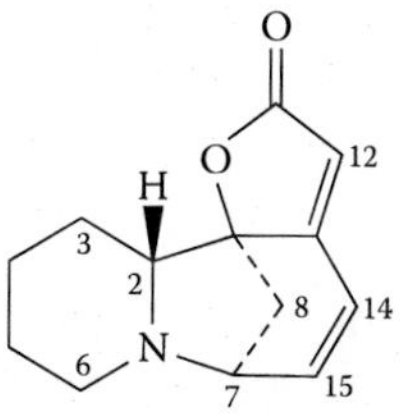

e.g. Securinine
Securinega alkaloids (56)
VX2100

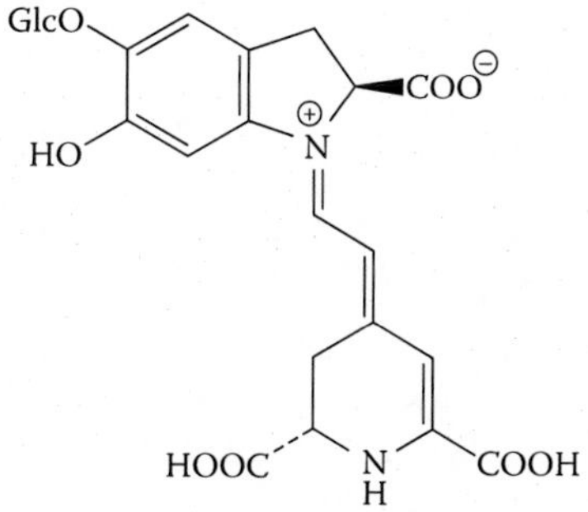

e.g. Betanin
Betalains (95)
VX2140

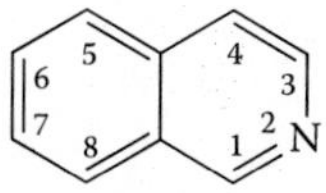

Simple isoquinolines (380)
VX2200, VX2310

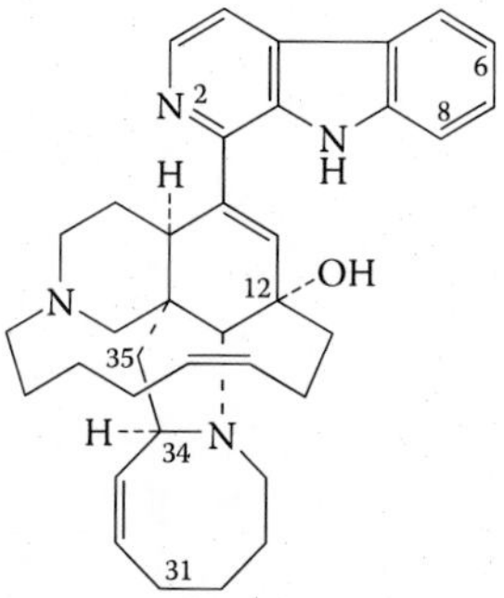

e.g. Manzamine A
Manzamines (60)
VX2250

(*Continued*)

TABLE 6.1 (*Continued*)
Natural Product Parent Skeletons

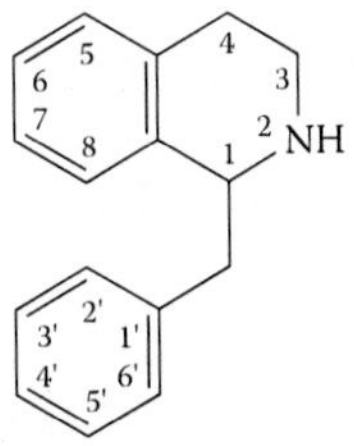

Benzylisoquinolines (179)
VX2320

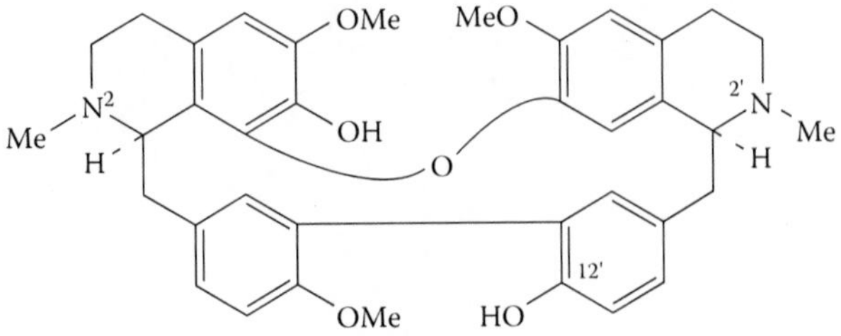

e.g. Cordobine
Bisbenzylisoquinolines (470)
VX2340-VX2400

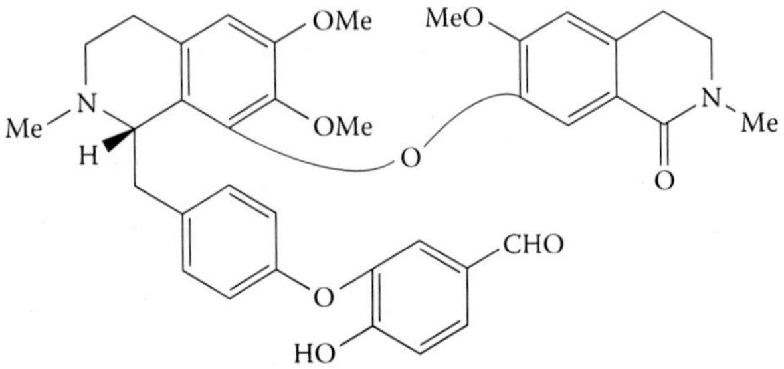

e.g. Baluchistanamine
Secobisbenzylisoquinolines (32)
VX2430

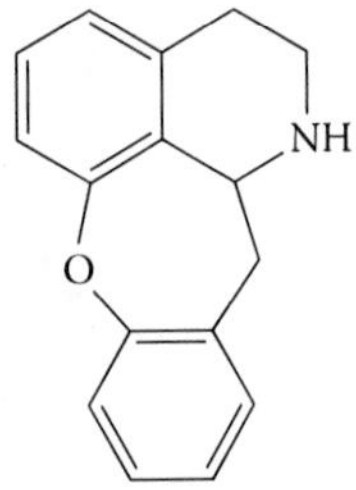

Cularines (32)
VX2440

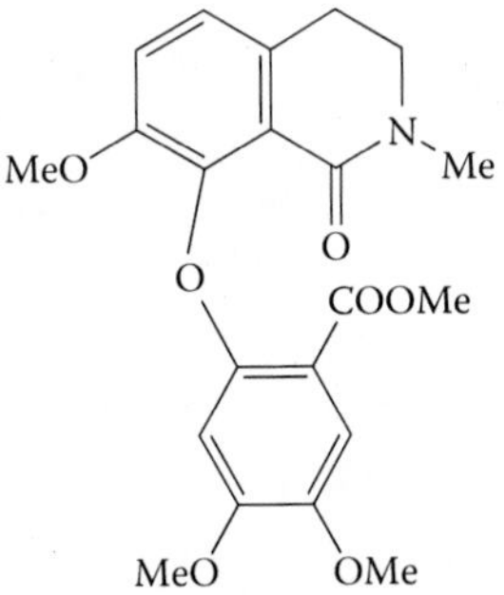

e.g. Noyaine
Secoisoquinolines (15)
VX2450

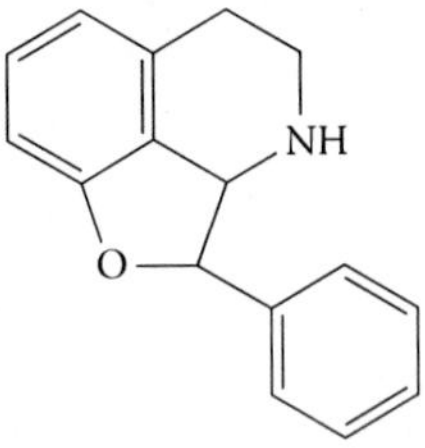

Quettamines (3)
VX2470

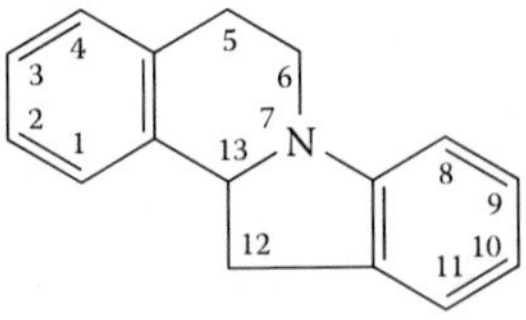

Dibenzopyrrocolines (8)
VX2480

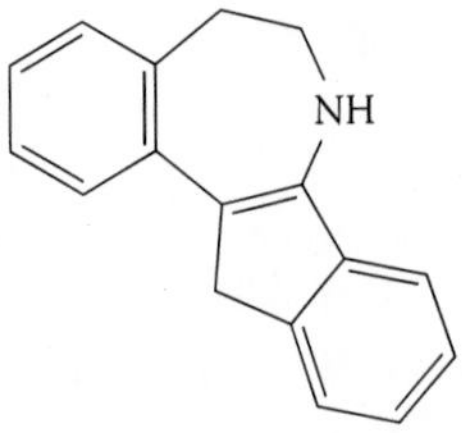

Indenobenzazepines (10)
VX2500

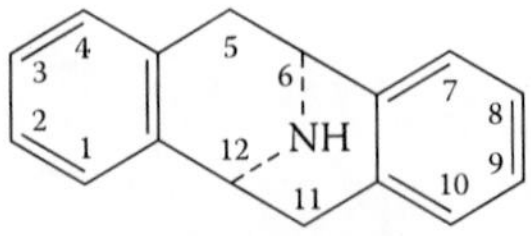

Pavines (46)
VX2520

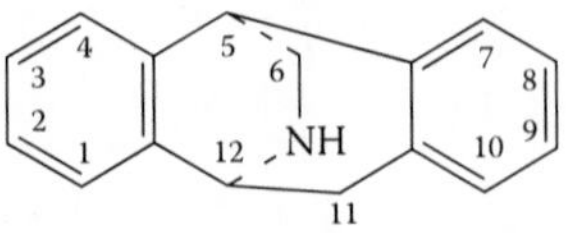

Isopavines (21)
VX2540

(*Continued*)

TABLE 6.1 (*Continued*)
Natural Product Parent Skeletons

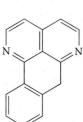

Proaporphines (85)
VX2600

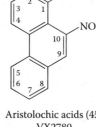

Aporphines (665)
VX2640

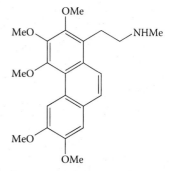

e.g. Thalicarpine
Aporphine-benzylisoquinoline dimers (70)
VX2700

Oxoisoaporphines (13)
VX2750

Aristolochic acids (45)
VX2780

Aristolactams (72)
(numbering systems vary)
VX2800

e.g. Thalicpureine
Phenanthrene alkaloids (54)
VX2820

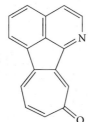

Azaaporphines (14)
VX2830

Azafluoranthrenes (21)
VX2840

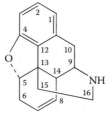

Morphines (119)
VX2900

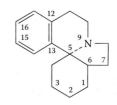

Erythrina alkaloids (134)
VX2940

(*Continued*)

TABLE 6.1 (*Continued*)
Natural Product Parent Skeletons

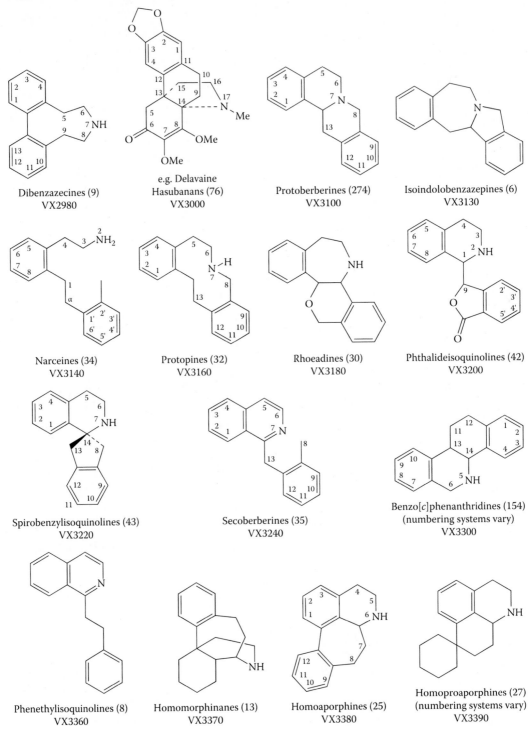

Dibenzazecines (9)
VX2980

e.g. Delavaine
Hasubanans (76)
VX3000

Protoberberines (274)
VX3100

Isoindolobenzazepines (6)
VX3130

Narceines (34)
VX3140

Protopines (32)
VX3160

Rhoeadines (30)
VX3180

Phthalideisoquinolines (42)
VX3200

Spirobenzylisoquinolines (43)
VX3220

Secoberberines (35)
VX3240

Benzo[*c*]phenanthridines (154)
(numbering systems vary)
VX3300

Phenethylisoquinolines (8)
VX3360

Homomorphinanes (13)
VX3370

Homoaporphines (25)
VX3380

Homoproaporphines (27)
(numbering systems vary)
VX3390

(*Continued*)

TABLE 6.1 (*Continued*)
Natural Product Parent Skeletons

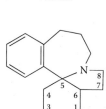

Colchicines (64)
VX3400

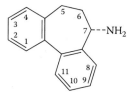

Dibenzocycloheptylamines (6)
VX3410

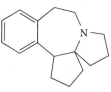

Cephalotaxus alkaloids (59)
(numbering systems vary)
VX3420

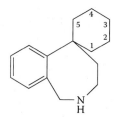

Homoerythrina alkaloids (67)
(numbering systems vary)
VX3440

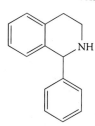

1-Aryltetrahydro-
isoquinolines (9)
VX3510

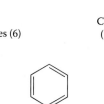

4-Aryltetrahydro-
isoquinolines (2)
VX3520

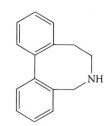

Apogalanthamines (4)
VX3530

Galanthamines (34)
VX3540

Haemanthidines (122)
VX3550

e.g. Graciline
Gracilines (6)
VX3555

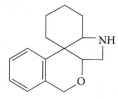

Tazettines (28)
VX3560

Pancracines (11)
VX3570

Lycorines (109)
VX3580

Lycorenines (75)
(numbering systems vary)
VX3585

Narciclasines (26)
VX3590

(*Continued*)

TABLE 6.1 (*Continued*)
Natural Product Parent Skeletons

e.g. Mesembrine
Mesembrenoid alkaloids (35)
VX3600

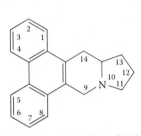

Emetines (91)
VX3690

Phenanthroindolizidines (101)
VX3700

Phenanthroquinolizidines (9)
VX3760

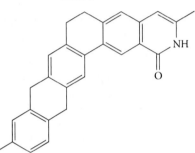

Ericamycins (74)
VX3900

Simple indoles (729)
VX4000

e.g. Indirubin
Biindoles (51)
VX4020

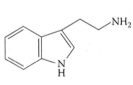

Tryptamines (203)
VX4040

Cyclotryptamines (117)
VX4100

(*Continued*)

TABLE 6.1 (*Continued*)
Natural Product Parent Skeletons

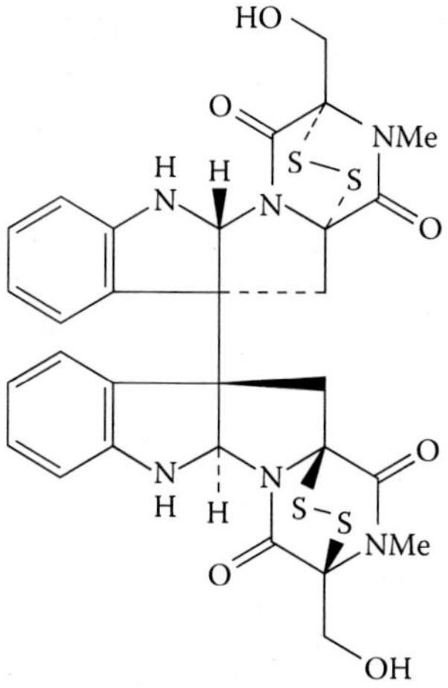

e.g. Chaetocin
Chaetocin alkaloids (205)
VX4110

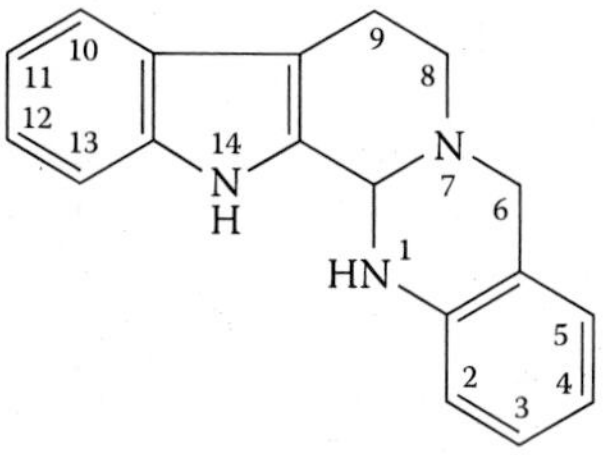

Evodia alkaloids (43)
VX4120

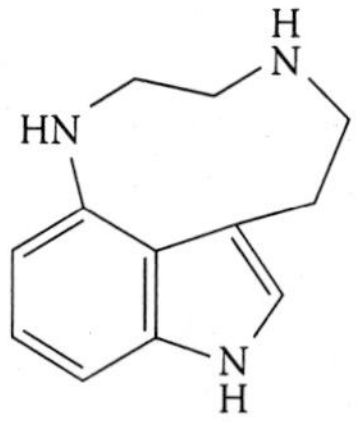

Indolactams (40)
VX4200

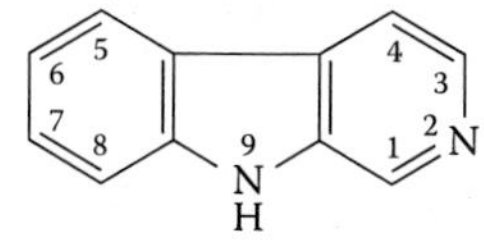

β-Carbolines (579)
VX4240

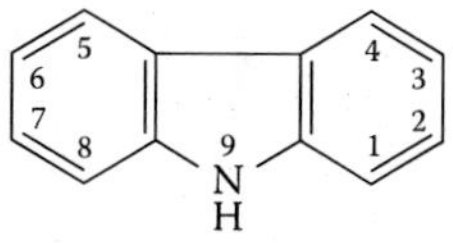

Carbazoles (342)
VX4300

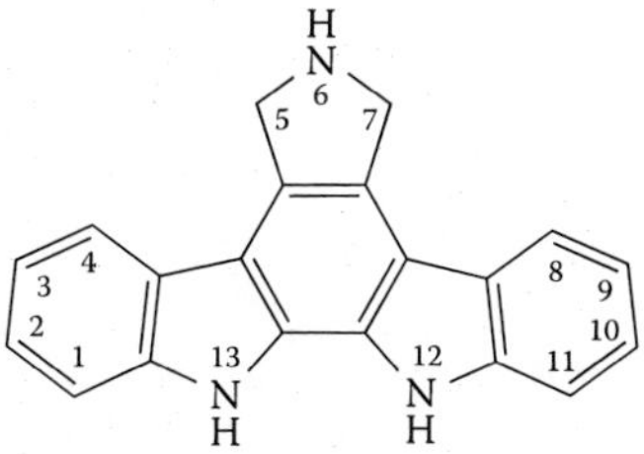

Indolo[2,3-*a*]carbazoles (118)
VX4350

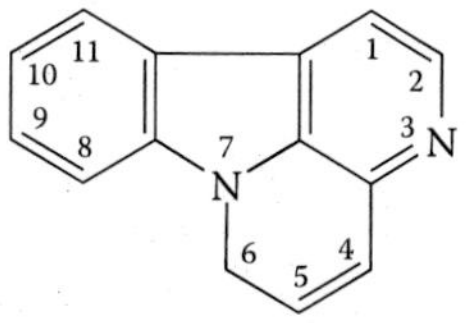

Canthines
Indolonaphthyridines (88)
VX4400

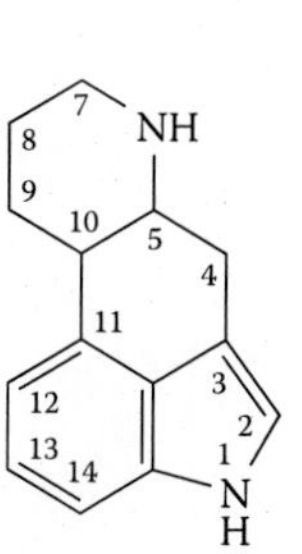

Ergot alkaloids (125)
VX4460

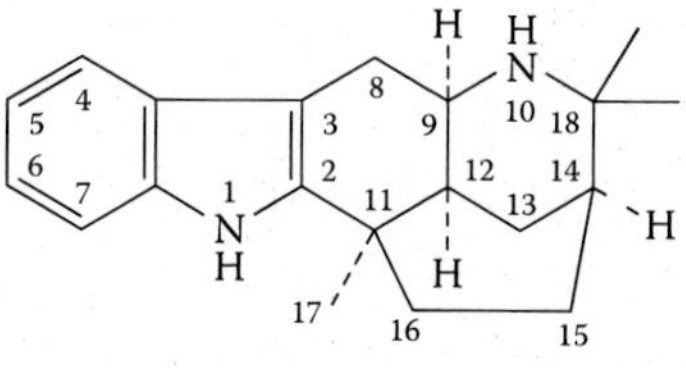

e.g. Aristoteline
Indoloterpenoid alkaloids (171)
VX4620

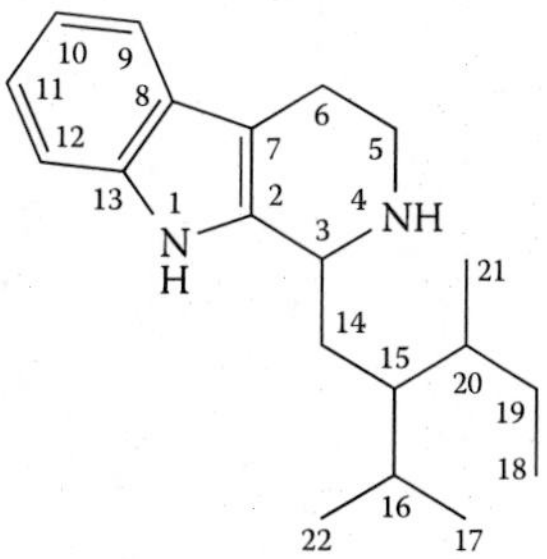

Strictosidines (107)
VX4640

(*Continued*)

TABLE 6.1 (*Continued*)
Natural Product Parent Skeletons

Camptothecins (44)
VX4700

Corynantheans (273)
VX4800

Akagerans (39)
VX4840

Ajmalicines (152)
VX4860

Macrolines (84)
VX4900

Malindans (20)
VX4920

Cadambans (8)
VX4940

Vallesiachotamans (36)
VX4960

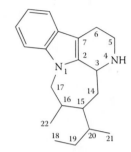

Gelsemium alkaloids (94)
VX5000

Yohimbanoid alkaloids (109)
VX5040

Sarpagines (152)
VX5100

Ajmalines (107)
VX5120

(*Continued*)

TABLE 6.1 (*Continued*)
Natural Product Parent Skeletons

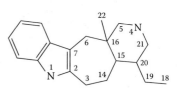

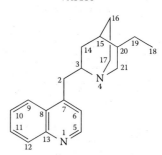

Vobasines (154)
VX5140

Koumines (7)
VX5150

Ervatamia alkaloids (23)
VX5180

Akuammilines (194)
VX5200

Pleiocarpamans (41)
VX5220

Cinchona alkaloids (50)
VX5240

Akuammicines (272)
VX5260

Strychnos alkaloids (98)
VX5280

Condylocarpans (45)
VX5320

Secodines (18)
VX5360

Aspidosperma alkaloids (349)
VX5400

Rhazinilams (11)
VX5450

(*Continued*)

TABLE 6.1 (*Continued*)
Natural Product Parent Skeletons

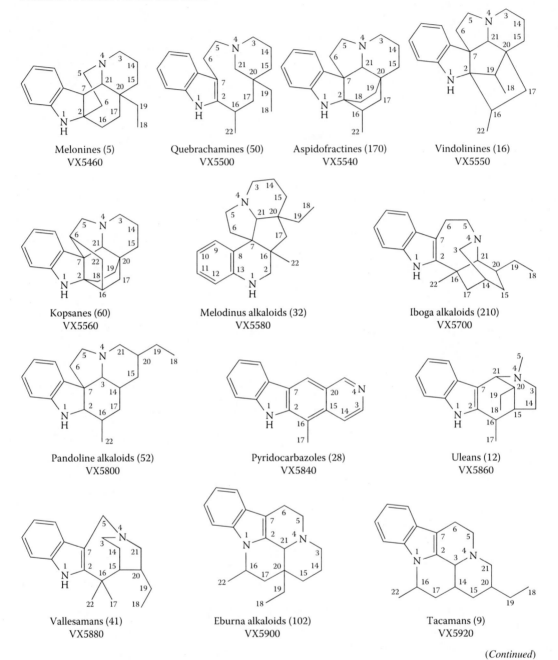

Melonines (5)
VX5460

Quebrachamines (50)
VX5500

Aspidofractines (170)
VX5540

Vindolinines (16)
VX5550

Kopsanes (60)
VX5560

Melodinus alkaloids (32)
VX5580

Iboga alkaloids (210)
VX5700

Pandoline alkaloids (52)
VX5800

Pyridocarbazoles (28)
VX5840

Uleans (12)
VX5860

Vallesamans (41)
VX5880

Eburna alkaloids (102)
VX5900

Tacamans (9)
VX5920

(*Continued*)

TABLE 6.1 (*Continued*)
Natural Product Parent Skeletons

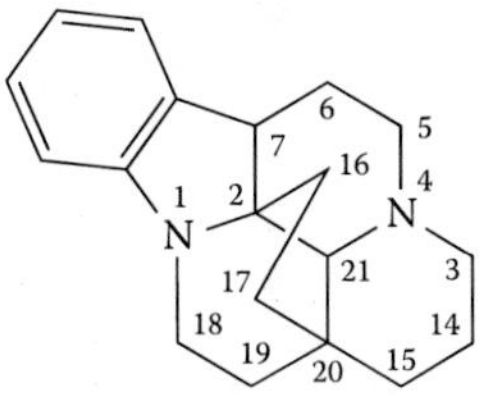

Schizozygines (11)
VX5930

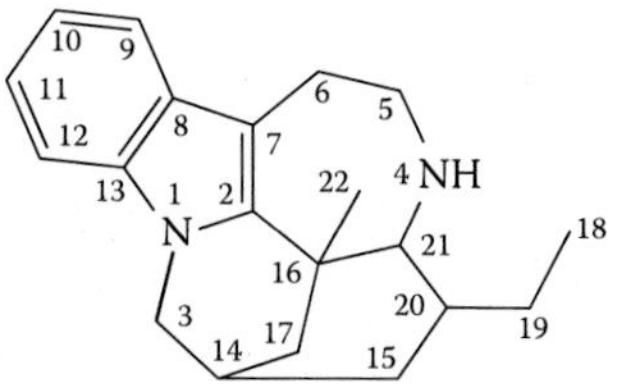

Chippiines (6)
VX5940

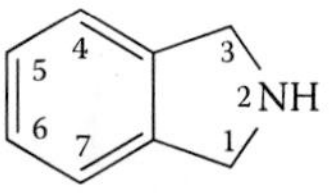

Isoindoles (118)
VX6000

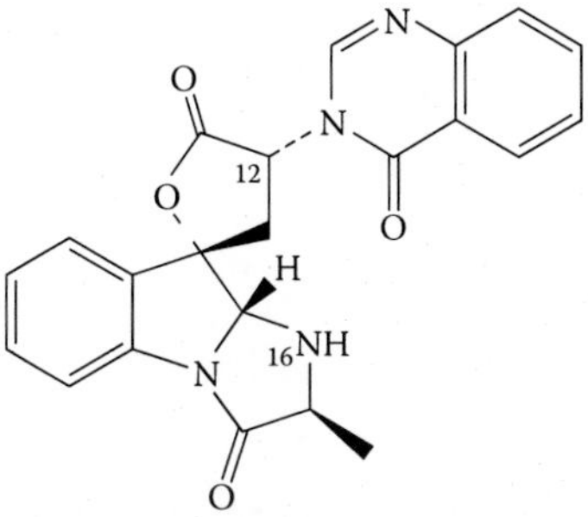

e.g. Tryptoquivaline F
Tryptoquivalines (92)
VX6030

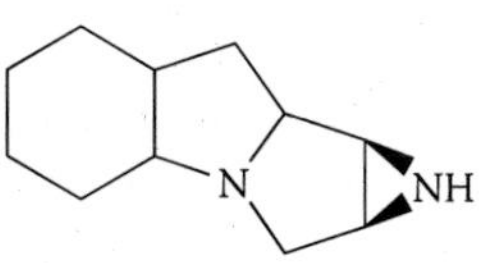

Mitomycins (23)
VX6050

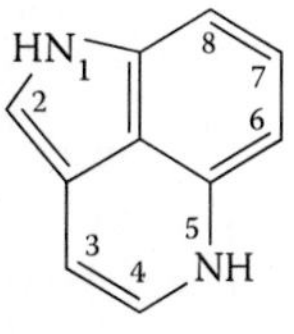

Pyrrolo[4,3,2-*de*]-
quinolines (118)
VX6070

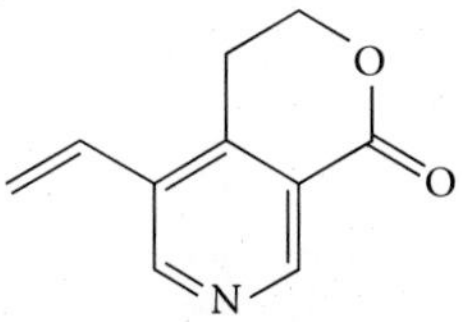

e.g. Gentianine
Monoterpenoid alkaloids (208)
VX6240

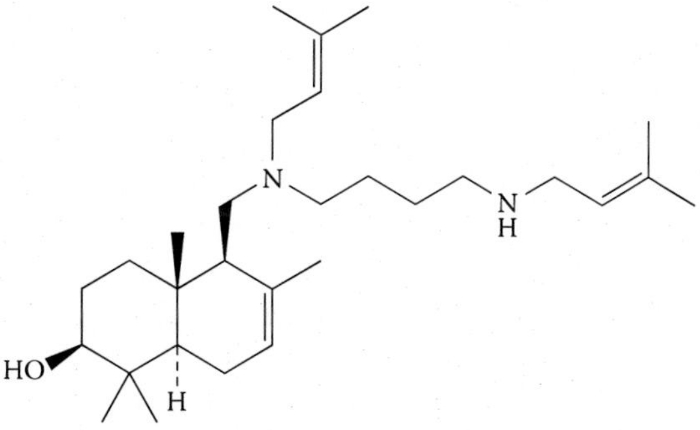

e.g. Halichonine A
Sesquiterpene alkaloids (330)
VX6300

(*Continued*)

TABLE 6.1 (*Continued*)
Natural Product Parent Skeletons

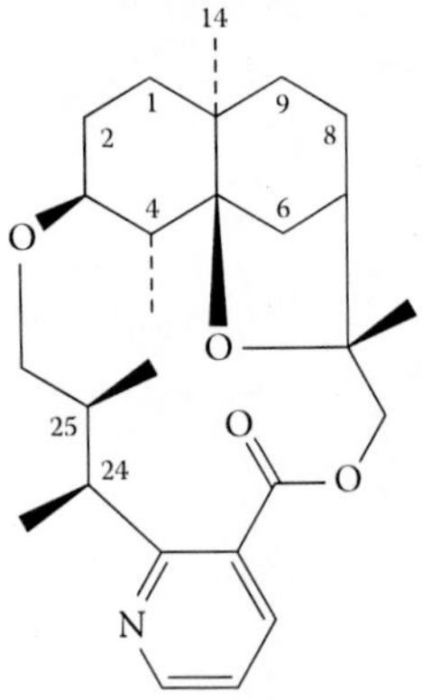

Macrocyclic sesquiterpene
alkaloids (186)
VX6320

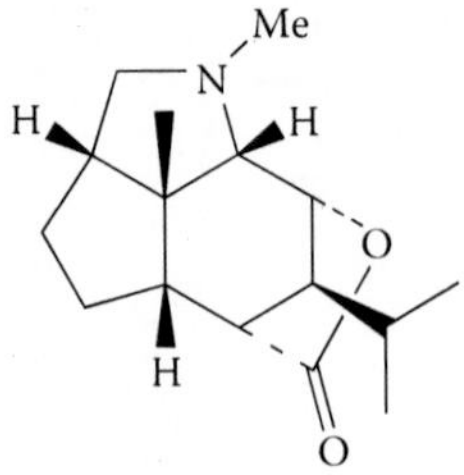

e.g. Dendrobine
Dendrobium alkaloids (26)
VX6340

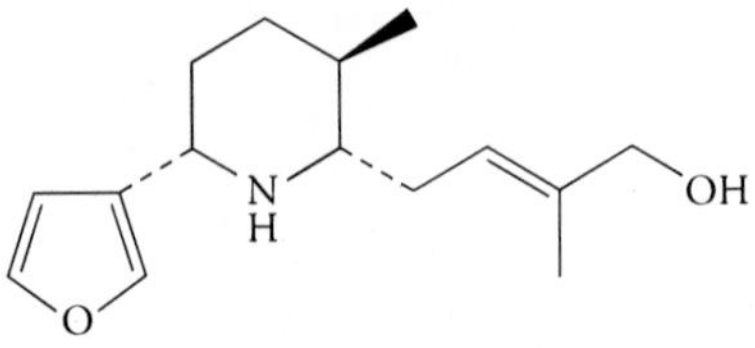

e.g. Nuphamine
Nuphar alkaloids (61)
VX6360

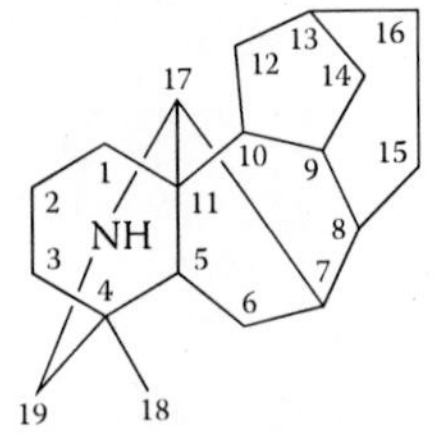

C_{19} Diterpenoid alkaloids (922)
VX6400

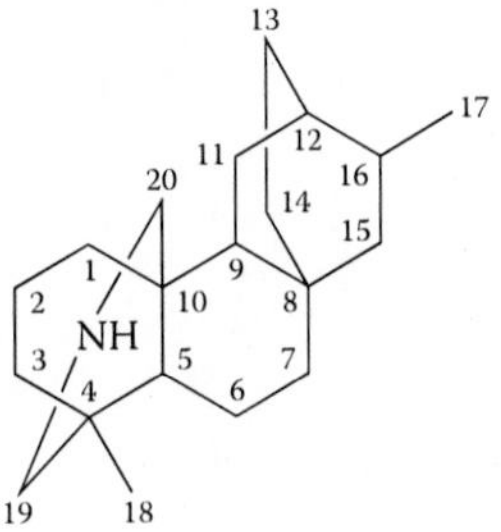

C_{20} Diterpenoid alkaloids (410)
VX6420

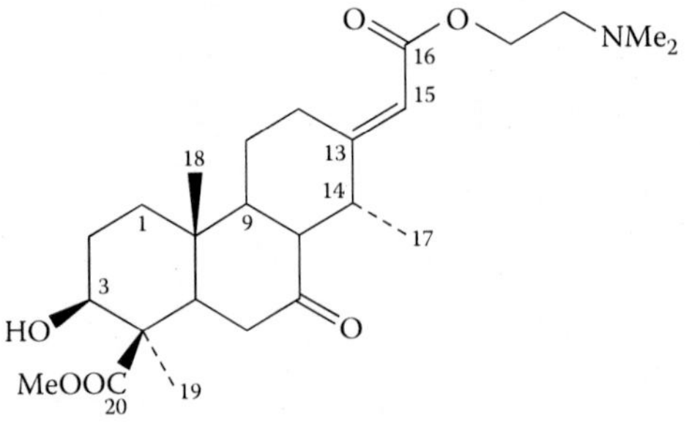

e.g. Erythrophlamine
Erythrophleum alkaloids (42)
VX6460

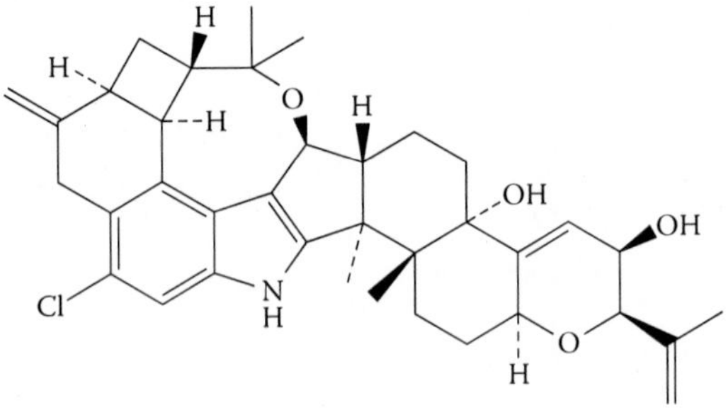

e.g. Penitrem C
Penitrems (148)
VX6470

(*Continued*)

TABLE 6.1 (*Continued*)
Natural Product Parent Skeletons

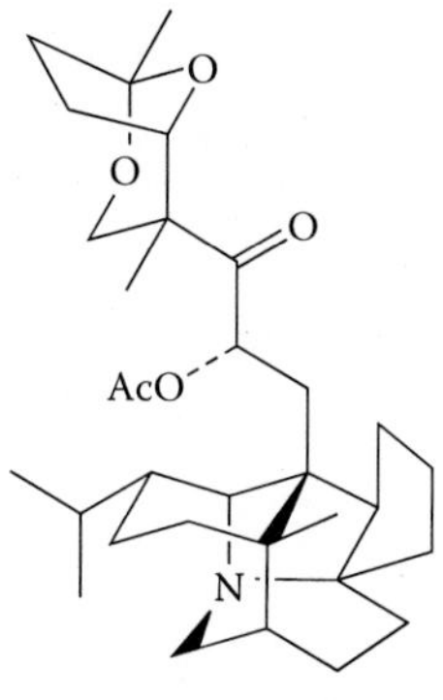

e.g. Daphniphylline
Daphniphylline alkaloids (270)
VX6500

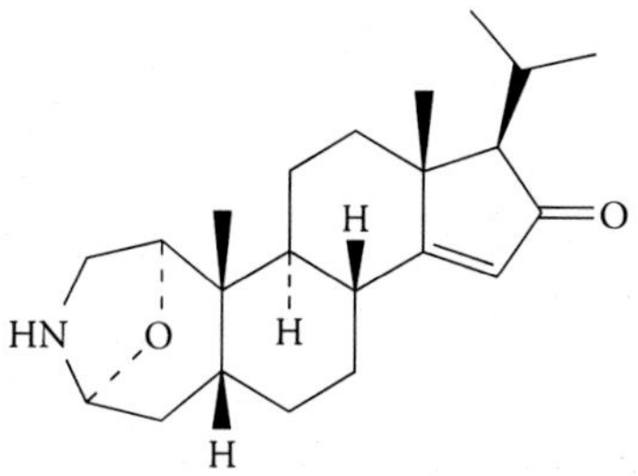

e.g. Samandenone
Salamandra alkaloids (10)
VX6640

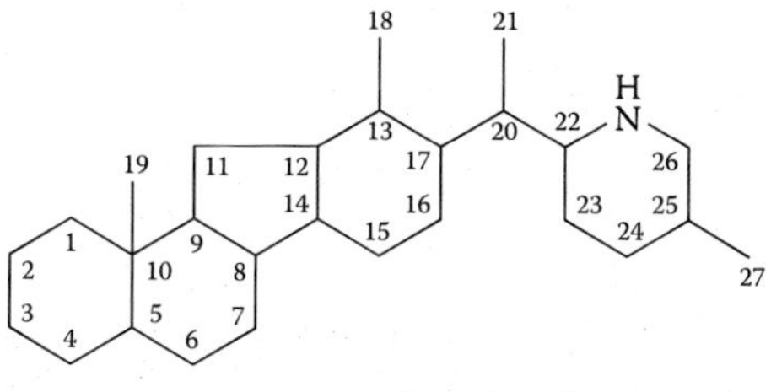

Jerveratrum alkaloids (40)
VX6660

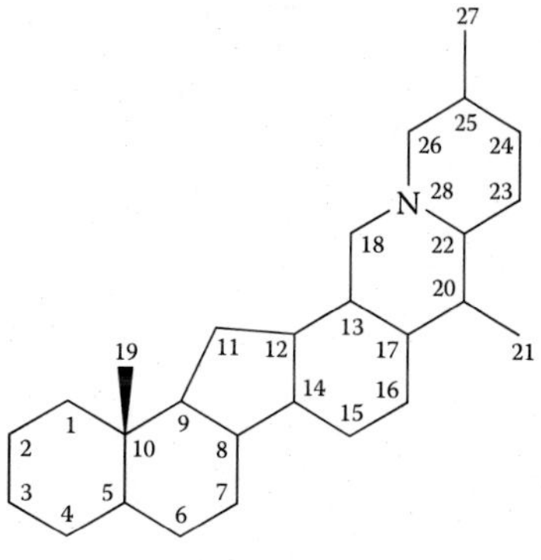

Cerveratrum alkaloids (171)
VX6680

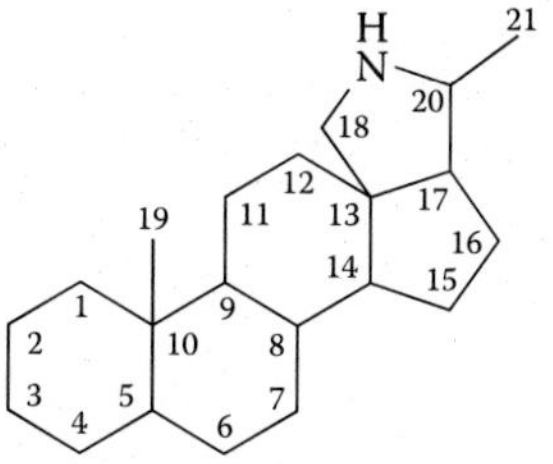

Conanines (69)
VX6700

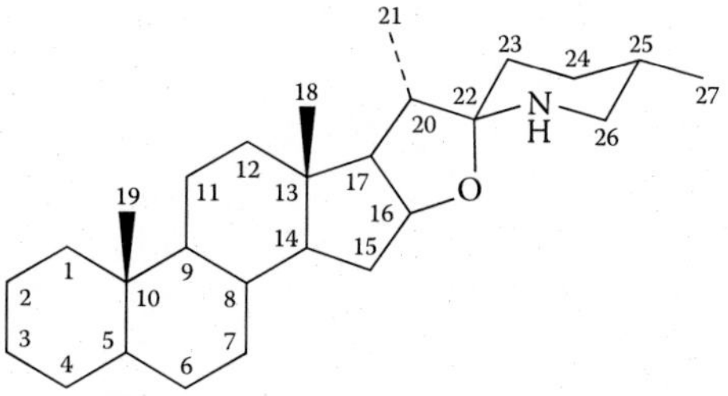

Spirosolanes (117)
VX6720

(*Continued*)

TABLE 6.1 (*Continued*)
Natural Product Parent Skeletons

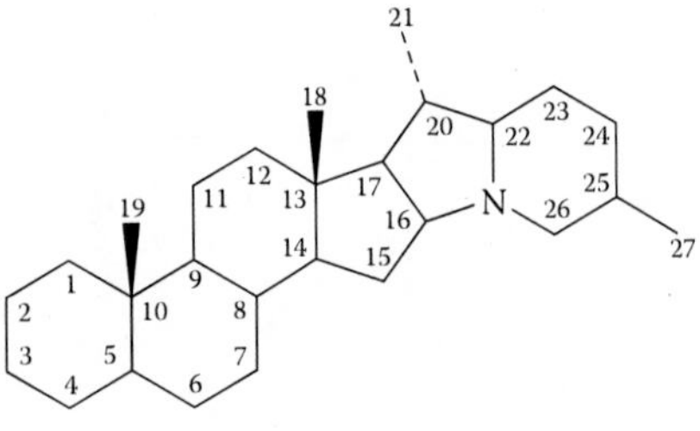

Solanidines (163)
VX6740

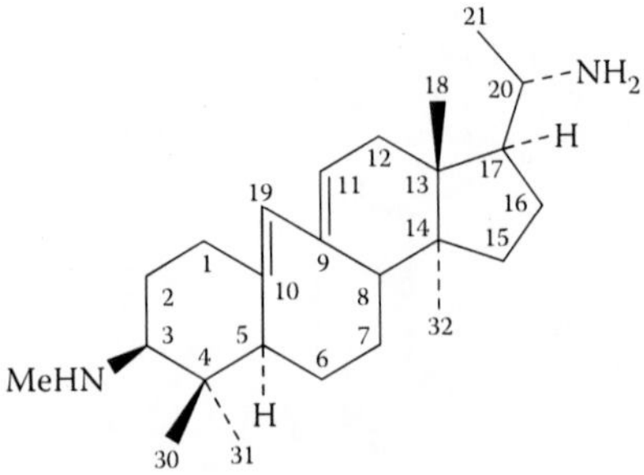

e.g. Buxamine G
Buxus alkaloids (279)
VX6760

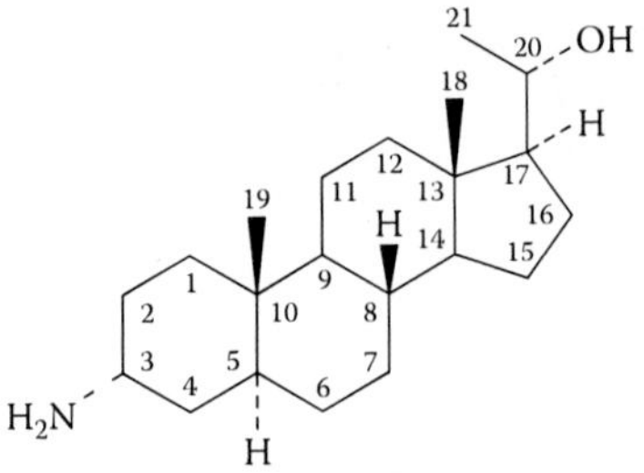

e.g. Funtumidine
Pregnane alkaloids (250)
VX6780

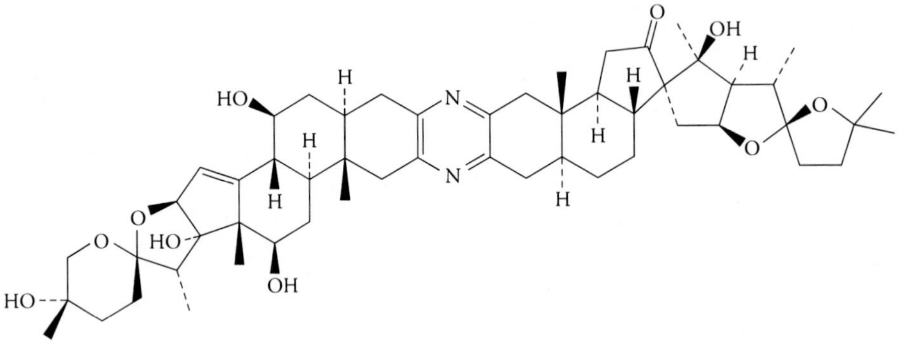

e.g. Ritterazine A
Cephalostatins/Ritterazines (45)
VX6785

Steroidal Alkaloids (VX6640-VX6785)

(*Continued*)

TABLE 6.1 (*Continued*)
Natural Product Parent Skeletons

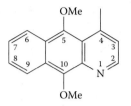

e.g. Annopholine
Azaanthracene alkaloids (36)
VX6820

Azafluorene alkaloids (26)
Indeno[1,2-*b*]pyridines
VX6840

Pyrazoles (23)
VX6900

Imidazoles (480)
VX6920

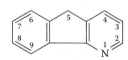

e.g. Oroidin
Pyrrole-imidazole alkaloids (163)
VX6922

Cycloheptadiimidazoles (32)
VX6925

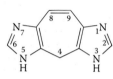

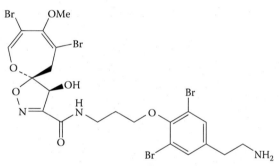

Oxazoles and benzoxazoles (494)
VX6930

Isoxazoles (34)
VX6932

e.g. Psammaplysin A
Spirobenzoxazolines (135)
VX6934

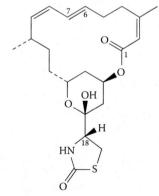

Thiazoles and benzothiazoles (623)
VX6935, VX6937

e.g. Latrunculin A
Latrunculins (20)
VX6936

(Continued)

TABLE 6.1 (*Continued*)
Natural Product Parent Skeletons

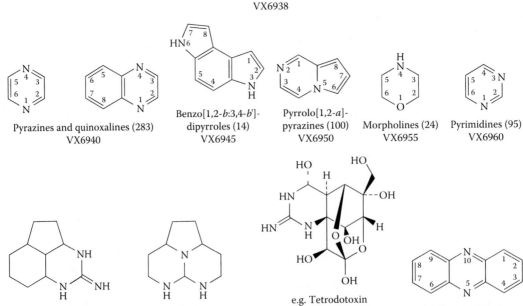

e.g. Zorbamycin
Bleomycins (77)
VX6938

Pyrazines and quinoxalines (283)
VX6940

Benzo[1,2-*b*:3,4-*b'*]-
dipyrroles (14)
VX6945

Pyrrolo[1,2-*a*]-
pyrazines (100)
VX6950

Morpholines (24)
VX6955

Pyrimidines (95)
VX6960

Ptilocaulins (24)
VX6970

Triazaacenaphthylenes (53)
VX6980

e.g. Tetrodotoxin
Tetrodotoxins (24)
VX6990

Phenazines (145)
VX7000

(*Continued*)

TABLE 6.1 (*Continued*)
Natural Product Parent Skeletons

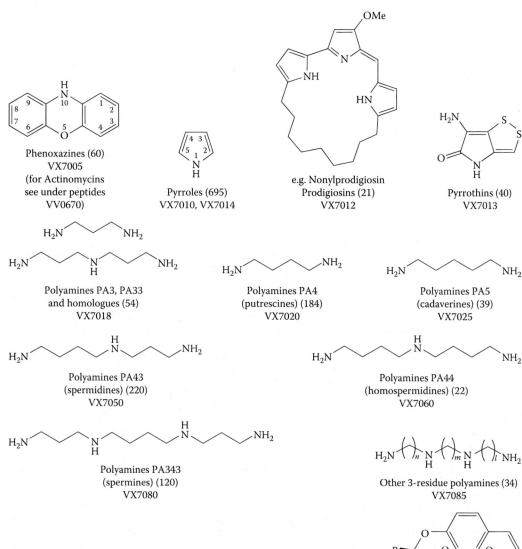

Phenoxazines (60)
VX7005
(for Actinomycins
see under peptides
VV0670)

Pyrroles (695)
VX7010, VX7014

e.g. Nonylprodigiosin
Prodigiosins (21)
VX7012

Pyrrothins (40)
VX7013

Polyamines PA3, PA33
and homologues (54)
VX7018

Polyamines PA4
(putrescines) (184)
VX7020

Polyamines PA5
(cadaverines) (39)
VX7025

Polyamines PA43
(spermidines) (220)
VX7050

Polyamines PA44
(homospermidines) (22)
VX7060

Polyamines PA343
(spermines) (120)
VX7080

Other 3-residue polyamines (34)
VX7085

Polyamines
with more than 3 residues (98)
VX7090

Ansa-peptide alkaloids (227)
VX7100

(*Continued*)

TABLE 6.1 (*Continued*)
Natural Product Parent Skeletons

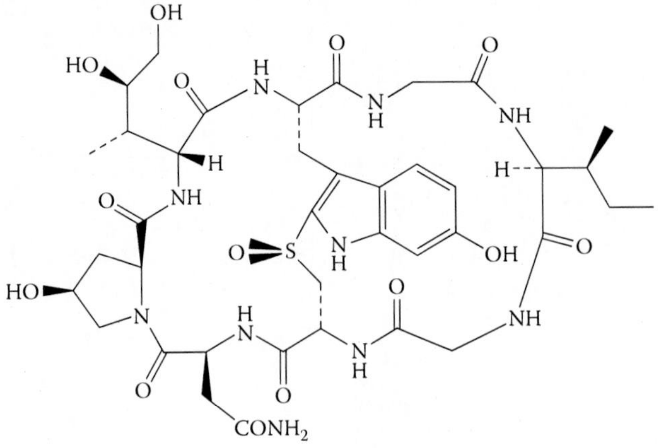

e.g. α-Amanitin
Amanita alkaloids (22)
VX7120

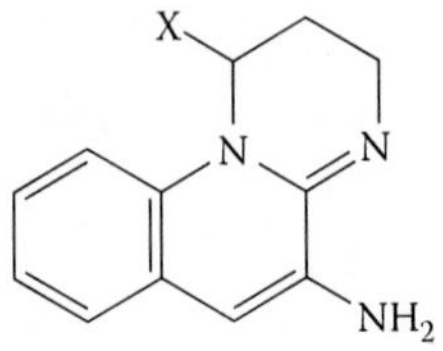

Pyoverdins (134)
VX7170

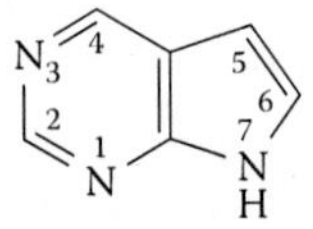

Pyrrolo[2,3-*d*]-
pyrimidines (34)
VX7200

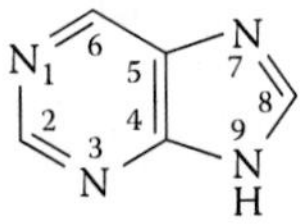

Purines (207)
VX7300

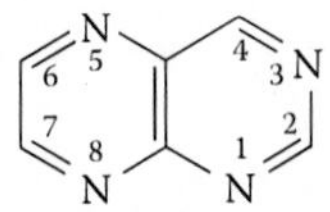

Pteridines and
analogues (128)
VX7350

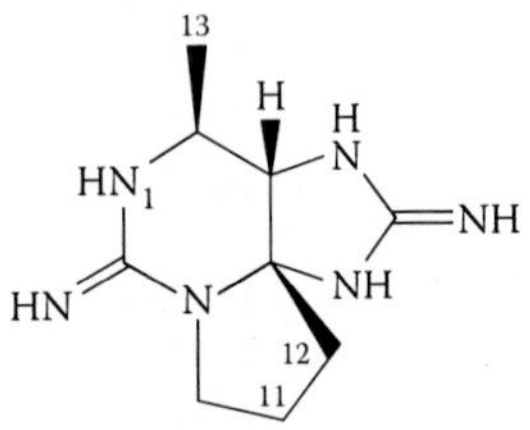

Saxitoxins (49)
VX7400

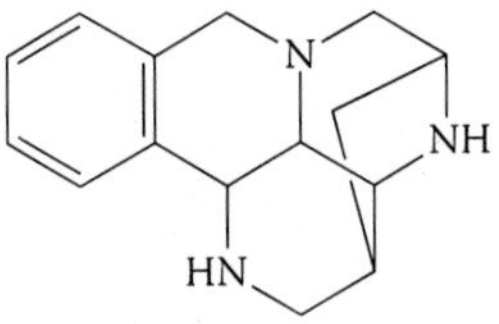

Naphthyridinomycins (14)
VX7700

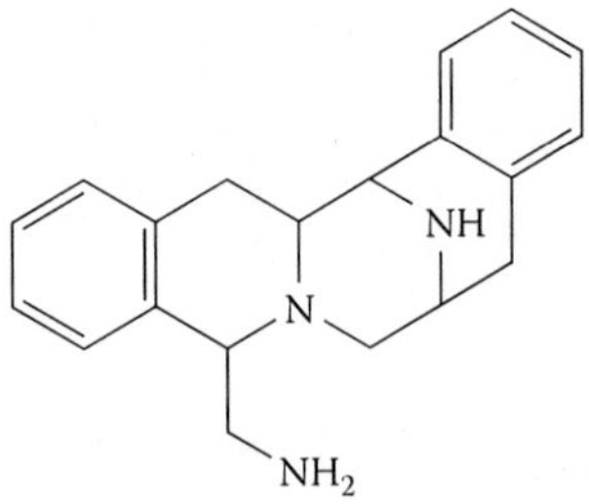

Saframycins (67)
VX7800

(*Continued*)

TABLE 6.1 (*Continued*)
Natural Product Parent Skeletons

Polypyrroles (VY)

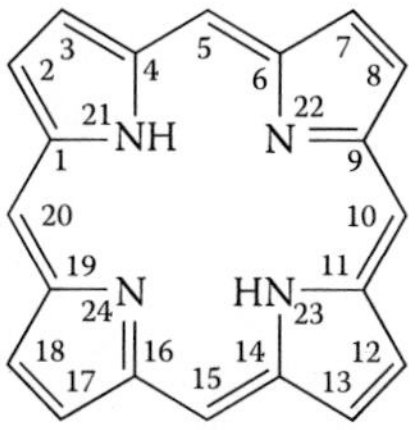

Porphyrins and
porphyrinogens (61)
VY0905

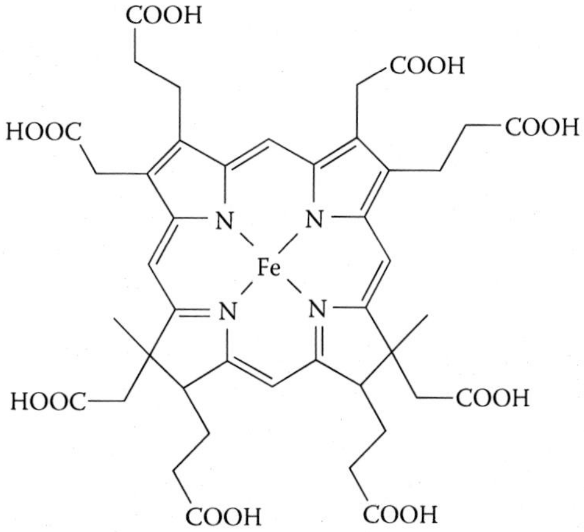

e.g. Siroheme
Haems (15)
VY0910

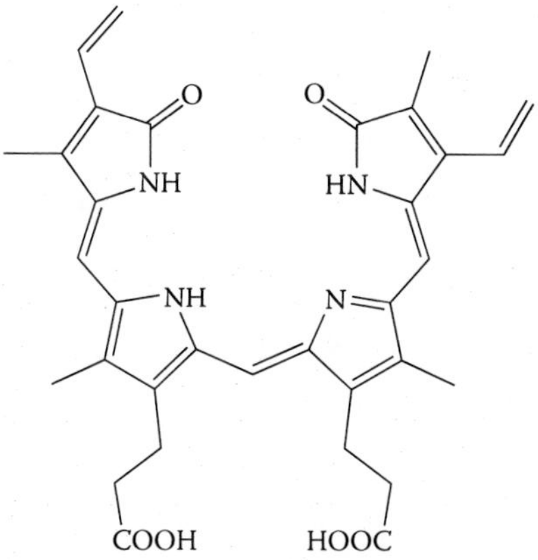

e.g. Biliverdin
Bilins (37)
VY0915

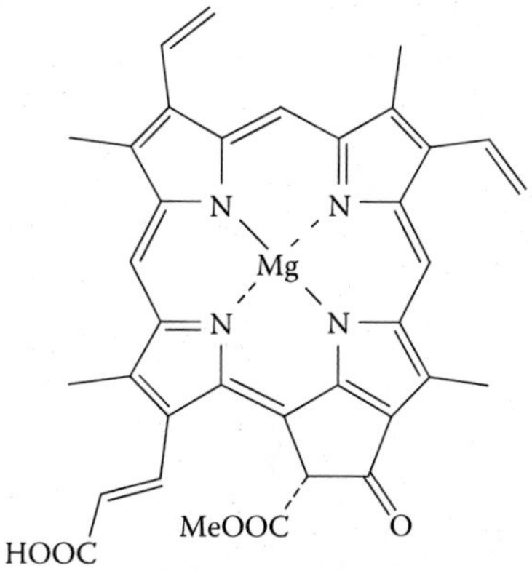

e.g. Chlorophyll c_1
Chlorophylls (114)
VY0920

(*Continued*)

TABLE 6.1 (*Continued*)
Natural Product Parent Skeletons

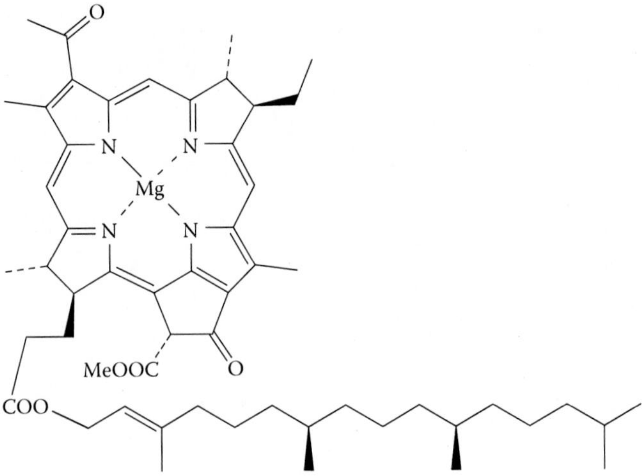

e.g. Bacteriochlorophyll a
Bacteriochlorophylls (17)
VY0925

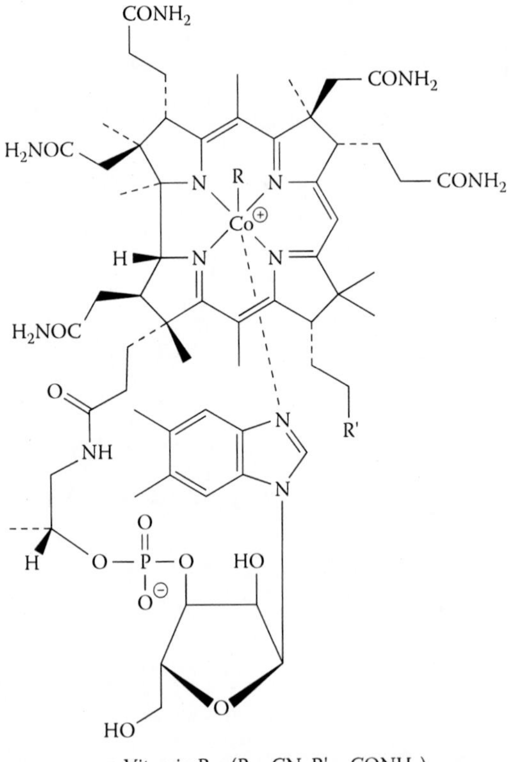

e.g. Vitamin B_{12} (R = CN, R' = $CONH_2$)
Vitamin B_{12} precursors and variants (42)
VY0930, VY0935

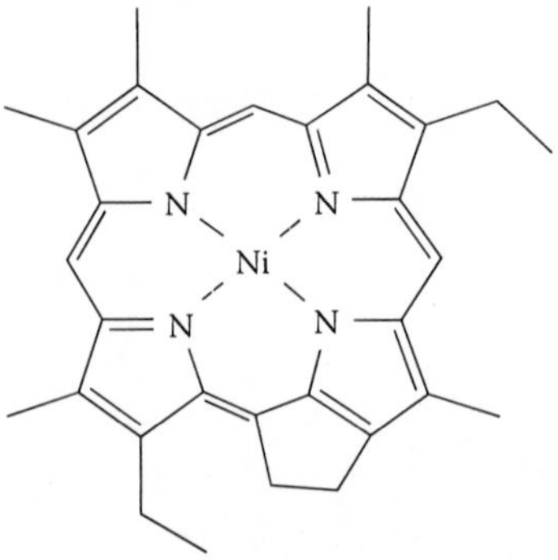

e.g. Abelsonite
Geoporphyrins (51)
VY0940

7 Structure and Nomenclature of Some Specialised Types of Natural Product

7.1 CARBOHYDRATES (VE)

Carbohydrate nomenclature impacts on stereochemistry, and on the nomenclature of compounds other than mainstream carbohydrates (e.g. hydroxylactones), often named as modified carbohydrates in CAS and elsewhere. (*Pure Appl. Chem.*, 1996, **68**, 1919–2008.)

7.1.1 FUNDAMENTAL ALDOSES

The fundamental carbohydrates are polyhydroxyaldehydes (aldoses) and -ketones (ketoses). Of these, the most important for nomenclature are the aldoses. An aldose, $HOCH_2(CHOH)_{n-2}CHO$, has $(n - 2)$ chiral centres. The stereochemical designation of a fundamental aldose is arrived at by assigning it to the D- or L-series depending on the absolute configuration of the *highest-numbered chiral centre (penultimate carbon atom)* of the chain, together with the aldose name which defines the relative configuration of all the chiral centres, thus D-Glucose. This system of stereodescription is used extensively in organic chemistry to specify the absolute configurations of compounds that can be related to carbohydrates. When applied in this general sense, the descriptors are italicised, e.g. L-*erythro*-, D-*gluco*-.

The *Dictionary of Carbohydrates*, Ed. P.M. Collins, is recommended for an overview of all the fundamental types of carbohydrate. Each compound is classified under one or more Type of Compound code (e.g. AF2700, 1,6-Anhydrosugars, 323 listed) and perusal of the printed and/or electronic versions can often resolve uncertainties of nomenclature.

Carbohydrates may be represented as *Fischer, Haworth* or *Planar (Mills)* diagrams, as well as zigzag diagrams as used for non-carbohydrates. Figure 7.1 shows how these representations are related and how to go from one to another.

In a *Fischer projection* of an open-chain carbohydrate, the chain is written vertically with carbon number 1 at the top. The OH group on the highest-numbered chiral carbon atom is depicted on the right in monosaccharides of the D-series and on the left in the L-series. To go from a Fischer projection to the correct absolute configuration, the groups attached to the *horizontal* bonds are pulled *above* the plane of the paper. Rotation of a Fischer diagram by 180° in the plane of the paper is an allowed operation which leaves the configuration unchanged. *Caution.* Rotating a Fischer projection by 90° inverts the stereochemistry. Occasionally, Fischer diagrams are drawn horizontally to save space. This should never be done!

The configuration of a group of consecutive asymmetric carbon atoms (such as >CHOH) containing one to four centres of chirality is designated by one of the following configurational prefixes (Table 7.1).

Each prefix is preceded by the D- or L- depending on the configuration of the highest-numbered chiral carbon atom in the Fischer projection of the prefix.

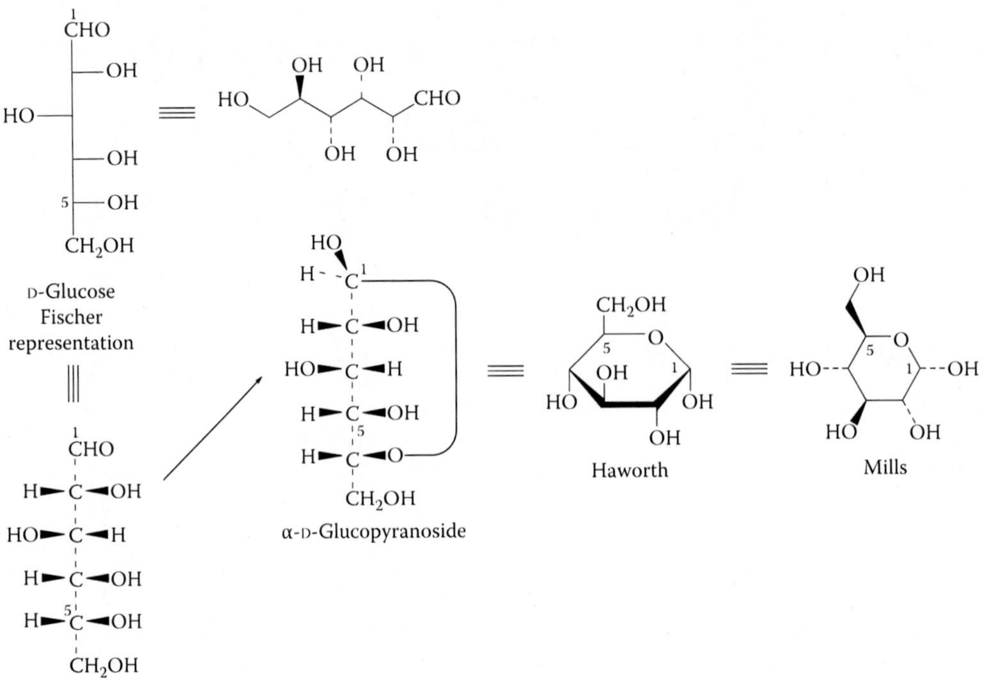

FIGURE 7.1 Fischer, Haworth and Mills diagrams and the relationship between them.

TABLE 7.1
Configurational Prefixes

No. of Carbon Atoms	Prefixes
1	*glycero-*
2	*erythro-, threo-*
3	*arabino-, lyxo-, ribo-, xylo-*
4	*allo-, altro-, galacto-, gluco-, gulo-, ido-, manno-, talo-*

The names of the aldoses and their formulae are

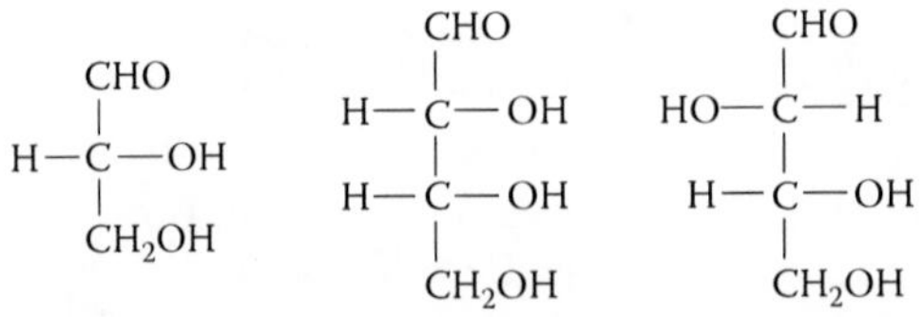

D-glycerose D-erythrose D-threose

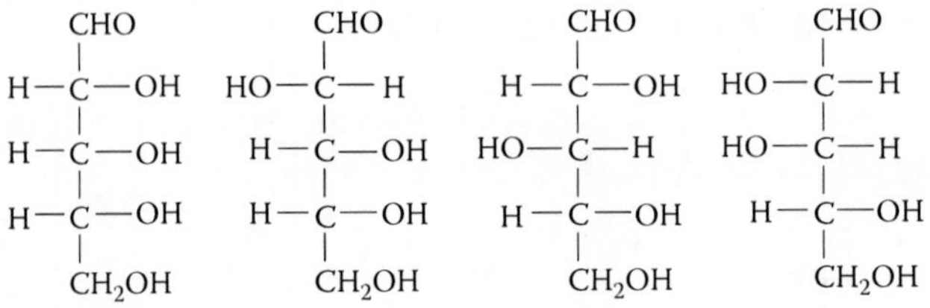

D-ribose D-arabinose D-xylose D-lyxose

```
     CHO            CHO             CHO            CHO
      |              |               |              |
 H—C—OH       HO—C—H          H—C—OH       HO—C—H
      |              |               |              |
 H—C—OH        H—C—OH        HO—C—H       HO—C—H
      |              |               |              |
 H—C—OH        H—C—OH         H—C—OH       H—C—OH
      |              |               |              |
 H—C—OH        H—C—OH         H—C—OH       H—C—OH
      |              |               |              |
   CH₂OH         CH₂OH          CH₂OH         CH₂OH
   D-allose       D-altrose      D-glucose      D-mannose
```

```
     CHO            CHO             CHO            CHO
      |              |               |              |
 H—C—OH       HO—C—H          H—C—OH       HO—C—H
      |              |               |              |
 H—C—OH        H—C—OH        HO—C—H       HO—C—H
      |              |               |              |
HO—C—H        HO—C—H        HO—C—H       HO—C—H
      |              |               |              |
 H—C—OH        H—C—OH         H—C—OH       H—C—OH
      |              |               |              |
   CH₂OH         CH₂OH          CH₂OH         CH₂OH
   D-gulose       D-idose       D-galactose     D-talose
```

Strictly, carbohydrates containing one chiral centre should have their configuration specified as D- or L-*glycero*-. In practice this is often omitted, and such compounds can often be named equally well as aliphatics.

2,3-Dideoxy-D-*glycero*-pentose ≡ (S)-4,5-Dihydroxypentanal (note different numbering)

2,3-Dideoxy-D-*glycero*-pentofuranose
= Tetrahydro-5-(hydroxymethyl)-2-furanol
(note different numbering)
or Tetrahydro-5-hydroxy-2-furanmethanol
(note different numbering)

The consecutive asymmetric carbon atoms need not be contiguous. Thus, the following four arrangements are all L-*erythro*- ('X' is attached to the lowest-numbered carbon atoms).

```
                                                   X
                                                   |
                              X           HO—C—H          X
                              |               |            |
                  X       HO—C—H          CH₂       HO—C—H
                  |           |               |            |
           HO—C—H         CH₂          CH₂        C=O
                  |           |               |            |
           HO—C—H      HO—C—H      HO—C—H     HO—C—H
                  |           |               |            |
                  Y           Y               Y            Y
```
L-*erythro*-

7.1.2 Fundamental Ketoses

The most important ketoses are the hexos-2-uloses $HOCH_2(CH_2)_{n-3}COCH_2OH$ such as Fructose. They have one less chiral centre than the aldoses of the same chain length, i.e. there are only four diastereomerically different hexos-2-uloses.

Trivial names for the 2-hexuloses and their formulae are as follows:

D-psicose   D-fructose   D-sorbose   D-tagatose

7.1.3 Modified Aldoses and Ketoses

Suffixes are employed to denote modification of functional groups in an aldose or ketose, e.g. by oxidation of an OH group (Table 7.2).

7.1.4 Higher Sugars

Sugars having more than six carbon atoms are named using two prefixes, one defining the configuration at C(2)–C(5) as in a hexose, and the other, which appears first in the name, defining the configuration at the remaining chiral centres.

Examples of the use of configurational prefixes are as follows:

D-*arabino*-3-hexulose   D-*glycero*-D-*gluco*-heptose

7.1.5 Cyclic Forms; Anomers

When a monosaccharide exists in the heterocyclic intramolecular hemiacetal form, the size of the ring is indicated by the suffixes '-furanose', '-pyranose' and '-septanose' for five-, six- and seven-membered rings, respectively.

TABLE 7.2

Suffixes Used in Carbohydrate Nomenclature

-ose	Aldose	$X=CHO, Y=CH_2OH$	
-odialdose	Dialdose	$X=Y=CHO$	
-onic acid	Aldonic acid	$X=COOH, Y=CH_2OH$	
-uronic acid	Uronic acid	$X=CHO, Y=COOH$	
-aric acid	Aldaric acid	$X=Y=COOH$	
-itol	Alditol	$X=Y=CH_2OH$	
-ulose	Ketose	$X=Y=CH_2OH$	
-osulose	Ketoaldose	$X=CHO, Y=CH_2OH$	
-ulosonic acid	Ulosonic acid	$X=COOH, Y=CH_2OH$	
-ulosuronic acid	Ulosuronic acid	$X=CHO, Y=COOH$	
-ulosaric acid	Ulosaric acid	$X=Y=COOH$	
-odiulose	Diketose		

$$\begin{array}{c} X \\ | \\ (CHOH)_x \\ | \\ Y \end{array}$$

$$\begin{array}{c} X \\ | \\ C{=}O \\ | \quad \text{(2-hexulose series)} \\ (CHOH)_2 \\ | \\ Y \end{array}$$

Two configurations known as anomers, may result from the formation of the ring. These are distinguished by the anomeric prefixes 'α-' and 'β-', which relate the configuration of the anomeric carbon atom to the configuration at a reference chiral carbon atom (normally the highest-numbered chiral carbon atom). For example, consider the glucopyranoses:

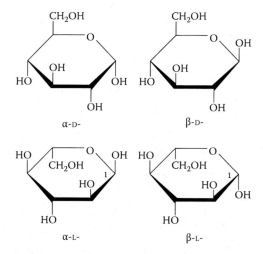

In the D-series, the CH₂OH is projected above the ring.

- In the D-series, the CH_2OH is projected above the ring.
- In the L-series, the CH_2OH is projected below the ring.
- In the α-series, the anomeric OH (at position 1) is on the opposite side of the ring to the CH_2OH group.
- In the β-series, the anomeric OH (at position 1) is on the same side of the ring as the CH_2OH group.

Suffixes used in carbohydrate nomenclature to indicate cyclic forms are as follows:

-ose (acyclic form) → -ofuranose (5-membered ring), -opyranose (6-membered ring), -oheptanose (7-membered ring).

Similar suffixes can be constructed for dicarbonyl sugars and other modifications, e.g.

-ulose	-ulopyranose
-osulose	-opyranosulose or -osulopyranose
-odialdose	-odialdopyranose

The suffixes for the acids can be modified to indicate the corresponding amide, nitriles, acid halides, etc., e.g. '-uronamide', '-ononitrile', '-ulosonyl chloride'.

7.1.6 GLYCOSIDES

These are mixed acetals resulting from the replacement of the hydrogen atom on the anomeric (glycosidic) OH of the cyclic form of a sugar by a radical R derived from an alcohol or phenol (ROH). They are named by changing the terminal '-e' of the name of the corresponding cyclic form of the saccharide by '-ide'; the name of the R radical is put at the front of the name followed by a space.

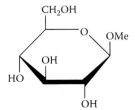

Methyl β-D-glucopyranoside

7.1.7 DISACCHARIDES AND OLIGOSACCHARIDES

These are sugars produced where the alcohol forming the glycoside of a sugar is another sugar. Where the resulting sugar has a (potentially) free aldehyde function, it is called a reducing disaccharide, and where both aldehyde functions are involved in the linkage (1→1) glycoside, it is a non-reducing disaccharide.

Maltose (4-O-α-D-glucopyranosyl-
D-glucose), a reducing disaccharide

α-D-galactopyranosyl α-D-galactopyranoside,
a non-reducing disaccharide

The *Dictionary of Natural Products* style for Maltose is α-D-Glucopyranosyl-(1→4)-D-glucose, which is easier to extend to tri- and higher saccharides, and can also cope with complex natural

TABLE 7.3
Abbreviations for Use in Representing Oligosaccharides

Hexoses	All	Allose
	Alt	Altrose
	Gal	Galactose
	Glc	Glucose
	Gul	Gulose
	Ido	Idose
	Man	Mannose
	Tal	Talose
Pentoses	Ara	Arabinose
	Lyx	Lyxose
	Rib	Ribose
	Xyl	Xylose
Other	Rha	Rhamnose
	Fuc	Fucose
	Fru	Fructose
Suffixes	*f*	Furanose
	p	Pyranose
	A	Uronic acid
	N	2-Deoxy-2-amino sugar
Prefixes	D-	Configurational descriptor
	L-	Configurational descriptor
	An	Anhydro

glycosides in which saccharides alternate in the chain with non-saccharide moieties. Chain branching is shown by nesting brackets, e.g. β-D-Glucopyranosyl-(1→2)-[β-D-glucopyranosyl-(1→4)]-D-glucose.

Abbreviations for use in representing oligosaccharides are shown in Table 7.3. (*See Pure Appl. Chem.*, 1982, **54**, 1517.)

Examples are

Ara*f*	Arabinofuranose
Glc*p*	Glucopyranose
Gal*p*A	Galactopyranuronic acid
D-Glc*p*N	2-Amino-2-deoxy-D-glucopyranose
3,6-AnGal	3,6-Anhydrogalactose

7.1.8 TRIVIALLY NAMED SUGARS

A number of names for modified sugars, which occur frequently in natural glycosides, are in common use.

The following are allowed descriptors in *Dictionary of Natural Products* (DNP):

Cymarose: 2,6-Dideoxy-3-*O*-methyl-*ribo*-hexose
Diginose: 2,6-Dideoxy-3-*O*-methyl-*lyxo*-hexose
Digitoxose: 2,6-Dideoxy-*ribo*-hexose
Fucose: 6-Deoxygalactose
Rhamnose: 6-Deoxymannose
Olandrose: 2,6-Dideoxy-3-*O*-methyl-*arabino*-hexose
Thevetose: 6-Deoxy-3-*O*-methylglucose

The following are also of common occurrence but are named systematically in DNP:

Allomethylose: 6-Deoxyallose
Quinovose: 6-Deoxyglucose

In addition, all disaccharides are named systematically in DNP, for example Rutinose $= \alpha$-L-Rhamnopyranosyl-$(1 \rightarrow 6)$-D-glucose.

Note that if the absolute configuration of a sugar is not clear from the literature, CAS makes certain assumptions, e.g. rhamnose is assumed to be L-.

7.2   ALDITOLS AND CYCLITOLS

7.2.1   ALDITOLS

Reduction of the carbonyl group of an aldose (or of the oxo group in a ketose) gives the series of alditols (called tetritols, pentitols, hexitols, etc. with 4, 5, 6, ... carbon atoms).

Because of their higher symmetry compared to the aldoses, the number of possible isomers is lower and some isomers are *meso*-forms or, in the C_7 series, some isomers show pseudoasymmetry.

The alditols derived from the C_4, C_5 and C_6 monosaccharides in the D-series. Degenerate symmetry means that there are only three pentitols and six hexitols.

Some isomers can, therefore, be named in more than one way. A choice is made according to a special carbohydrate rule which says that allocation to the D-series take precedence over alphabetical assignment to the parent carbohydrate diastereoiomer.

7.2.2 CYCLITOLS

The most important cyclitols are the inositols (1,2,3,4,5,6-cyclohexanehexols). The relative arrangement of the six hydroxyl groups below or above the plane of the cyclohexane ring is denoted by an italicised configurational prefix in the eight inositol stereoparents (the numerical locants indicate OH groups which are on the same side of the ring) (Hudlicky, T. and Cebulak, M., *Cyclitols and Their Derivatives*, VCH, New York, 1993; Posternak, T., *The Cyclitols*, Holden-Day, San Francisco, CA, 1965.)

cis-Inositol	(1,2,3,4,5,6)
epi-Inositol	(1,2,3,4,5)
allo-Inositol	(1,2,3,4)
myo-Inositol	(1,2,3,5)
muco-Inositol	(1,2,4,5)
neo-Inositol	(1,2,3)
chiro-Inositol	(1,2,4)
scyllo-Inositol	(1,3,5)

Six of these isomers (*scyllo*-, *myo*-, *epi*-, *neo*-, *cis*- and *muco*-) have one or more planes of symmetry and are *meso*-compounds; *chiro*-inositol lacks a plane of symmetry and exists as the D- and L-forms. In *myo*-inositol, the plane of symmetry is C-2/C-5; unsymmetrically substituted derivatives on C-1, C-3, C-4, C-6 are chiral. Substitution at C-2 and/or C-5 gives a *meso*-product. *allo*-Inositol appears to have a plane of symmetry, but, at room temperature, it is actually a racemate formed of two enantiomeric conformers in rapid equilibrium.

7.2.2.1 Assignment of Locants for Inositols

From the IUPAC–IUB 1973 Recommendations for the Nomenclature of Cyclitols; *Biochem. J.*, 1976, **153**, 23–31; based upon proposals first issued in 1967; *Biochem. J.*, 1969, **112**, 17–28.

1. The lowest locants are assigned to the set (above or below the plane) which has the most OH groups.
2. For *meso*-inositols only, the C-1 locant is assigned to the (prochiral) carbon atom which has the L-configuration.

7.2.2.2 Absolute Configuration

Using a horizontal projection of the inositol ring, if the substituent on the lowest numbered asymmetric carbon is above the plane of the ring and the numbering is counterclockwise, the configuration is assigned D-, and if clockwise, the configuration is L- (illustrated in Figure 7.2 for *myo*-inositol 1-phosphate enantiomers).

Note that 1D-*myo*-inositol 1-phosphate is the same as 1L-*myo*-inositol 3-phosphate (and 1L-*myo*-inositol 1-phosphate is the same as 1D-*myo*-inositol 3-phosphate), but the lower locant has precedence over the stereochemical prefix (D- or L-) in naming the derivative. A consequence of applying the 1973 IUPAC–IUB rules to *myo*-inositol is that the numbering of C-2 and C-5 remains invariant.

Before 1968, the nomenclature for inositols assigned the symbols D- and L- to the *highest*-numbered chiral centre, C-6. This convention was based on the rules for naming carbohydrates. For substituted *myo*-inositols, in particular, where C-1 and C-6 hydroxyl groups are *trans*, compounds identified in the literature before 1968 as D- are now assigned as 1L-.

Locants for unsubstituted inositols other than *myo*-inositol are shown in Figure 7.3.

In order to clarify the metabolic pathways for substituted *myo*-inositols (in practice *myo*-inositol phosphates), the lowest-locant rule, which gives priority to a 1L-locant has been relaxed, and

OPO₃H₂

1ᴅ-*myo*-Inositol 1-(dihydrogen phosphate)

H₂O₃PO

1ʟ-*myo*-Inositol 1-(dihydrogen phosphate)

FIGURE 7.2   *myo*-inositol 1-(dihydrogen phosphate) enantiomers.

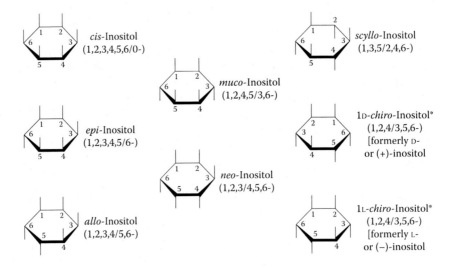

cis-Inositol
(1,2,3,4,5,6/0-)

scyllo-Inositol
(1,3,5/2,4,6-)

muco-Inositol
(1,2,4,5/3,6-)

epi-Inositol
(1,2,3,4,5/6-)

1ᴅ-*chiro*-Inositol*
(1,2,4/3,5,6-)
[formerly ᴅ-
or (+)-inositol

neo-Inositol
(1,2,3/4,5,6-)

allo-Inositol
(1,2,3,4/5,6-)

1ʟ-*chiro*-Inositol*
(1,2,4/3,5,6-)
[formerly ʟ-
or (−)-inositol

FIGURE 7.3   Unsubstituted inositols.

numbering based on the 1ᴅ-series is now allowed. (*Biochem. J.*, 1989, **258**, 1–2.) This is to allow substances related by simple chemical or biochemical transformations to carry the same labels. Thus 1ʟ-*myo*-inositol 1-phosphate may now be called 1ᴅ-*myo*-inositol-3-phosphate.

In a further simplification, the symbol Ins may be used to denote *myo*-inositol, with the numbering of the 1ᴅ-configuration implied (unless the prefix ʟ is explicitly added).

7.3 NUCLEOTIDES AND NUCLEOSIDES (VE)

Nucleic acids, e.g. DNA, are ladder polymers of *nucleotides*, the heterocyclic bases of which (purines and pyrimidines), encode the genetic information. The nucleosides are linked together through the carbohydrate residues of the nucleotides to form the polymer.

Nucleotides consist of *nucleosides* O-phosphorylated in the sugar residues. The nucleosides, e.g. Thymidine, consist of a heterocyclic base (in this case Thymine) attached to a carbohydrate residue as an *N*-glycoside. The most common bases found in nucleosides are shown in Table 7.4 and the commonest nucleosides in Table 7.5.

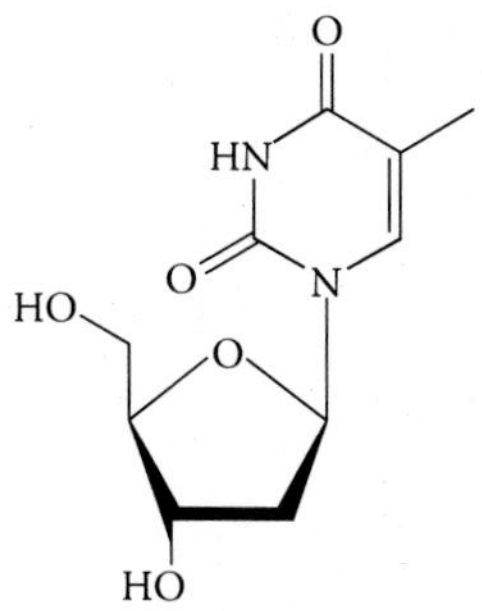

Thymidine

TABLE 7.4
Common Nucleoside Bases and Their Abbreviated Forms

Ade	Adenine
Cyt	Cytosine
Gua	Guanine
Hyp	Hypoxanthine
Oro	Orotate
Pur	Unknown purine
Pyr	Unknown pyrimidine
Shy	Thiohypoxanthine
Sur	Thiouracil
Thy	Thymine
Ura	Uracil
Xan	Xanthine

TABLE 7.5
Abbreviations for Nucleosides

Ado	A	Adenosine
BrUrd	B	5-Bromouridine
Cyd	C	Cytidine
	D or hU	5,6-Dihydrouridine
Guo	G	Guanosine
Ino	I	Inosine
Nuc	N	Unspecified nucleoside
Oro	O	Orotidine
Puo	R	Unspecified purine nucleoside
Pyd	Y	Unspecified pyrimidine nucleoside
Ψrd	ψ or Q	Pseudouridine
Sno	M or sI	Thiouridine
Srd	S or sU	6-Thioinosine
Thd	T	Ribosylthymine (not thymidine)
Urd	U	Uridine
Xao	X	Xanthosine
D		2-Deoxy
dThd	dT	Thymidine
Nir		Ribosylnicotinamide
-P		Phosphoric residue

7.4 AMINO ACIDS AND PEPTIDES (VV)

7.4.1 AMINO ACIDS

In α-amino acids, the L-compounds are those in which the NH_2 group is on the left-hand side of the Fischer projection in which the COOH group appears at the top. (See *Pure Appl. Chem.*, 1984, **56**, 595.)

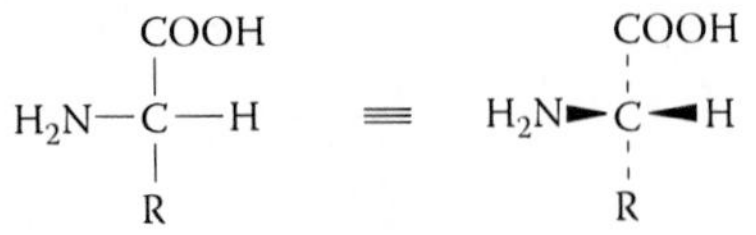

Table 7.6 lists the common α-amino acids. Most of these are found in proteins. Those marked * are nonproteinaceous but common in peptides and are also used as stem names in CAS. The list is in order of nomenclatural priority according to current CAS practice (2006), e.g. CAS name alanylarginine not argininylalanine. The order pre-2006 was different.

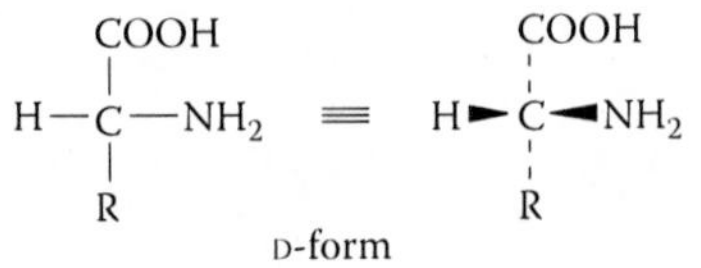

For all the amino acids in the table, except for cysteine, the L-form has the *S*-configuration. For cysteine, the L-form has the *R*-configuration, because the $-CH_2SH$ group has higher priority than $-COOH$ according to the Sequence Rule.

Other one-letter abbreviations are as follows:

B: asparagine or aspartic acid
X: unspecified amino acid
Z: glutamine or glutamic acid

Other abbreviations that may be encountered in the literature include those listed in Table 7.7.

7.4.2 PEPTIDES

Peptides are oligomers notionally derived from amino acids by condensation to produce amide linkages. They are named either systematically or trivially.

The *primary structure* of a peptide is the amino acid sequence. The *secondary structure* is that resulting from modifying bonds, especially hydrogen bonds and cysteine/cysteine dimerisation, and the *tertiary structure* is the three-dimensional organisation resulting from the folding of a peptide chain into helices, sheets, etc. The *quaternary structure* of a protein comprises the macrostructure formed by the coming together of two or more smaller subunits. Peptide chains are numbered from the *N*-terminal end as shown below.

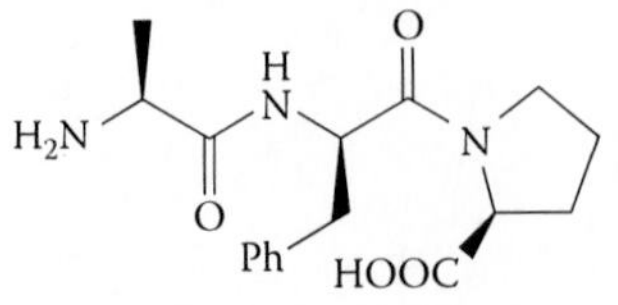

Alanyl-D-phenylalanylproline

TABLE 7.6

α-Amino Acids Listed in the Order of Precedence in CAS Nomenclature (2006 Revision)

Name	Abbreviations		R Group (Side-Chain)	Molecular Formula
Glutamic acid	Glu	E	$-CH_2CH_2COOH$	$C_5H_9NO_4$
Aspartic acid	Asp	D	$-CH_2COOH$	$C_4H_7NO_4$
Tryptophan	Trp	W		$C_{11}H_{12}N_2O_2$
Histidine	His	H		$C_6H_9N_3O_2$
Proline	Pro	P		$C_5H_9NO_2$
Tyrosine	Tyr	Y		$C_9H_{11}NO_3$
Phenylalanine	Phe	F	$-CH_2Ph$	$C_9H_{11}NO_2$
Lysine	Lys	K	$-(CH_2)_3CH_2NH_2$	$C_6H_{14}N_2O_2$
Norleucine*	Nle		$-(CH_2)_3CH_3$	$C_6H_{13}NO_2$
Glutamine	Gln (or Glu(NH$_2$))	Q	$-CH_2CH_2CONH_2$	$C_5H_{10}N_2O_3$
Arginine	Arg	R	$-(CH_2)_3NHC(NH_2)=NH$	$C_6H_{14}N_4O_2$
Ornithine*	Orn		$-(CH_2)_3NH_2$	$C_5H_{12}N_2O_2$
Isoleucine	Ile	I	$-CH(CH_3)CH_2CH_3$ ($R*,R*-$)	$C_6H_{13}NO_2$
Alloisoleucine	aIle		$-CH(CH_3)CH_2CH_3$ ($R*,S*-$)	$C_6H_{13}NO_2$
Leucine	Leu	L	$-CH_2CH(CH_3)_2$	$C_6H_{13}NO_2$
Norvaline*	Nva (or Avl)		$-CH_2CH_2CH_3$	$C_5H_{11}NO_2$
Asparagines	Asn		$-CH_2CONH_2$	$C_4H_8N_2O_2$
Threonine	Thr	T (also stands for thymine)	$-CH(OH)CH_3$ ($R*,S*-$)	$C_4H_9NO_3$
Allothreonine	aThr		$-CH(OH)CH_3$ ($R*,R*-$)	$C_4H_9NO_3$
Homoserine*	Hse		$-CH_2CH_2OH$	$C_4H_9NO_3$
Methionine	Met	M	$-CH_2CH_2SMe$	$C_5H_{11}NO_2S$
Homocysteine	Hcy		$-CH_2CH_2SH$	$C_4H_9NO_2S$
Valine	Val	V	$-CH(CH_3)_2$	$C_5H_{11}NO_2$
Isovaline	Iva		$-CH_2CH_3 + \alpha CH_3$	$C_5H_{11}NO_2$
Serine	Ser	S	$-CH_2OH$	$C_3H_7NO_3$
Cystine				$C_6H_{12}N_2O_4S_2$
Cysteine	Cys	C (also stands for cytosine)	$-CH_2SH$	$C_3H_7NO_2S$
Alanine	Ala	A (also stands for adenine)	$-CH_3$	$C_3H_7NO_2$
β-Alanine*	βAla		$-CH_2NH_2$ (replaces α-NH$_2$)	$C_3H_7NO_2$
Glycine	Gly	G (also stands for guanine)	$-H$	$C_2H_5NO_2$

Notes: Non-protein amino acids are marked *. Three-letter codes in brackets are non-recommended.

TABLE 7.7

Other Amino Acid–Related Abbreviations Found in the Literature

βAad	3-Aminoadipic acid	Hsl	Homoserine Lactone
Aad	2-Aminoadipic acid	Hyl	5-Hydroxylysine
A2bu	2,4-Diaminobutanoic acid	5Hyl	5-Hydroxylysine
Abu	2-Aminobutanoic acid	Hyp	4-Hydroxyproline
εAhx	6-Aminohexanoic acid	4Hyp	4-Hydroxyproline
Ahx	2-Aminohexanoic acid (norleucine)	J (one-letter code)	Indistinguishable leucine or isoleucine
2-MeAla	2-Methylalanine	MetO	Methionine *S*-oxide
Ape	2-Aminopentanoic acid (norvaline)	MetO$_2$	Methionine *S,S*-dioxide
Apm	2-Aminopimelic acid	Mur	Muramic acid
A2pr	2,3-Diaminopropionic acid	Neu	Neuraminic acid
Asp(NH$_2$)	Asparagine	Neu5Ac	*N*-acetylneuraminic acid
Asx	Asparagine or aspartic acid	5-oxoPro	5-Oxoproline (pyroglutamic acid)
Dpm	2,6-Diaminopimelic acid	Sar	Sarcosine
Dpr	2,3-Diaminopropionic acid	Sec or U	Selenocysteine (one-letter code U; also stands for uridine)
Gla	4-Carboxyglutamic acid	Ser(P)	Phosphoserine
Glp or pGlu or <Glu	5-Oxoproline (pyroglutamic acid)	Thx	Thyroxine
Glx	Glutamine or glutamic acid	Tyr(I$_2$)	3,5-Diiodotyrosine
Glx	Glutamine or glutamic acid	Tyr(SO$_3$H)	O^4-sulfotyrosine
		Xaa	Unspecified amino acid

Trivial names of peptides can be modified in the following ways to denote a change in the amino acid sequence.

Replacement. When a peptide with a trivial name has an amino acid replaced by another amino acid, the modified peptide can be named as a derivative of the parent peptide by citing the new amino acid as a replacement. The new amino acid is designated by the appropriate amino acid residue number.

H-^{1}Arg-^{2}Pro-^{3}Pro-^{4}Gly-^{5}Phe-^{6}Ser-^{7}Pro-^{8}Phe-^{9}Arg-OH = **bradykinin**
H-Arg-Pro-Pro-Gly-Phe-Ser-Phe-Phe-Arg-OH = **7-L-phenylalaninebradykinin**

Extension. Extension of a trivially named peptide at the N-terminal end is denoted by substitutive nomenclature. Extension at the C-terminal end is made by citing the new amino acid residues with locants derived by suffixing the highest locant with a, b, etc. Extension in the middle of the chain is denoted by use of the term *endo-*.

H-Lys-Arg-Pro-Pro-Gly-Phe-Ser-Pro-Phe-Arg-OH = N^2-**L-lysylbradykinin**
H-Arg-Pro-Pro-Gly-Phe-Ser-Pro-Phe-Arg-Arg-OH = **9a-L-argininebradykinin**
H-Arg-Pro-Pro-Gly-Phe-Ser-Ala-Pro-Phe-Arg-OH = **6a-*endo*-L-alaninebradykinin**

Removal. Removal of an amino acid residue is denoted using 'de-'.

H-Pro-Pro-Gly-Phe-Ser-Pro-Phe-Arg-OH = **1-de-L-argininebradykinin**

Various numbering methods have been used to indicate substitution or other modification in or of the residues of a peptide. The method now recommended by IUPAC and introduced into the *Dictionary of Natural Products* uses numerical locants of the type 3^2, where 3 is the locant of the substituent in the amino acid residue, and 2 is the amino acid position in the ring or chain, numbered from the N-terminal end, which is standard for all peptides. $N^{5.4}$-methyloxytocin would indicate a methyl substituent on N-5 of the glutamine residue at position 4 of oxytocin.

7.4.3 CYCLOPEPTIDES

IUPAC Nomenclature of Cyclic Peptides Recommendations 2004. IUPAC recommends the term anhydro- for indicating cyclic peptide formation, but the term cyclo- is more common. A cyclopeptide could be named and numbered starting at any position on the ring, but the choice is made using alphabetical priority. *Cyclo-* should be italicised and round brackets used.

cyclo-(^{1}Ala-^{2}Ala-^{3}Lys-^{4}Ala-^{5}His) = cyclo-(Alanylalanyllysylalanylhistidine)

See the IUPAC recommendations for nomenclature of heterodetic cyclopeptides (those containing bonds other than α- –CONH-), Also see Spengler, J. et al., *J. Peptide Res.*, 2005, **65**, 550–555 for extension/improvement of linear representations to facilitate error-free searching of linear representations.

Specification of locants in peptide side chains. Various systems have been used to number aminoacid side chains in peptides. IUPAC now recommends the use of superscripts to indicate the residue, e.g. the NH_2 group in the lysine residue of the above peptide occupies position 6^3.

7.4.4 RECENT CAS PEPTIDE NOMENCLATURE REVISIONS

For the 14th Collective index period, several revisions in CAS peptide nomenclature have been made, and further changes were made as part of the 2006 revisions.

- No structure is assigned a peptide name unless it contains at least two standard amino acid residues. A standard amino acid residue is any amino acid that can stand alone as a stereoparent (e.g. glycine or tryptophan), plus asparagine and glutamine. 2-Aminobutanoic acid and 2,4-diaminobutanoic acid are now considered nonstandard amino acids.
- The 2006 changes to CAS nomenclature have reduced the number of peptide parent names from about 3000 to fewer than 100 of the most studied examples, e.g. bradykinin. All others are now named systematically.
- The nomenclature of linear peptides has been simplified. The C-terminal residue is the index heading parent, and the other residues are cited in the substituent, beginning with the N-terminal residue and continuing from left to right in the sequence, e.g. L-Lysine, D-alanylglycyl-L-leucyl-. This replaces the many locants and enclosing marks that have been used in the past, e.g. L-Lysine, N^2-[N-(N-D-alanylglycyl)-L-leucyl]-.
- *S*-oxides of sulfur-containing amino acids are now named at the peptide e.g.

$$MeS(O)CH_2CH_2CH(NH_2)COOH$$

Butanoic acid, 2-amino-4-(methylsulfinyl), 9CI → Methionine *S*-oxide

7.5 ENZYMES (VV)

Enzymes are catalysts that change the rate of a reaction without undergoing change themselves. They are highly specific and their activity can be regulated. Most enzymes are proteins although some catalytically active RNAs are known.

The *active site* is the region of the enzyme that binds the substrate to form an enzyme–substrate complex and changes it into product. It is a 3D structure, often a cleft or a crevice on the surface of the protein in which the substrate is bound by multiple weak interactions. Two models are used to explain how the enzyme binds to its substrate: the 'lock and key' model and the 'induced fit' model. The substrate specificity of an enzyme is determined by the properties and spatial arrangement of the amino acid residues forming the active site.

Many enzymes need the presence of small non-protein units or *cofactors* to carry out their particular reaction. Cofactors may be either inorganic ions (e.g. Zn^{2+} or Fe^{2+}) or a complex organic molecule called a *coenzyme*. A metal or coenzyme that is covalently attached to the enzyme is called a *prosthetic group* (e.g. Heme in Haemoglobin). A complete catalytically active enzyme with its coenzyme or metal ion is called a *holoenzyme*. The protein part of the enzyme without the cofactor is called an *apoenzyme*. Nicotinamide adenine dinucleotide (NAD^+) and Nicotinamide adenine dinucleotide phosphate ($NADP^+$) are coenzymes having a common function acting as carriers of electrons and are involved in oxidation–reduction reactions. *Isoenzymes* (isozymes) are different forms of an enzyme, which catalyse the same reaction but exhibit different physical or kinetic properties. Different isoenzyme forms of a given enzyme are usually derived from different genes and mostly occur in different tissues of the body.

An enzyme is readily assayed by measuring the rate of appearance of product or the rate of disappearance of substrate. If they absorb light at a specific wavelength, then changes in the concentration of these molecules is measured by following the change of absorbance at this wavelength. Two common molecules used in such studies are the coenzymes NADH and NADPH which each absorb in the UV region at 340 nm.

Most enzymes are named by adding the suffix '-ase' to the name of their substrates. For example, urease catalyses hydrolysis of urea and fructose-1,6-bisphosphatase hydrolyses fructose-1,6-bisphosphate. Certain enzymes like trypsin and chymotrypsin have names that do not denote their substrates. Some have several alternative names.

A system of enzyme nomenclature has been created according to the classification of the nomenclature committee of the International Union of Biochemistry and Molecular Biology (IUBMB). All enzymes are placed into one of six major classes based on the type of reactions they catalyse.

1. **Oxidoreductases**: Catalyse reactions involving transfer of electrons. Typical enzyme name: dehydrogenase, oxidase.
2. **Transferases**: Catalyse reactions involving transfer of functional groups such as methyl, acyl, amino or a phosphate group. Typical enzyme name: transferase, kinase.
3. **Hydrolases**: Catalyse hydrolysis reactions. Typical enzyme name: lipase, amylase, peptidase.
4. **Lyases**: Catalyse reactions involving non-hydrolytic cleavage of C–C, C–O, C–N or C–S bonds, mostly forming a double bond. Typical enzyme name: decarboxylase, lyase.
5. **Isomerases**: Catalyse reactions with transfer of groups within a molecule. Typical enzyme name: isomerase, mutase.
6. **Ligases** or **Synthases**: Catalyse reactions with new C–O, C–S, C–N or C–C bond formation by breaking down ATP. Typical enzyme name: synthetase.

Each enzyme has a unique four digit classification number. The first digit denotes class of the enzyme, the second and third digits denote sub-class and sub sub-class. The fourth digit denotes the serial number of the enzyme. Trypsin has the enzyme commission (EC) number 3.4.21.4, where the first number (3) denotes a hydrolase; the second number (4) shows a protease that hydrolyses peptide bonds; the third number (21) denotes a serine protease with a critical serine residue at the active site; and the fourth number (4) indicates that it was the fourth enzyme assigned to this class.

Table 7.8 gives the recommendations of the Nomenclature Committee of the IUBMB on the classification of enzymes showing the first and second sets of numbers with the appropriate subclass. It is available in a web version prepared by G.P. Moss.

TABLE 7.8

Classification of Enzymes

Subclass	Name
EC 1	*Oxidoreductases*
EC 1.1	Acting on the CH–OH group of donors
EC 1.2	Acting on the aldehyde or oxo group of donors
EC 1.3	Acting on the CH–CH group of donors
EC 1.4	Acting on the CH–NH$_2$ group of donors
EC 1.5	Acting on the CH–NH group of donors
EC 1.6	Acting on NADH or NADPH
EC 1.7	Acting on other nitrogenous compounds as donors
EC 1.8	Acting on a sulphur group of donors
EC 1.9	Acting on a heme group of donors
EC 1.10	Acting on diphenols and related substances as donors
EC 1.11	Acting on a peroxide as acceptor
EC 1.12	Acting on hydrogen as donor
EC 1.13	Acting on single donors with incorporation of molecular oxygen (oxygenases)
EC 1.14	Acting on paired donors with incorporation or reduction of molecular oxygen
EC 1.15	Acting on superoxide radicals as acceptor
EC 1.16	Oxidising metal ions
EC 1.17	Acting on CH or CH$_2$ groups
EC 1.18	Acting on iron–sulphur proteins as donors
EC 1.19	Acting on reduced flavodoxin as donor
EC 1.20	Acting on phosphorus or arsenic donors
EC 1.21	Acting on X–H and Y–H to form an X–Y bond
EC 1.97	Other oxidoreductases
EC 2	*Transferases*
EC 2.1	Transferring one-carbon groups
EC 2.2	Transferring aldehyde or ketonic groups
EC 2.3	Acyltransferases
EC 2.4	Glycosyltransferases
EC 2.5	Transferring alkyl or aryl groups, other than methyl groups
EC 2.6	Transferring nitrogenous groups
EC 2.7	Transferring phosphorus-containing groups
EC 2.8	Transferring sulphur-containing groups
EC 2.9	Transferring selenium-containing groups
EC 3	*Hydrolases*
EC 3.1	Acting on ester bonds
EC 3.2	Glycosylases
EC 3.3	Acting on ether bonds
EC 3.4	Acting on peptide bonds (peptidases)
EC 3.5	Acting on carbon–nitrogen bonds, other than peptide bonds
EC 3.6	Acting on acid anhydrides
EC 3.7	Acting on carbon–carbon bonds
EC 3.8	Acting on halide bonds
EC 3.9	Acting on phosphorus–nitrogen bonds
EC 3.10	Acting on sulphur–nitrogen bonds
EC 3.11	Acting on carbon–phosphorus bonds
EC 3.12	Acting on sulphur–sulphur bonds
EC 3.13	Acting on carbon–sulphur bonds

(Continued)

TABLE 7.8 (*Continued*)
Classification of Enzymes

Subclass	Name
EC 4	*Lyases*
EC 4.1	Carbon–carbon lyases
EC 4.2	Carbon–oxygen lyases
EC 4.3	Carbon–nitrogen lyases
EC 4.4	Carbon–sulfur lyases
EC 4.5	Carbon–halide lyases
EC 4.6	Phosphorus–oxygen lyases
EC 4.99	Other lyases
EC 5	*Isomerases*
EC 5.1	Racemases and epimerases
EC 5.2	*cis-trans*-Isomerases
EC 5.3	Intramolecular isomerases
EC 5.4	Intramolecular transferases (mutases)
EC 5.5	Intramolecular lyases
EC 5.99	Other isomerases
EC 6	*Ligases*
EC 6.1	Forming carbon–oxygen bonds
EC 6.2	Forming carbon–sulphur bonds
EC 6.3	Forming carbon–nitrogen bonds
EC 6.4	Forming carbon–carbon bonds
EC 6.5	Forming phosphoric ester bonds
EC 6.6	Forming nitrogen–metal bonds

FURTHER INFORMATION

ELECTRONIC DATABASES

The most useful is the electronic database **BRENDA** (**BR**aunschweig **EN**zyme **DA**tabase) founded in 1987 and reviewed by Schomberg, I. et al., *Trends Biochem. Sci.*, 2002, **27**, 54–56.

Other useful electronic databases are **EXPASY** (**EX**pert **P**rotein **A**nalysis **SY**stem) and **KEGG** (**K**yoto **E**ncyclopedia of **G**enes and **G**enomes).

PRINT SOURCES

Methods in Enzymology, Vol. 1 (1955)—Vol. 545 (2014), Academic Press/Elsevier. Online version http://www.sciencedirect.com/science/bookseries/00766879

The Enzyme Reference: Purich, D.L. and Allison, R.D., *A Comprehensive Guide to Enzyme Nomenclature, Reactions and Methods*, Academic Press, Waltham, MA, 2002.

8 Chemical Hazard Information for Natural Products

8.1 INTRODUCTION

The general precautions and advice for handling chemicals in a laboratory environment are also relevant to natural product research. In addition, two particular groups of hazard need to be recognised during the isolation, manipulation and biological testing of natural products: hazards resulting from the acute toxicological and pharmacological properties of some classes of natural products and hazards associated with the use of organic solvents. Toxins exemplify the hazards of working with natural products of high acute oral toxicity, and hazard information for some selected toxins is tabulated in Section 8.4. The fire and health hazards associated with handling organic solvents used to extract natural products are summarised in Section 8.5.

The safety and assessment procedures required for handling chemicals in the laboratory (and in the workplace, in general) apply to natural product research:

- Identify the risks of handling hazardous substances and inform employees.
- Prevent, minimise or control exposure.
- Ensure that control measures are correctly used and maintained and that personal protection equipment is available.
- Monitor exposure in the workplace and comply with national occupational exposure limits.
- Provide information, training and instruction of the risks involved.
- Keep records of risk assessments, records of the maintenance and testing of engineering controls and occupational health records.

The framework for enacting these procedures depends on an awareness of the relevant *hazard information* and current national *health and safety legislation*.

8.2 HAZARD AND RISK ASSESSMENT

8.2.1 DEFINITIONS

It is helpful to appreciate the distinction between 'hazard' and 'risk' in the context of extrapolating *hazard information* to *hazard and risk assessments* in the chemical laboratory, in general, and in natural product research, in particular.

Hazard is the set of *inherent properties* of a chemical substance that make it capable of causing adverse effects in people or the environment when a particular degree of exposure occurs. *Risk* is the predicted or actual *frequency of occurrence of an adverse effect* of a chemical substance from a given exposure to humans or the environment.

Risk assessment, therefore, requires knowledge of both the hazard of a chemical and the purpose for which it is being used. A highly hazardous substance presents a very low risk if it is securely contained with no likely exposure. Conversely, a substance of relatively low hazard may present unacceptable risks if extensive exposure can occur. Both hazard and exposure must be considered before the risk can be adequately assessed.

8.2.2 HEALTH AND SAFETY LEGISLATION

In the United Kingdom, laboratory managers and supervisors (and employers, in the wider sense, in general) have a legal obligation under the *Control of Substances Hazardous to Health (COSHH) Regulations 2002* to assess the risks to health from hazardous substances used in or created by workplace activities. Also in the United Kingdom, the *Classification, Labelling and Packaging Regulation (CLP Regulation)* requires manufacturers and suppliers of chemicals to provide end-users (laboratory managers and personnel) with information about the physical and health hazards of chemicals by labelling their products with relevant hazard information. Where appropriate, labels for laboratory chemicals and reagents should incorporate

- A pictogram(s) depicting the hazard (chosen from a set of nine pictograms)
- A hazard statement(s) summarising the nature and degree of hazard of a substance
- A precautionary statement(s) with information on minimising or preventing the physical, health or environmental effects of a substance, including first aid measures

The UK CLP Regulation adopts the *United Nations Globally Harmonised System of Classification and Labelling of Chemicals* (UN-GHS). Other European Union (EU) countries have enacted equivalent legislation. Details of the nine GHS pictograms and a list of the hazard and the precautionary statements, respectively, are available: *Globally Harmonised System of Classification and Labelling of Chemicals*, 4th edn., United Nations, New York and Geneva may be obtained as a downloadable PDF from the United Nations Economic Commission for Europe (UNECE) website.

Chemicals produced in the research laboratory are exempted from the obligations of the CLP Regulation, provided they are not placed on the market and they are used under controlled conditions in accordance with workplace and environmental legislation. However, if substances used in research and development are physically made available or supplied to another establishment, for example by sending samples from a university to another research institute or by importing such samples, then the CLP Regulation will apply.

The CLP Regulation replaced the Chemicals (Hazard Information and Packaging for Supply) (CHIP) Regulations in the United Kingdom in 2015. Under the EU's *REACH* legislation, which complements the CLP Regulation, suppliers of substances or mixtures meeting the criteria for classification as hazardous according to CLP have to compile and supply end-users with Safety Data Sheets (SDS) for use, for example in *COSHH Assessments*. [REACH (Registration, Evaluation, Authorisation and restriction of CHemicals)].

8.2.3 WORKPLACE EXPOSURE LIMITS

The U.K. adopted Workplace Exposure Limits (WELs) in 2005 to replace Maximum Exposure Limits (MELs) and Occupational Exposure Standards (OESs). Workplace Exposure Limits – long-term exposure limits (8 h time-weighted average exposures) and short-term exposure limits (15 min time-weighted average exposures) – are set by the U.K. Health and Safety Executive (HSE). *EH40/2005 Workplace Exposure Limits*, 2nd edn., revised and corrected in 2011, is available as a downloadable PDF from the HSE website.

Recommendations for controlling and monitoring substances assigned WELs are part of the COSHH Regulations 2002. Exposure limits are also set by other regulatory and advisory bodies, e.g. Threshold Limit Values (TLVs) by the American Conference of Governmental Industrial Hygienists (ACGIH) and Maximale Arbeitsplatzkonzentrationen (MAK) by German authorities. In EH40, the route of exposure is mainly by inhalation, but exposure limits are also assigned to some substances that are easily absorbed by the skin or are skin sensitisers.

8.3   HEALTH HAZARDS OF CHEMICALS

Natural products are included in the following groups of chemicals, which present health hazards for laboratory workers and others in the working environment. These groups are differentiated for the purposes of risk assessment and product labelling under U.K. and EU health and safety regulations, and by the UN-GHS:

- Human carcinogens and probable human carcinogens according to the International Agency for Research on Cancer (IARC) classifications. (*IARC Monographs on the Evaluation of Carcinogenic Risks to Humans*, Supplement 7 (1987) Overall Evaluations of Carcinogenicity: An Updating of *IARC Monographs*, Volumes 1–42; available online).
- Human teratogens and chemicals that may have an effect on human reproduction or an effect on or via lactation. Two groups of chemicals are distinguished: substances which may damage development of an unborn child and substances which may impair fertility (data from human exposure or animal tests).
- Chemicals that are irritants to the skin, eyes and respiratory system (data from human exposure or animal tests).
- Chemicals that are corrosive to the skin, eyes and respiratory system (data from human exposure or animal tests).
- Sensitisers (respiratory allergens; dermal allergens) (data from human exposure or animal tests).
- Chemicals with known specific target organ toxicity or toxicity due to a specific pharmacological mechanism (data from human exposure or animal tests).
- Mutagenic chemicals.
- Chemicals classified as fatal (… if swallowed; … in contact with skin; … if inhaled); toxic (… if swallowed; … in contact with skin; … if inhaled); harmful (… if swallowed; … in contact with skin; … if inhaled).

The toxicological criteria, which are used for these classifications, are described, for example by Bender, H. F. et al., *Hazardous Chemicals: Control and Regulation in the European Market*, Wiley-VCH, Weinheim, Germany, 2007. The health and other hazards associated with solvents are described in the 'Health Hazards of Solvents'.

Acute toxicity classifications for chemicals are based on acute toxicity data from animals. For example, for the oral route of exposure, six toxicity categories are allocated from oral LD_{50} data:

Category 1: Fatal if swallowed ($LD_{50} \leq 5$ mg/kg)
Category 2: Fatal if swallowed ($5 < LD_{50} \leq 50$ mg/kg)
Category 3: Toxic if swallowed ($50 < LD_{50} \leq 300$ mg/kg)
Category 4: Harmful if swallowed ($300 < LD_{50} \leq 2000$ mg/kg)
Category 5: May be harmful if swallowed ($2000 < LD_{50} \leq 5000$ mg/kg)
Category 6: Non-toxic ($LD_{50} > 5000$ mg/kg)

8.4   HAZARD INFORMATION FOR SELECTED TOXINS

Handling toxins in a laboratory environment requires an assessment of their physical and biological properties and the provision of control measures to minimise and prevent exposure, including an adequate disposal method for inactivating contaminated residues and solvents, aqueous waste discharges and apparatus. In addition, control of exposure to certain classes of toxins requires specialised and mandatory ventilation equipment and is regulated under specific legislation, including, in the United Kingdom, the Anti-Terrorism, Crime and Security Act 2001. Hazard information and validated disposal methods for selected toxins are shown in Table 8.1.

TABLE 8.1
Hazard Information and Disposal Methods for Experimental Procedures with Selected Toxins[a]

Toxin (CAS Registry No.)	Biological Source (Occurrence)[b]	Acute Toxicity and Related Information[b,c] (Selected UN-GHS Acute Toxicity Hazard Statement)	Selected Chronic Toxicity Information[b,c]	Selected Pharmacological Activity[b]	Disposal Method[d]
Aflatoxin B$_1$ (1162-65-8)	Prod. by *Aspergillus flavus* and *Aspergillus parasiticus*	LD$_{50}$ (rat, orl) 4.8 mg/kg. Crystalline material is electrostatic and can present an inhalation hazard. (*Fatal if swallowed*) [and fatal by other routes of administration]	Human and exp. carcinogen. Exp. reprod. and teratogenic effects. Hepatotoxic. One of the most potent carcinogens known in animals		Includes (1) aq. KMnO$_4$ in 2M NaOH then addn. of sod. metabisulfite and (2) aq. NaOCl followed by addition of acetone.
Anatoxin a (64285-06-9)	Alkaloid from *Anabaena flos-aquae* NRC-44h (*Euphorbiaceae*). Also isol. from *Oscillatoria* spp., *Aphanizomenon flos-aquae* and *Cylindrospermum* sp.	Very toxic by intraperitoneal route; LD$_0$ (mus, ipr) 0.2 mg/kg. Skin and eye irritant. (*Fatal if swallowed*) [and fatal by other routes of administration]	Gastrointestinal and other effects in humans. Exp. reprod. effects	Neurotoxin. Depolarising agent. Potent agonist for the nicotinic acetylcholine receptor	(1) Ozone and (2) aq. KMnO$_4$
Botulinum toxins	Single chain polypeptides. Synth. by *Clostridium botulinum*	Acutely lethal by all routes of administration. (*Fatal if swallowed*) [and fatal by other routes of administration]	Causative agents of human botulism food poisoning	Potent neurotoxin	Includes (1) steam autoclave, (2) heat 100°C 10 min and (3) aq. NaOCl
Brevetoxin A (98112-41-5)	Prod. by *Gymnodinium breve* (*Ptychodiscus brevis*)	Acutely lethal by all routes of administration. LD$_{50}$ (mus, ipr) 0.5 mg/kg. (*Fatal if swallowed*) [and fatal by other routes of administration]	Teratogen	Potent neuro- and ichthyotoxin	Aq. NaOCl
Citrinin (518-75-2)	Isol. from *Penicillium citrinum*, *Guanomyces polythrix*, *Aspergillus terreus*, *Monascus* spp. and other microorganisms	Severe skin irritant. LD$_{50}$ (mus, orl) 112 mg/kg. (*Toxic if swallowed*)	Exp. reprod. and teratogenic effects. Exp. hepatotoxic and nephrotoxic effects		Includes (1) aq. KMnO$_4$ in 2M NaOH then addn. of sod. metabisulfite and (2) aq. NaOCl (*Continued*)

TABLE 8.1 (Continued)
Hazard Information and Disposal Methods for Experimental Procedures with Selected Toxins[a]

Toxin (CAS Registry No.)	Biological Source (Occurrence)[b]	Acute Toxicity and Related Information[b,c] (Selected UN-GHS Acute Toxicity Hazard Statement)	Selected Chronic Toxicity Information[b,c]	Selected Pharmacological Activity[b]	Disposal Method[d]
Cylindrospermopsin (143545-90-8)	Alkaloid from the cyanobacteria *Aphanizomenon ovalisporum*, *Cylindrospermopsis raciborskii*, *Oscillatoria* sp. and *Umezikia natans*	LD_{50} (mus, ipr) 2.1 mg/kg. *(Fatal if swallowed)* [and fatal by other routes of administration]	Potent hepatotoxin. Exp. carcinogen and teratogen. Mutagen	Glutathione synthesis inhibitor	Ozone
Microcystins (77238-39-2)	Peptide toxin complex. Prod. by *Microcystis aeruginosa* and other cyanobacteria	LD_{50} (mus, ipr) 0.47 mg/kg. Irritant *(Fatal if swallowed)* [and fatal by other routes of administration]	Hepatotoxin	Serine-threonine phosphatase inhibitor	Aq. NaOCl
Ochratoxin A (303-47-9)	Prod. by *Aspergillus melleus*, *Aspergillus carbonarius*, *Penicillium ostianus* and other spp.	LD_{50} (rat, orl) 20–25 mg/kg. Crystalline material is electrostatic. *(Fatal if swallowed)* [and fatal by other routes of administration]	Possible human carcinogen. Exp. carcinogen and neoplastic agent. Nephrotoxic. Exp. reprod. and teratogenic effects		(1) Aq. NaOCl and (2) aq. $KMnO_4$ in 2M NaOH then addn. of sod. metabisulfite
Palytoxin (77734-91-9)	Prod. by *Palythoa tuberculosa*, *Palythoa toxica* and *Radianthus macrodactylus*	LD_{50} (rat, ivn) 89 ng/kg, LD_{50} (rat, ims) 240 ng/kg. Skin and severe eye irritant. *(Fatal if swallowed)* [and fatal by other routes of administration]		Na/K-ATPase inhibitor, sperm motility inhibitor, cardiotoxic and haemolytic agent, coronary vasoconstrictor	Aq. NaOCl
Patulin (149-29-1)	Produced by several fungi, e.g. *Aspergillus clavatus*, *Aspergillus terreus*, a marine *Aspergillus varians*, *Penicillium patulum*, *Penicillium griseofulvum* and *Byssochlamys nivea*	LD_{50} (rat, orl) 27.8 mg/kg. Crystalline material is electrostatic.	Exp. neoplastic agent and teratogen. Exp. kidney and gastrointestinal effects		(1) Aq. $KMnO_4$ in 2M NaOH then addn. of sod. metabisulfite and (2) aq. NH_3 + autoclave at 120°C

(Continued)

TABLE 8.1 (Continued)
Hazard Information and Disposal Methods for Experimental Procedures with Selected Toxins[a]

Toxin (CAS Registry No.)	Biological Source (Occurrence)[b]	Acute Toxicity and Related Information[b,c] (Selected UN-GHS Acute Toxicity Hazard Statement)	Selected Chronic Toxicity Information[b,c]	Selected Pharmacological Activity[b]	Disposal Method[d]
T-2 Toxin (21259-20-1)	Trichothecene mycotoxin. Isol. from *Fusarium* spp. and *Trichoderma lignorum*.	LD_{50} (rat, orl) 2.7 mg/kg; LD_{50} (rat, ipr) 0.9 mg/kg. Irritant. (*Fatal if swallowed*) [and fatal by other routes of administration]	Exp. effects incl. emesis, anorexia, ataxia. Exp. neoplastigenic data. Pulmonary toxin, haemorrhagic agent	Immunosuppressant. Inhibits protein synth. at the ribosomal level	Aq. NaOCl + NaOH
Tetrodotoxin (4368-28-9)	Isol. from the ovaries and liver of Japanese puffer fish (*Sphoeroides rubripes*, *Sphoeroides vermicularis*, *Sphoeroides phyreus*) and other spp.	LD_{50} (mus, orl) 435 µg/kg; LD_{50} (mus, ivn) 9 µg/kg; LD_{50} (mus, ipr) 8 µg/kg. (*Fatal if swallowed*) [and fatal by other routes of administration]	Initial symptoms are neurologic and gastrointestinal. In severe poisoning, dysrhythmias, hypotension, fatalities reported	Highly potent neurotoxin. Sodium channel blocker	Aq. NaOCl

a   Duffus, J.H. et al., A toxin is a poisonous substance produced by a biological organism such as a microbe, animal, plant or fungus. (*IUPAC Glossary of Terms Used in Toxicology*, 2nd edn., IUPAC Recommendations, 2007; *Pure Appl. Chem.*, 2007, 79(7), 1153–1344.)

b   *The Combined Chemical Dictionary on DVD*, Version 18:1, CRC Press, Boca Raton, FL, 2014.

c   Lewis, R.J. Sr., *Sax's Dangerous Properties of Industrial Materials*, 12th edn., Wiley, Hoboken, NJ, 2012.

d   Lunn, G. et al., *Destruction of Hazardous Chemicals in the Laboratory*, 3rd edn., Wiley, Hoboken, NJ, 2012.

8.5 SOLVENTS

Although alternative techniques for the isolation of natural products have been developed, which minimise the use of solvents, for example supercritical fluid extraction, microwave-assisted extraction and pressurised solvent extraction, traditional solvent-based procedures using relatively large quantities of organic solvents are still prevalent in the literature.

Solvents are fire, health and environmental hazards, and caution is necessary when using these substances in a laboratory environment. The flammable properties and flammability classifications of many solvents impose restrictions on their handling and storage. Particular concerns are vapour leaks of solvents as sources of ignition and the inappropriate storage of Winchester bottles containing solvents (especially if exposed to sunlight, which can result in peroxidation) in laboratories. The peroxidation of solvents during storage as a reactive hazard is described in more detail in Section 8.5.3. Both acute and chronic low-level exposures contribute to the recognised health hazards of solvents.

8.5.1 Flammability Classifications and Legislation

The UN Globally Harmonised System of Classification and Labelling of Chemicals divides flammable liquids into four categories:

Category 1: Extremely flammable liquid and vapour (flash point (fl. p.) <23°C and initial Bp ≤ 35°C)

Category 2: Highly flammable liquid and vapour (fl. p. <23°C and initial Bp > 35°C)

Category 3: Flammable liquid and vapour (fl. p. ≥ 23°C and ≤60°C)

Category 4: Combustible liquid (fl. p. > 60°C and ≤93°C)

Flammable substances used and stored in the laboratory are also subject to further risk assessment and control in U.K. law under the Health and Safety at Work Act 1974, the Management of Health and Safety at Work Regulations 1999, the COSHH Regulations 2002, the Dangerous Substances and Explosive Atmospheres Regulations 2002 (DSEAR) and The Regulatory Reform (Fire Safety) Order 2005.

Flammability classifications for a selection of solvents (and some other substances) are given in Table 8.2. The chemicals are listed in order of increasing boiling point to the nearest 1°C. Solvents in Table 8.2 that are also peroxidation hazards may be identified from data in Tables 8.6 and 8.7.

8.5.2 Health Hazards of Solvents

Apart from the acute toxic effects of high concentrations of the more volatile solvents, there are health hazards from the long-term (chronic) exposure to low levels of solvents. Reproductive effects that are associated with chronic exposure to some solvents used in laboratories are shown in Table 8.3. Evidence from animal studies suggests there are reproductive hazards from handling other solvents. For example,

- 2-ethoxyethanol and 2-butanone are teratogenic (in animal models)
- Dichloromethane, styrene, 1,1,1-trichloroethane, tetrachloroethylene and xylene isomers have foetotoxic properties (in animal models)

The IARC classifications for the carcinogenic risk from exposure to some laboratory solvents (and other selected reagents) are summarised in Table 8.4. Other toxic effects for classes of solvents categorised by functional group are given in Table 8.5.

Solvents which are currently assigned a Workplace Exposure Limit (8 h long-term exposure limit) which is less than or equal to 100 ppm are marked in Table 8.2 with an arrow head (►). (Data from EH40/2005, 2nd edn.)

TABLE 8.2
Fire Hazards of Some Common Laboratory Solvents and Other Substances

Bp (°C)	Mp (°C)	Name[a]	Flash Point (°C)[b]	UN-GHS Flammability Classification[c]
30-60		Petrol[d]		Extremely flammable liquid and vapour
30	−161	2-Methylbutane	<−51	Extremely flammable liquid and vapour
32	−99	Methyl formate	<−19	Extremely flammable liquid and vapour
35	−116	► Diethyl ether	−45	Extremely flammable liquid and vapour
36	−129	Pentane	−49	Extremely flammable liquid and vapour
38	−98	Dimethyl sulfide	−34	Highly flammable liquid and vapour
40	−97	► Dichloromethane		Conc. 12%–19% in air, flammable liquid and vapour
46	−112	► Carbon disulfide	−30	Flammable liquid and vapour
46	−14	1,1,1-Trichloro-2,2,2-trifluoroethane		Non-flammable
47	−111	1,2-Dibromo-1,1,2,2-tetrafluoroethane		Non-flammable
50	−94	Cyclopentane	−37	Flammable liquid and vapour
54	−109	► 2-Methoxy-2-methylpropane	−28	Highly flammable liquid and vapour
56	−94	Acetone	−17	Highly flammable liquid and vapour
56	−98	Methyl acetate	−9	Highly flammable liquid and vapour
61	−63	► Chloroform		Non-flammable
65	−108	► Tetrahydrofuran	−14	Highly flammable liquid and vapour
65	−98	Methanol	10	Highly flammable liquid and vapour
69	−87	Diisopropyl ether	−28	Highly flammable liquid and vapour
69	−94	► Hexane	−23	Highly flammable liquid and vapour
72	−15	Trifluoroacetic acid		Non-flammable
74	−32	► 1,1,1-Trichloroethane		Non-flammable
75	−95	1,3-Dioxolane	2 (oc)	Highly flammable liquid and vapour
77	−84	Ethyl acetate	−4	Highly flammable liquid and vapour
77	−21	► Carbon tetrachloride		Non-flammable
78	−117	Ethanol	12	Highly flammable liquid and vapour
78	−123	1-Chlorobutane	−12	Highly flammable liquid and vapour
80	−86	2-Butanone	−1	Highly flammable liquid and vapour
80	6	► Benzene	−11	Highly flammable liquid and vapour
81	6	► Cyclohexane	−20	Highly flammable liquid and vapour
82	−45	► Acetonitrile	6 (oc)	Highly flammable liquid and vapour
82	−90	2-Propanol	12	Highly flammable liquid and vapour
83	26	2-Methyl-2-propanol	11	Highly flammable liquid and vapour
84	−35	► 1,2-Dichloroethane	13	Highly flammable liquid and vapour
85	−58	1,2-Dimethoxyethane	1	Highly flammable liquid and vapour
87	−85	► Trichloroethylene		Non-flammable
88	−45	Tetrahydropyran	−20	Highly flammable liquid and vapour
97	−127	1-Propanol	15	Highly flammable liquid and vapour
98	−92	Heptane	−4	Highly flammable liquid and vapour
99	−108	2,2,4-Trimethylpentane	−12	Highly flammable liquid and vapour
100	−115	► 2-Butanol	24	Highly flammable liquid and vapour
100	0	Water		Non-flammable
101	11	► 1,4-Dioxane	11	Highly flammable liquid and vapour
101	−127	Methylcyclohexane	−4	Highly flammable liquid and vapour
101	−29	► Nitromethane	35	Flammable liquid and vapour
101	8	► Formic acid	69	

(Continued)

TABLE 8.2 (*Continued*)
Fire Hazards of Some Common Laboratory Solvents and Other Substances

Bp (°C)	Mp (°C)	Name[a]	Flash Point (°C)[b]	UN-GHS Flammability Classification[c]
102	−42	3-Pentanone	13	Highly flammable liquid and vapour
103	15	Trimethyl orthoformate	15	Highly flammable liquid and vapour
104	−6	Bromotrichloromethane		Non-flammable
108	−108	2-Methyl-1-propanol	28	Flammable liquid and vapour
111	−95	► Toluene	4	Highly flammable liquid and vapour
114	−36	► 1,1,2-Trichloroethane		Non-flammable
116	−42	► Pyridine	20	Highly flammable liquid and vapour
117	−80	► 4-Methyl-2-pentanone	17	Highly flammable liquid and vapour
118	−90	1-Butanol	29	Flammable liquid and vapour
118	17	Acetic acid	39	Flammable liquid and vapour
121	−19	► Tetrachloroethylene		Non-flammable
125	−86	► 2-Methoxyethanol	43	Flammable liquid and vapour
126	−77	Butyl acetate	22	Flammable liquid and vapour
126	−57	Octane	13	Highly flammable liquid and vapour
132	−117	3-Methyl-1-butanol	43	Flammable liquid and vapour
132	10	► 1,2-Dibromoethane		Non-flammable
132	−45	► Chlorobenzene	24	Flammable liquid and vapour
135	−70	► 2-Ethoxyethanol	44	Flammable liquid and vapour
136	−94	► Ethylbenzene	15	Highly flammable liquid and vapour
138	14	► 1,4-Dimethylbenzene	25	Flammable liquid and vapour
139	−47	► 1,3-Dimethylbenzene	25	Flammable liquid and vapour
142	−79	Isopentyl acetate	25	Flammable liquid and vapour
142	−98	Dibutyl ether	25	Flammable liquid and vapour
144	−25	► 1,2-Dimethylbenzene	17	Highly flammable liquid and vapour
146	30	Triethyl orthoformate	30	Flammable liquid and vapour
150	−51	Nonane	30	Flammable liquid and vapour
153	−61	► Dimethylformamide	55	Flammable liquid and vapour
155	−38	Methoxybenzene	52 (oc)	Flammable liquid and vapour
155	−45	► Cyclohexanone	44	Flammable liquid and vapour
156	−31	Bromobenzene	51	Flammable liquid and vapour
161	−68	Diglyme	67	
166	−20	► *N,N*-Dimethylacetamide	67 (oc)	
172	−75	► 2-Butoxyethanol	61	
175	−42	2,4,6-Trimethylpyridine	57	Flammable liquid and vapour
180	−17	► 1,2-Dichlorobenzene	66	
185	−31	*trans*-Decahydronaphthalene	54	Flammable liquid and vapour
189	18	Dimethyl sulfoxide	95 (oc)	
191	−13	Benzonitrile	72	
195	−17	1-Octanol	81	
196	−46	Trimethyl phosphate	107	
196	−43	*cis*-Decahydronaphthalene	54	Flammable liquid and vapour
197	−13	► 1,2-Ethanediol	111	
202	20	Acetophenone	77	
202	−2	Hexachloro-2-propanone		Non-flammable
202	−24	► 1-Methyl-2-pyrrolidinone	96 (oc)	
205	−15	Benzyl alcohol	93	

(*Continued*)

TABLE 8.2 (*Continued*)
Fire Hazards of Some Common Laboratory Solvents and Other Substances

Bp (°C)	Mp (°C)	Name[a]	Flash Point (°C)[b]	UN-GHS Flammability Classification[c]
207	<−50	1,3-Butanediol	109	
207	−35	1,2,3,4-Tetrahydronaphthalene	71	
210	3	► Formamide	>77	
211	6	► Nitrobenzene	88	
214	17	► 1,2,4-Trichlorobenzene	105	
215	−12	Dodecane	74	
216	−45	1,2-*bis*(2-Methoxyethoxy)ethane	111	
222	82	Acetamide	>104	
235	7	Hexamethylphosphoric triamide	>55	
240	−55	4-Methyl-1,3-dioxolan-2-one	135	
255	72	Biphenyl	113	
279	96	Acenaphthene	>66	
285	27	Tetrahydrothiophene 1,1-dioxide	177	
290	18	Glycerol	160	
328	−6	Tetraethylene glycol	174 (oc)	

Note: Carbon-, sulphur-, nitrogen- and phosphorus-containing solvents will evolve oxides of their constituent elements, including CO, on combustion, and these gases are toxic and probably irritants if a fire involving such materials is encountered. Some chlorinated solvents can form phosgene (carbonyl chloride) in fires.

[a] ►Indicates a substance currently assigned a Workplace Exposure Limit (8-h long-term exposure limit), which is less than or equal to 100 ppm (data from EH40/2005 2nd edn., revised and corrected in 2011).

[b] Flash point measurements from the closed-cup method are quoted unless only data from the open-cup (oc) method are available. Data from Stephenson, R.M., *Flash Points of Organic and Organometallic Compounds*, Elsevier, New York, 1987; Bond, J., *Sources of Ignition*, Butterworth, Oxford, U.K., 1991.

[c] Substances having flash points above 60°C are considered non-flammable, but may ignite if brought to a high temperature.

[d] Mixture of hydrocarbons, typically 73% *n*-pentane, 23% branched pentanes, 3% cyclopentane. Higher boiling petrol have correspondingly decreasing flammability hazards.

TABLE 8.3
Reproductive Effects of Some Solvents

Reproductive Effects	Solvent(s)
Menstrual disorders	Toluene, styrene, benzene
Abortion or infertility	Formaldehyde, benzene
Testicular atrophy	2-Ethoxyethanol
Decreased foetal growth, low birth weight	Toluene, formaldehyde, vinyl chloride

Source: Reproduced with permission from Winder, C. et al. (eds.), *Occupational Toxicology*, 2nd edn., CRC Press, Boca Raton, FL, 2004.

TABLE 8.4

IARC Classifications[a] for Some Laboratory Solvents

Solvent	IARC Group
Benzene	Group 1
Carbon tetrachloride	Group 2B
Chloroform	Group 2B
Cyclohexanone	Group 3
1,2-Dichloroethane	Group 2B
Dichloromethane	Group 2B
Dimethylformamide	Group 3
1,4-Dioxane	Group 2B
Mineral oils (untreated)	Group 1
Tetrachloroethylene	Group 2A
Toluene	Group 3
1,1,1-Trichloroethane	Group 3
1,1,2-Trichloroethane	Group 3
Trichloroethylene	Group 2A
Xylene-	Group 3

Source: Adapted with permission from Winder, C. et al. (eds.), *Occupational Toxicology*, 2nd edn., CRC Press, Boca Raton, FL, 2004.

[a] IARC Classifications – Group 1, agents which are carcinogenic to humans; Group 2A, agents which are probably carcinogenic to humans; Group 2B, agents which are possibly carcinogenic to humans; Group 3, agents which are not classifiable; Group 4, agents which are probably not carcinogenic to humans. From *IARC Monographs on the Evaluation of Carcinogenic Risks to Humans*, Supplement 7 (1987) Overall Evaluations of Carcinogenicity: An Updating of *IARC Monographs* Volumes 1–42; available online.

8.5.3 PEROXIDE-FORMING CHEMICALS

Peroxide-forming solvents and reagents should be dated at the time they are first opened and should be either discarded or tested for peroxides within a fixed period of time after their first use. Peroxides can be detected with NaI/AcOH, though dialkyl peroxides may need treatment with conc. HCl or 50% H_2SO_4 before detection with iodide is possible. A commercially available test paper, which contains a peroxidase, can detect hydroperoxides and dialkyl peroxides, as well as oxidising anions, in organic and aqueous solvents.

The types of structures that have been identified as likely to produce peroxides are listed in Table 8.6, and some common peroxidisable chemicals are quoted in Table 8.7.

Hydroperoxides, but not dialkyl peroxides, can be removed from peroxide-forming solvents by passage through basic activated alumina, by treatment with a self-indicating activated molecular sieve (type 4A) under nitrogen, or by treatment with Fe^{2+}/H^+, CuCl or other reductants.

The following references provide further information:

Detection and removal of peroxides from solvents; Riddick, J.A. et al. (eds.), *Organic Solvents: Physical Properties and Methods of Purification*, 4th edn., Chichester, U.K., Wiley, 1986.

Deperoxidation of ethers with molecular sieves; Burfield, D.R., *J. Org. Chem.*, 1982, **47**, 3821–3824.

Determination of organic peroxides; Mair, R.D. et al., in *Treatise on Analytical Chemistry*, eds. Kolthoff, I.M. et al., Vol. 14, Part II, New York, Interscience, 1971, p. 295.

TABLE 8.5
Toxic Effects of Groups of Solvents

Solvent Group	Examples	Effects	
		Acute	Chronic
Aliphatic hydrocarbons	Petrol, kerosene, diesel, *n*-hexane	Nausea, pulmonary irritation, ventricular arrhythmia	Weight loss, anaemia, proteinuria, haematuria, bone-marrow hypoplasia
Aromatic hydrocarbons	Toluene, xylene, benzene	Nausea, ventricular arrhythmia, respiratory	Headache, anorexia, lassitude
Halogenated hydrocarbons	Carbon tetrachloride, dichloromethane, trichloroethane, trichloroethylene, tetrachloroethylene	Irritant, liver, kidney, heart	Fatigue, anorexia, liver, kidney, cancer[a]
Ketones	Acetone, methyl ethyl ketone, methyl *n*-butyl ketone	Irritant, respiratory depression	
Alcohols	Methanol, ethanol, isopropanol	Irritant, gastrointestinal	Liver, immune function
Esters	Methyl formate, methyl acetate, amyl acetate	Irritant, liver, palpitations	
Glycols	Ethylene glycol, diethylene glycol, propylene glycol	Kidney	Kidney
Ethers	Diethyl ether, isopropyl ether	Irritant, nausea	
Glycols ethers	Ethylene glycol monomethyl ether, ethylene glycol, monoethyl ether, propylene glycol monomethyl ether	Irritant, nausea, anaemia, liver, kidney, reproductive system	

Source: Reproduced with permission from Winder, C. et al. (eds.), *Occupational Toxicology*, 2nd edn., CRC Press, Boca Raton, FL, 2004.

[a] In experimental animals, nongenotoxic.

TABLE 8.6
Types of Chemicals That May Form Peroxides

Organic structures
Ethers and acetals with α-hydrogen atoms
Olefins with allylic hydrogen atoms
Chloroolefins and fluoroolefins
Vinyl halides, esters and ethers
Dienes
Vinylacetylenes with α-hydrogen atoms
Alkylacetylenes with α-hydrogen atoms
Alkylarenes that contain tertiary hydrogen atoms
Alkanes and cycloalkanes that contain tertiary hydrogen atoms
Acrylates and methacrylates
Secondary alcohols
Ketones that contain α-hydrogen atoms
Aldehydes
Ureas, amides and lactams that have a H atom linked to a C attached to a N
Inorganic substances
Alkali metals, especially potassium, rubidium and caesium
Metal amides
Organometallic compounds with a metal atom bonded to carbon
Metal alkoxides

Source: Reproduced with permission from IUPAC-IPCS, *Chemical Safety Matters*, Cambridge University Press, Cambridge, U.K., 1992.

TABLE 8.7
Common Peroxide-Forming Chemicals

Severe peroxide hazard on storage with exposure to air. *Discard within 3 months*

Diisopropyl ether	Sodium amide (sodamide)
Divinylacetylene[a]	Vinylidene chloride (1,1-dichloroethylene)[a]
Potassium metal	Potassium amide

Peroxide hazard on concentration: Do not distil or evaporate without first testing for the presence of peroxides. *Discard or test for peroxides after 6 months*

Acetaldehyde diethyl acetal (1,1-diethoxyethane)	Ethylene glycol dimethyl ether (glyme)
Cumene (isopropylbenzene)	Ethylene glycol ether acetates
Cyclohexene	Ethylene glycol monoethers (cellosolves)
Cyclopentene	Furan
Decalin (decahydronaphthalene)	Methylacetylene
Diacetylene (1,2-butadiyne)	Methylcyclopentane
Dicyclopentadiene	Methyl isobutyl ketone
Diethyl ether (ether)	Tetrahydrofuran
Diethylene glycol dimethyl ether (diglyme)	Tetralin (tetrahydronaphthalene)
Dioxan/dioxolan (dioxane)	Vinyl ethers[a]

Hazard or rapid polymerisation **initiated by internally formed peroxides**[a]

a. Normal liquids. *Discard or test for peroxides after 6 months*[b]

Chloroprene (2-chloro-1,3-butadiene)[c]	Vinyl acetate
Styrene	Vinylpyridine

b. Normal gases. *Discard after 12 months*[d]

Butadiene[c]	
Tetrafluoroethylene[c]	Vinylacetylene[c]
	Vinyl chloride

Source: Reproduced with permission from IUPAC-IPCS, *Chemical Safety Matters*, Cambridge University Press, Cambridge, U.K., 1992.

[a] Monomers may polymerise and should be stored with a polymerisation inhibitor from which the monomer can be separated by distillation just before use.

[b] Although common acrylic monomers such as acrylonitrile, acrylic acid, ethyl acrylate and methyl methacrylate can form peroxides, they have not been reported to develop hazardous levels in normal use and storage.

[c] The hazard from peroxide formation in these compounds is substantially greater when they are stored in the liquid phase.

[d] Although air cannot enter a gas cylinder in which gases are stored under pressure, these gases are sometimes transferred from the original cylinder to another in the laboratory, and it is difficult to be sure that there is no residual air in the receiving cylinder. An inhibitor should be put into any secondary cylinder before transfer. The supplier can suggest an appropriate inhibitor to be used. The hazard posed by these gases is much greater if there is a liquid phase in the secondary container. Even inhibited gases that have been put into a secondary container under conditions that create a liquid phase should be discarded within 12 months.

FURTHER LITERATURE SOURCES

RISK AND HAZARD ASSESSMENT (GENERAL)

DiBerardinis, L.J. (ed.), *Handbook of Occupational Safety and Health*, 2nd edn., Wiley, New York, 1999.

Koradecka, D. (ed.), *Handbook of Occupational Safety and Health*, CRC Press, Boca Raton, FL, 2010.

Richardson, M.L. (ed.), *Toxic Hazard Assessment of Chemicals*, Royal Society of Chemistry, London, U.K., 1986 (definitions of risk and hazard).

Van Leeuwen, C.J. et al. (eds.), *Risk Assessment of Chemicals*, 2nd edn., Springer, Dordrecht, the Netherlands, 2007.

PHYSICAL PROPERTIES RELATED TO HAZARD

Bond, J., *Sources of Ignition*, Butterworth, Oxford, U.K., 1991 (flash points, explosive limits and auto-ignition temperatures).
Lide, D.R., *Handbook of Organic Solvents*, CRC Press, Boca Raton, FL, 1995.
Riddick, J.A. et al., *Organic Solvents: Physical Properties and Methods of Purification*, 4th edn., Wiley-Interscience, New York, 1986.
Stephenson, R.M., *Flash Points of Organic and Organometallic Compounds*, Elsevier, New York, 1987.
Verschueren, K., *Handbook of Environmental Data on Organic Chemicals*, 4th edn., Wiley, Chichester, U.K., 2001.

OCCUPATIONAL EXPOSURE LIMITS

2015 TLVs® and BEIs®, American Conference of Governmental Industrial Hygienists, Cincinnati, OH, 2015.
EH40/2005 Workplace Exposure Limits, 2nd edn., revised and corrected in 2011, available as a downloadable pdf from the HSE website. http://www.hse.gov.uk/pubns/priced/eh40.pdf. Accessed 31 July, 2015.
Deutsche Forschungsgemeinschaft, *List of MAK and BAT Values 2014*, Wiley-VCH, Weinheim, 2014.
Occupational Exposure Limits for Airborne Toxic Substances, 3rd edn., ILO, Geneva, Switzerland, 1991 (data from 16 countries).

REACTIVE HAZARDS

Clark, D.E., *Chem. Health Saf.*, 2001, **8**(5), 12–22 (peroxidizable organic compounds).
IUPAC-IPCS, *Chemical Safety Matters*, Cambridge University Press, Cambridge, U.K., 1992.
Jackson, H.L., et al., *J. Chem. Ed.*, 1970, **47**, A175 (peroxidizable compounds).
Kelly, R.J., *Chem. Health Saf.*, 1996, **3**(5), 28–36 (peroxidizable organic compounds).
Luxon, S.G. (ed.), *Hazards in the Chemical Laboratory*, 5th edn., Royal Society of Chemistry, Cambridge, U.K., 1992.
Pohanish, R.P. et al., *Wiley Guide to Chemical Incompatibilities*, 3rd edn., Wiley, Hoboken, NJ, 2009.
Urben, P.G. (ed.), *Bretherick's Handbook of Reactive Chemical Hazards*, 7th edn., Elsevier, Oxford, U.K., 2007.

TOXICOLOGY

General

Grandjean, P., *Skin Penetration: Hazardous Chemicals at Work*, Taylor & Francis, London, U.K., 1990 (300 chemicals which are toxic by skin absorption).
Hathaway, G.J. et al. (eds.), *Proctor and Hughes' Chemical Hazards of the Workplace*, 5th edn., Wiley, Hoboken, NJ, 2004.
IARC Monographs on the Evaluation of the Carcinogenic Risk of Chemicals to Humans, International Agency for Research on Cancer, Lyon, France, 1971–.
Lewis, R.J. Sr., *Sax's Dangerous Properties of Industrial Materials*, 12th edn., Wiley, Hoboken, NJ, 2012.
Maibach, H. et al., *Applied Dermatotoxicology: Clinical Aspects*, Academic Press/Elsevier, Amsterdam, the Netherlands, 2014.
McQueen, C.A. (ed.), *Comprehensive Toxicology*, 2nd edn., Elsevier, Oxfordshire, U.K., 2010.
Rose, V.E. and Cohrssen, B. (eds.), *Patty's Industrial Hygiene and Toxicology*, 6th edn., Wiley, Hoboken, NJ, 2012.
Wexler, P. et al. (eds.), *Information Resources in Toxicology*, 4th edn., Elsevier/Academic Press, Amsterdam, the Netherlands, 2009.
Winder, C. et al. (eds.), *Occupational Toxicology*, 2nd edn., CRC Press, Boca Raton, FL, 2004.

Reproductive Toxicology

Kolb, V.M. (ed.), *Teratogens*, 2nd edn., Elsevier, Amsterdam, the Netherlands, 1993.
Lewis, R.J., *Reproductively Active Chemicals: A Reference Guide*, Van Nostrand-Reinhold, New York, 1991.
Shepard, T.H., *Catalog of Teratogenic Agents*, 9th edn., The John Hopkins University Press, Baltimore, MD, 1998.

Solvent Toxicology

Chemical Safety Data Sheets, Vol. 1, Solvents, Royal Society of Chemistry, Cambridge, U.K., 1989.

Commission of the European Communities, *Long-Term Neurotoxic Effects of Paint Solvents*, Royal Society of Chemistry, London, U.K., 1993 (neurotoxicity).

Henning, H. (ed.), *Solvent Safety Sheets: A Compendium for the Working Chemist,* Royal Society of Chemistry, Cambridge, U.K.,1993.

McParland, M. et al. (ed.), *Toxicology of Solvents*, RAPRA Technology Ltd., Shropshire, U.K., 2002 (toxicity and treatment of solvent exposure for all the commonly-used laboratory solvents).

Sarker, S.D. et al. (ed.), *Natural Products Isolation: Methods and Protocols*, Methods in Molecular Biology, Vol. 864, Humana Press, New York, 2012.

Snyder, R. (ed.), *Ethel Browning's Toxicity and Metabolism of Industrial Solvents*, 2nd edn., Vols. 1/2/3, Elsevier, Amsterdam, the Netherlands, 1987/1990/1992.

Solvents in Common Use: Health Risks to Workers, Royal Society of Chemistry, London, U.K., 1988.

SAFETY DATA SHEETS

Chemical Safety Sheets, Kluwer Academic, Dordrecht, the Netherlands, 1991 (includes a section on the prediction of chemical handling properties from physical data).

International Chemical Safety Cards, Commission of the European Communities, Luxembourg (produced for the International Programme on Chemical Safety), available online: http://www.inchem.org/. Accessed 31 July, 2015.

Keith, L.H. (ed.), *Compendium of Safety Data Sheets for Research and Industrial Chemicals*, Parts I–VI, VCH, Deerfield Park, MI, 1985–1987.

Lenga, R.E. (ed.), *The Sigma-Aldrich Library of Chemical Safety Data*, 2nd edn., Sigma-Aldrich Corp., Milwaukee, WI, 1988.

LABORATORY SAFETY

Alaimo, R.J. (ed.), *Handbook of Chemical Health and Safety*, American Chemical Society/Oxford University Press, New York, 2001.

Chosewood, L.C. and Wilson, D.E. (eds.), *Biosafety in Microbiological and Biomedical Laboratories*, 5th edn., US Department of Health and Human Services, 2009, available as a downloadable pdf from http://www.cdc.gov/biosafety/publications/bmbl5/

Fleming, D.O. et al. (eds.), *Biological Safety: Principles and Practices*, 4th edn., ASM Press, Washington, DC, 2006.

Furniss, B.S. et al., *Vogel's Textbook of Practical Organic Chemistry*, 5th edn., Longman Scientific & Technical, Essex, U.K., 1989, pp. 35–51 (hazards in organic chemistry laboratories).

Furr, A.K., *CRC Handbook of Laboratory Safety*, 5th edn., CRC Press, Boca Raton, FL, 2000.

Fuscaldo, A.A. (ed.), *Laboratory Safety: Theory and Practice*, Academic Press, New York, 1980.

Gottschall, W. C. et al., *Laboratory Health and Safety Dictionary*, Wiley, New York, 2001.

IUPAC-IPCS, *Chemical Safety Matters*, Cambridge University Press, Cambridge, U.K., 1992 (useful laboratory safety advice including storage and disposal of waste chemicals).

J. Chem. Health Safety. Published by Elsevier, Amsterdam, the Netherlands.

Laboratory Biosafety Manual, 3rd edn., WHO, Geneva, Switzerland, 2004, available as a downloadable pdf from http://www.who.int/csr/resources/publications/biosafety/en/Biosafety7.pdf. Accessed 31 July, 2015.

Laboratory Hazards Bulletin. Published online by the Royal Society of Chemistry, Cambridge, U.K.

Lunn, G. et al., *Destruction of Hazardous Chemicals in the Laboratory*, 3rd edn., Wiley, Hoboken, NJ, 2012.

Luxon, S.G. (ed.), *Hazards in the Chemical Laboratory*, 5th edn., Royal Society of Chemistry, Cambridge, U.K., 1992.

National Research Council (ed.), *Safe Science: Promoting a Culture of Safety in Academic Chemical Research*, National Academies Press, Washington DC, 2014.

National Research Council (US) Committee on Laboratory Security and Personnel Reliability Assurance Systems for Laboratories Conducting Research on Biological Select Agents and Toxins, *Responsible Research with Biological Select Agents and Toxins*, National Academies Press, Washington DC, 2009, available as a downloadable pdf http://www.nap.edu/catalog/12774/responsible-research-with-biological-select-agents-and-toxins. Accessed 31 July, 2015.

Palluzi, R.P., *Pilot Plant and Laboratory Safety*, McGraw Hill, New York, 1994.

Pipitone, D.A. (ed.), *Safe Storage of Laboratory Chemicals*, 2nd edn., Wiley, New York, 1991.

Prudent Practices in the Laboratory, National Academies Press, Washington DC, updated 2011, available as a downloadable pdf, http://www.ncbi.nlm.nih.gov/books/NBK55878/pdf/TOC.pdf. Accessed 31 July, 2015.

Slein, M.W. et al. (eds.), *Degradation of Chemical Carcinogens: An Annotated Bibliography*, Van Nostrand-Reinhold, New York, 1980.

Stricoff, R.S. et al., *Handbook of Laboratory Health and Safety*, 2nd edn., Wiley, New York, 1995.

Young, J.A. (ed.), *Improving Safety in the Chemical laboratory: A Practical Guide*, 2nd edn., Wiley, New York, 1991.

HEALTH AND SAFETY LEGISLATION

Bender, H.F. et al., *Hazardous Chemicals: Control and Regulation in the European Market*, Wiley-VCH, Weinheim, Germany, 2007.

Globally Harmonised System of Classification and Labelling of Chemicals, 4th edn., United Nations, New York, 2011, available as a downloadable pdf from the United Nations Economic Commission for Europe (UNECE) website. Accessed 31 July, 2015.

Moore, R. et al., *Law of Health and Safety at Work 2015/16*, 24th edn., Croner, Kingston upon Thames, U.K., 2015.

The Control of Substances Hazardous to Health Regulations 2002. Approved Code of Practice and Guidance, 6th edn., 2013.

Tolley's Health and Safety at Work Handbook 2015, 27th edn., LexisNexis, London, U.K., 2014.

ELECTRONIC SOURCES FOR HAZARD INFORMATION

The web is a vast resource for hazard information and for advice on safe practices in the chemical laboratory. Many U.K. and U.S. university chemistry departments have posted their safety policies and guidance for laboratory workers on the web and added links to other health and safety websites. Websites of the following organisations are also useful sources of hazard information:

Organisation	Internet Address and Description of Content
Agency for Toxic Substances and Disease Registry	http://www.atsdr.cdc.gov/
American Conference of Governmental Industrial Hygienists	http://www.acgih.org
	Sources of information on TLVs, biological exposure indices, chemicals under study and revisions to TLVs
Health and Safety Executive (HSE)	http://www.hse.gov.uk/
COSHH homepage	http://www.hse.gov.uk/coshh/index.htm
The CLP Regulation	http://www.hse.gov.uk/chemical-classification/legal/clp-regulation.htm
International Agency for Research On Cancer	http://www.iarc.fr/
Agents Classified by the *IARC Monographs*, Volumes 1–111	http://monographs.iarc.fr/ENG/Classification/
International Programme on Chemical Safety (WHO)	http://www.inchem.org/
Full-text access to the following *WHO Environmental Health Criteria*: No. 11 Mycotoxins (1979); No. 37 Biotoxins, aquatic (marine and freshwater) (1984); No. 80 Pyrrolizidine alkaloids (1988); No. 105 Mycotoxins, selected (1990); No. 219 Fumonisin B_1 (2000).	http://www.who.int/ipcs/publications/ehc/en/
National Institute for Occupational Health and Safety	http://www.cdc.gov/niosh
	Research studies, health hazard evaluations, extensive links to occupational safety and health resources on the Internet
National Library of Medicine	http://www.nlm.nih.gov

(Continued)

Organisation	Internet Address and Description of Content
Databases include PubMed and TOXNET	http://www.ncbi.nlm.nih.gov/pubmed/ and http://toxnet.nlm.nih.gov/
National Toxicology Program	http://ntp.niehs.nih.gov/
	Extensive information on chemicals, reactivity, long-term and short-term effects
NIOSH Pocket Guide to Chemical Hazards	http://www.cdc.gov/niosh/npg/
	Also gives a link to The Registry of Toxic Effects of Chemical Substances (RTECS) database
Royal Society of Chemistry	http://www.rsc.org/
RSC Environment Health & Safety Committee Notes	Includes: *COSHH in Laboratories, Version 5*, 2013; *Fire Safety in Chemical Laboratories* http://www.rsc.org/membership/ehsc/ehsc-guidance.asp

Index

A

Abelsonite, 184
Abeoabietanes, 135
19(4→3)-Abeoabietanes, 135
4(5→10)-Abeoeremophilane, 35
Abeopentalenanes, 131
11(15→1)-Abeotaxanes, 143
Abietanes, 134
Acalyphin, 99
Acarids (ZX0600), 69
Acenaphthylenes, 120
Acerogenin A, 101
Acetogenins, 87
ACGIH, *see* American Conference of Governmental
    Industrial Hygienists (ACGIH)
Acoranes, 130
Acorn worms, 70
Acridones, 164
Acronycine, 164
Actinomyces, 61
Actinomycetes (ZB5000), 60–61
Actinomycin D, 160
Actinomycins, 160
Actinorhodin, 118
Aculeatin C, 99
Acyclic sesterterpenoids, 146
Acylphloroglucinols, 100
Adenaflorin A, 117
Adenochromines, 68
Adenosine, 99
Adiananes, 152
Adocianes, 143
Adouétines, 31
Adriamycin, 119
Affixes, semisystematic names, 34–36
Aflatoxin B₁, 206
Aflatoxins, 93
Africananes, 128
Agarofuranoids, 125
Aiphanol, 116
Ajmalicines, 172
Ajmalines, 172
Akagerans, 172
Akuammicines, 34, 173
Akuammilines, 173
Alanyl-D-phenylalanylproline, 196
Albaspidins, 65
Alborixin, 92
Alditols, 192
Aldoses, carbohydrates, 185–187
Algae, 61
Algarobin, 110
Aliphatic-related skeletons (VA), 86
Aliphatics, 65, 69–70
Alkaloids (VX), 61–71, 161–182
Alliacanes, 126
Allitol, 192

Alloimperatorin, 104
Allomethylose, 192
Allomones, 64
D-Allose, 187
Almazole A, 31
Alpinumisoflavone, 108
D-Altritol, 192
D-Altrose, 187
α-Amanitin, 182
Amentoflavone, 106
American Conference of Governmental Industrial
    Hygienists (ACGIH), 204
Amino acids (VV), 158–161, 196–199
2-Aminobutanoic acid, 199
Aminocoumarins, 161
2-Amino-3,3-dimethylbutanoic acid, 30
Amorphanes, 126
Amphibians (ZZ2000), 71
Amphilectanes, 143
Amphotericin B, 90
Amylopectin, 61
Amylose, 61
Anabaenopeptins, 158
Anabasine alkaloids, 161
Anatoxin a, 206
Ancistrocladine, 163
Andirobin, 149
Androstanes, 35, 155
A′-Neogammacerane, 151
Angiosperms (dicots, ZQ; monocots, ZR), 65–66
Angucyclines, 91
Annelids (ZU3000), 67
Annopholine, 179
Anomers, 188–190
Ansamycins, 89
Ansa-peptide alkaloids, 181
Anthocerodiazonin, 64
Anthoceros agrestis, 64
Anthocyanidins, 105
Anthozoa, 66–67
Anthracenes, 118
Anthracyclinones, 119
Anthramycin, 164
Anthraquinones, 29, 64
1,2-Anthraquinones, 118
1,4-Anthraquinones, 118
9,10-Anthraquinones, 118
Antimycin A₁ₐ, 95
Antimycins, 95
Anti-Terrorism, Crime and Security Act 2001 (UK), 205
Aphidicolanes, 138
Aphis, 70
Aplysiatoxins, 93
Apocynaceae, 66
Apoenzyme, 200
Apogalanthamines, 169
Aporphine-benzylisoquinoline dimers, 167